Extrapolation Practice for Ecotoxicological Effect Characterization of Chemicals

Other Titles from the Society of Environmental Toxicology and Chemistry (SETAC)

Valuation of Ecological Resources: Integration of Ecology and Socioeconomics in Environmental Decision Making
Stahl, Kapustka, Munns, Bruins, editors
2007

Genomics in Regulatory Ecotoxicology: Applications and Challenges
Ankley, Miracle, Perkins, Daston, editors
2007

Population-Level Ecological Risk Assessment
Barnthouse, Munns, Sorensen, editors
2007

Effects of Water Chemistry on Bioavailability and Toxicity of Waterborne Cadmium, Copper, Nickel, Lead, and Zinc on Freshwater Organisms
Meyer, Clearwater, Doser, Rogaczewski, Hansen
2007

Ecosystem Responses to Mercury Contamination: Indicators of Change
Harris, Krabbenhoft, Mason, Murray, Reash, Saltman, editors
2007

Genomic Approaches for Cross-Species Extrapolation in Toxicology
Benson and Di Giulio, editors
2007

New Improvements in the Aquatic Ecological Risk Assessment of Fungicidal Pesticides and Biocides
Van den Brink, Maltby, Wendt-Rasch, Heimbach, Peeters, editors
2007

Freshwater Bivalve Ecotoxicology
Farris and Van Hassel, editors
2006

Estrogens and Xenoestrogens in the Aquatic Environment: An Integrated Approach for Field Monitoring and Effect Assessment
Vethaak, Schrap, de Voogt, editors
2006

For information about SETAC publications, including SETAC's international journals, Environmental Toxicology and Chemistry and Integrated Environmental Assessment and Management, contact the SETAC Administratice Office nearest you:

SETAC Office
1010 North 12th Avenue
Pensacola, FL 32501-3367 USA
T 850 469 1500   F 850 469 9778
E setac@setac.org

SETAC Office
Avenue de la Toison d'Or 67
B-1060 Brussells, Belguim
T 32 2 772 72 81   F32 2 770 53 86
E setac@setaceu.org

www.setac.org
Environmental Quality Through Science®

Extrapolation Practice for Ecotoxicological Effect Characterization of Chemicals

Edited by

Keith R. Solomon

Theo C.M. Brock

Dick de Zwart

Scott D. Dyer

Leo Posthuma

Sean M. Richards

Hans Sanderson

Paul K. Sibley

Paul J. van den Brink

Coordinating Editor of SETAC Books
Joseph W. Gorsuch
Gorsuch Environmental Management Services, Inc.
Webster, New York, USA

SETAC®

CRC Press
Taylor & Francis Group
Boca Raton London New York

CRC Press is an imprint of the
Taylor & Francis Group, an **informa** business

Published in collaboration with the Society of Environmental Toxicology and Chemistry (SETAC)

ISBN 13: 978-1-880611-45-6 (SETAC Press)

© 2008 by the Society of Environmental Toxicology and Chemistry (SETAC)
CRC Press is an imprint of Taylor & Francis Group, an Informa business
SETAC Press is an imprint of the Society of Environmental Toxicology and Chemistry (SETAC)

First issued in paperback 2019

No claim to original U.S. Government works

ISBN-13: 978-0-367-45264-3 (pbk)
ISBN-13: 978-1-4200-7390-4 (hbk)

International Standard Book Number-13: 978-1-880611-45-6 (Hardcover; SETAC)

Library of Congress Cataloging-in-Publication Data

Extrapolation practice for ecotoxicological effect characterization of chemicals / editor(s), Keith R. Solomon ... [et al.] ; SETAC.
    p. ; cm.
Includes bibliographical references and index.
ISBN 978-1-4200-7390-4 (hardback : alk. paper)
    1. Environmental toxicology--Mathematical models. 2. Health risk assessment--Mathematical models. 3. Extrapolation. I. Solomon, Keith R. II. SETAC (Society)
    [DNLM: 1. Environmental Monitoring--methods. 2. Models, Biological. 3. Risk Assessment--methods. WA 670 E984 2008]

RA1226.E97 2008
571.9'5--dc22

2008011861

Visit the Taylor & Francis Web site at
http://www.taylorandfrancis.com

and the CRC Press Web site at
http://www.crcpress.com

and the SETAC Web site at
www.setac.org

SETAC Publications

Books published by the Society of Environmental Toxicology and Chemistry (SETAC) provide in-depth reviews and critical appraisals on scientific subjects relevant to understanding the impacts of chemicals and technology on the environment. The books explore topics reviewed and recommended by the Publications Advisory Council and approved by the SETAC North America, Latin America, or Asia/Pacific Board of Directors; the SETAC Europe Council; or the SETAC World Council for their importance, timeliness, and contribution to multidisciplinary approaches to solving environmental problems. The diversity and breadth of subjects covered in the series reflect the wide range of disciplines encompassed by environmental toxicology, environmental chemistry, and hazard and risk assessment, and life-cycle assessment. SETAC books attempt to present the reader with authoritative coverage of the literature, as well as paradigms, methodologies, and controversies; research needs; and new developments specific to the featured topics. The books are generally peer reviewed for SETAC by acknowledged experts.

SETAC publications, which include Technical Issue Papers (TIPs), workshops summaries, newsletter (SETAC Globe), and journals (Environmental Toxicology and Chemistry and Integrated Environmental Assessment and Management), are useful to environmental scientists in research, research management, chemical manufacturing and regulation, risk assessment, and education, as well as to students considering or preparing for careers in these areas. The publications provide information for keeping abreast of recent developments in familiar subject areas and for rapid introduction to principles and approaches in new subject areas.

SETAC recognizes and thanks the past coordinating editors of SETAC books:

A.S. Green, International Zinc Association
Durham, North Carolina, USA

C.G. Ingersoll, Columbia Environmental Research Center
US Geological Survey, Columbia, Missouri, USA

T.W. La Point, Institute of Applied Sciences
University of North Texas, Denton, Texas, USA

B.T. Walton, US Environmental Protection Agency
Research Triangle Park, North Carolina, USA

C.H. Ward, Department of Environmental Sciences and Engineering
Rice University, Houston, Texas, USA

SETAC Publications

Table of Contents

List of Illustrations

List of Tables

About the Editors

Keith R. Solomon graduated from Rhodes University with a BSc (Hons) and holds MSc degrees from Rhodes University and the University of Illinois as well as a PhD from Illinois.  He is director of the Centre for Toxicology and a professor in the Department of Environmental Biology at the University of Guelph, where he teaches courses in toxicology and pesticides. He conducts research into the fate and effects of pesticides and other substances in the environment, the exposure of humans to pesticides and industrial chemicals, and risk assessment. He has supervised more than 54 MSc and PhD students. He is a fellow of the Academy of Toxicological Sciences and received the 2002 American Chemical Society International Award for Research in Agrochemicals. In 2006, he was awarded the SETAC-Europe Environmental Education Award and the Society of Environmental Toxicology and Chemistry (SETAC) Founders Award. He has more than 30 years of experience in research and teaching in pesticide science and environmental toxicology and has contributed to more than 300 scientific publications and reports in the scientific literature.

Theo C. M. Brock, after finishing his PhD in 1985 at the University of Nijmegen, had several positions as an aquatic and wetland ecologist at the University of Amsterdam, the Wageningen Agricultural University, and the University of Utrecht, all in The Netherlands.  Since 1991, he has been employed at Alterra (Wageningen University and Research Centre; formerly SC-DLO) to scientifically underpin the ecological risk assessment procedures for pesticides and other chemicals, with reference to freshwater ecosystems. He is coauthor of more than 140 scientific papers and reports, covering subjects on the ecology of macrophyte-dominated freshwater ecosystems, the vegetation dynamics of wetlands and wetland forests, aquatic ecotoxicology, and ecological risk assessment and risk management of pesticides.

Theo Brock has been secretary and president of the Netherlands Society of Aquatic Ecology, and council member and president of SETAC-Europe. Currently he is associate editor of the SETAC journal *Integrated Environmental Assessment and Management*. He participated in various ecotoxicological workgroups of the Health Council of the Netherlands, in the organizing committee of European Union–Organization for Economic Cooperation and Development–SETAC workshops on higher tier risk assessment (e.g., Higher-Tier Aquatic Risk Assessment for Pesticides [HARAP], Community-Level Aquatic System Studies Interpretation Criteria [CLASSIC], and Workshop Linking Aquatic Exposure and Effects in the Registration Procedure of Plant Protection Products [ELINK]), and in the European and Mediterranean Plant Protection Organization (EPPO) workgroup on aquatic organisms. He is regularly invited to participate as a lecturer in (post)university courses on ecological risk assessment in the Netherlands and abroad, and asked to advise governmental authorities, the European Food Safety Authority (EFSA), the Dutch Board for Registration of Pesticides, and the chemical industry.

Dick De Zwart holds a BSc, MSc (hons), and PhD in biology (aquatic ecology and ecotoxicology) from the University of Amsterdam, The Netherlands. Since 1980, he has worked as a senior scientist at the Laboratory for Ecological Risk Assessment of the Dutch National Institute for Public Health and the Environment (RIVM). Between 1988 and 1995 he was responsible for a large-scale international cooperation project with India. This project was concerned with teaching ecotoxicology and water quality monitoring, as well as the development of a methodology for ecological risk assessment for river water quality, based on chemical, toxicological, and ecological criteria. For a long period of time (1992–2002), he also provided a major input to the definition and execution of a pan-European integrated monitoring network on the ecological impacts of long-range transboundary air pollution. Since 2002, his main interest has been in the development and validation of probabilistic methods for ecological risk assessment.

Scott D. Dyer obtained his BS and MS degrees from Iowa State University in biology and toxicology, respectively. His PhD was awarded from the University of North Texas, where he studied the stress protein response in fish exposed to diverse contaminants via laboratory and field exposures. Since 1991, he has been employed by the Procter & Gamble Company (P&G), Cincinnati, Ohio, and is presently a principal scientist in Central Product Safety–Environment as an ecotoxicologist. His primary mission within P&G is researching methods that advance the predictions of potential environmental exposure and effects of chemicals found in consumer products. He currently has three major research programs, all collaborations with academia, government, and industry: 1) the ecoepidemiology of consumer product chemicals relative to other chemical and physical stressors, 2) the extrapolation of potential effects across species, and 3) the development of screening tools for the estimation of metabolism in fish, an important attribute for the prediction of bioaccumulation. He has authored more than 40 journal articles, book chapters, and technical reports, and currently participates with work groups within organizations such as the Soap and Detergent Association (SDA), the Water Environment Research Foundation (WERF), the ILSI Health and Environmental Sciences Institute (HESI), and the H. John Heinz III Center for Science, Economics and the Environment.

Leo Posthuma is currently research staff member in the Laboratory for Ecological Risk Assessment at the Dutch National Institute of Public Health and the Environment (RIVM) in the Netherlands, where he is involved in the development, validation, and practice testing of methods for ecological risk assessment. He holds a BSc and MSc (hons) in biology and a PhD in ecology and ecotoxicology from the Vrije Universiteit, Amsterdam, the Netherlands. Dr. Posthuma has authored and coauthored more than 75 open literature publications, reports, and book chapters, and has been book editor and  coeditor. His research experience includes phytopathological studies, studies on the evolutionary ecology and population genetics of contaminant adaptation of exposed soil arthropod populations, studies on community tolerance evolution, studies on the bioavailability of toxic compounds for terrestrial organisms, studies on the joint effects of compound mixtures, studies on the stability and resilience of soil ecosystems, and diagnostic (ecoepidemiological) studies. As acting project manager for RIVM's Strategic Project in ecology and ecotoxicology, he currently leads various project teams that address fundamental topics like ecological stability and stress. These include ecological epidemiology of multiple stressors, ecological side effects of genetically modified organisms, and also practical topics such as the establishment of ecotoxicity databases for risk assessment and a toolbox for site-specific risk assessment of soil contamination linked to Dutch soil legislation.

Sean M. Richards earned his MS from the University of Arkansas for Medical Sciences and his PhD from Texas Tech University. He is currently a UC Foundation Associate Professor of Environmental Toxicology in the Department of Biological and Environmental Sciences at the University of Tennessee of Chattanooga. Dr. Richards has more than 10 years of experience in researching and publishing on human and ecological toxicant effects and risk/ hazard assessment of environmental toxicants. He is associate editor for ecological risk assessment for the journal *Ecotoxicology* and is on  the editorial board of the journal *Environmental Toxicology and Chemistry*. Dr. Richards has been an ad hoc participant on US Environmental Protection Agency (USEPA) Scientific Advisory Panels for the purpose of evaluating and advising the USEPA on pesticide registration data. Dr. Richards directs MS and PhD students and teaches graduate courses on toxicology, environmental chemistry, environmental toxicology, risk assessment, and hazardous waste management.

Hans Sanderson completed his BSc (limnology), MSc (environmental science), and PhD (ecotoxicology) at Roskilde University in Denmark. He then held a research fellow position (ecotoxicology) at the University of Guelph, Ontario, Canada, before working as the director for environmental safety with the Soap and Detergent Association in Washington, D.C. Dr. Sanderson is currently a senior scientist with the Danish National Environmental Research Institute, Aarhus University, with

specialty within policy analysis and environmental toxicology. He has served on both a USEPA Science Advisory Board
and an OECD expert group regarding quantitative structure-activity relationships (QSARs). He conducts research
on industrial chemicals (e.g., hazard assessment of high-production-volume chemical categories) as well as on "emerging
contaminants" (pharmaceuticals, nanomaterials, and personal care product ingredients), with an emphasis on regulatory and policy aspects, hereunder risk assessment, including
modeling approaches to comprehend the potential challenges

that emerging contaminants may pose and to advise appropriate testing methods.
Under the EU 6th research framework, he is both a work package leader and a con-
tributor to Modelling of Ecological Risks Related to Sea-Dumped Chemical Weap-
ons (MERCW) and Novel Methods for Integrated Risk Assessment of Cumulative
Stressors in Europe (NoMIRACLE). Since 1999, he has contributed to more than 50
scientific publications.

Paul Sibley is an associate professor in environmental biology at the University of Guelph. He received his BSc (marine
biology, 1985) and MSc (environmental biology, 1989) from
the University of Guelph, and his PhD (environmental toxicology, 1994) from the University of Waterloo, Ontario, Canada.
He spent 3 years as a research scientist at the USEPA Mid-
Continent Ecology Division, where he directed a research
program on the development, application, and validation of
sediment toxicity tests. His current research interests focus
on issues of water quality and environmental management,

including invertebrate and sediment toxicology of novel compounds (with empha-
sis on testing, methods development, and validation), ecosystem-level responses to
contaminants using field-based systems (assessment of lab-to-field extrapolation
issues), the development of risk assessment methodologies, and disturbance ecology
(impacts of forest harvesting on boreal aquatic systems). He has published more than
70 peer-reviewed scientific papers and book chapters. He served as president of the
North American Geographical unit of the Society of Environmental Toxicology and
Chemistry from 2007 to 2008.

Paul J. van den Brink works at the research institute Alterra
and the Aquatic Ecology and Water Quality Management
Group of Wageningen University, both belonging to the
Wageningen University and Research Centre. He is involved
in supervising and executing international projects on the scientific underpinning of higher tier risk assessment procedures
for contaminants (e.g., for the registration of pesticides). He
is also experienced in the analysis of complex ecological data
sets. For this he improved multivariate ordination techniques
to assess the impact of chemical stress on freshwater (model)

ecosystems. Recent research topics are the development of effect models (e.g. food web, meta-population, and expert base models), the validation of risk assessment procedures (e.g., uniform principles and species sensitivity distribution concept), and human and ecological risk assessment of pesticide use in developing countries in Asia and Africa. Since 1994, Paul van den Brink has published more than 75 peer-reviewed papers, for 2 of which he won an international prize. In 2006 Paul won the Long-Range Research Institute (LRI)–SETAC Innovative Science Award. He also organized and took part in many international workshops and courses. Paul van den Brink is presently a member of the Socio-Economic and Natural Sciences of the Environment (SENSE) research school (www.sense.nl), president of SETAC-Europe (2008–2009), and editor of the journal *Environmental Toxicology and Chemistry*. Since May 2006, he is also an associate fellow of the Canadian River Institute.

Contributors and Workshop Participants

Rolf Altenburger
UFZ Helmholtz Centre for
  Environmental Research
Leipzig, Germany

Donald J. Baird
Environment Canada @ Canadian
  Rivers Institute
University of New Brunswick
Fredericton, New Brunswick, Canada

Theo C.M. Brock
Alterra Centre for Water and Climate
Wageningen, The Netherlands

Lars Carlsen
National Environmental Research
  Institute
Roskilde, Denmark

John Chapman
Department of Environment and
  Climate Change
Lidcombe, New South Wales, Australia

Dick De Zwart
National Institute of Public Health
  and the Enviornment
Bilthoven, The Netherlands

Scott D. Dyer
The Procter & Gamble Company
Cincinnati, Ohio, USA

Mark Ellersieck
University of Missouri
Columbia, Missouri, USA

Valery Forbes
Roskilde University
Roskilde, Denmark

Jeffrey Giddings
Compliance Services International
Rochester, Massachusetts, USA

Christopher W. Hickey
National Institute of Water and
  Atmospheric Research
Hamilton, New Zealand

Mike Hooper
Texas Tech University
Lubbock, Texas, USA

David Johnson
University of Guelph, Centre
  for Toxicology,
Guelph, Ontario, Canada

Thomas W. La Point
University of North Texas
Denton, Texas, USA

Hans Lökke
National Environmental
  Research Institute
Silkeborg, Denmark

Lorraine Maltby
The University of Sheffield
Sheffield, United Kingdom

Steve Maund
Syngenta Crop Protection AG
Basel, Switzerland

J. Vincent Nabholz
US Environmental Protection Agency
Washington DC, USA

Willie Peijnenburg
National Institute of Public Health
  and the Environment
Bilthoven, The Netherlands

Leo Posthuma
RIVM
Bilthoven, The Netherlands

Hans Toni Ratte
RWTH Aachen University
Aachen, Germany

Sean M. Richards
University of Tennessee at Chattanooga
Chattanooga, Tennessee, USA

Pete Robinson
Environment Canada
Ottawa, Ontario, Canada

Hans Sanderson
Danish National
  Environmental Research Institute
Roskilde, Denmark

Paul K. Sibley
University of Guelph
Guelph, Ontario, Canada

Keith R. Solomon
University of Guelph
Guelph, Ontario, Canada

Glenn Suter
US Environmental Protection Agency
Cincinnati, Ohio, USA

Henrik Tyle
Danish Environmental Protection
  Agency
Kovenhavn, Denmark

Dik van de Meent
National Institute of Public Health
  and the Environment
Bilthoven, The Netherlands

Paul J. van den Brink
Alterra and Wageningen University
Wageningen, Netherlands

Amanda Warne-Lorscheider
University of Guelph
Guelph, Ontario, Canada

Eva Bay Wedebye
National Food Institute, DTU
Soborg, Denmark

René van Wijngaarden
Alterra Centre for Water and
  Climate
Wageningen, The Netherlands

Maurice Zeeman
US Environmental Protection
  Agency
Washington DC, USA

Preface

PURPOSE OF THIS BOOK

This book is the culmination of a request to prepare a wide-ranging compilation of methods for extrapolation of ecological risks. In the context of ecotoxicology, extrapolation is the use of existing information for the prediction or forecasting of events in another situation that is biologically, temporally, or spatially different from that where the existing information was gathered. It is generally recognized that extrapolation is a key component of risk assessment for the protection of human health and the environment. It is also recognized that extrapolation is conducted in an environment of uncertainty that may be different for every situation. Thus it is not possible to develop generic methods of extrapolation that will address all or even a subset of situations in a completely satisfactory manner.

With this in mind, we have approached this book with a view to compiling and describing various methods of extrapolation in the framework of ecological risk assessment (ERA) as well as identifying the data needs and situations where these extrapolations can be most usefully applied. This book is directed to risk assessors and to the scientific community as a whole by reviewing the current state of extrapolation techniques and approaches. The major focus is on extrapolation of effects, but it also deals with extrapolation of exposures in the context of interactions between the toxicant and the matrix. It also contains a practical guide to the application of these extrapolation procedures that are designed to be useful to regulators and risk managers at several levels and in the education of students in these disciplines.

As extrapolation is intertwined with risk assessment, the chapters in the book follow the generalized framework for assessing risks. The chapters themselves also follow a similar framework, starting with a problem formulation and ending with a discussion of uncertainty and data needs. Chapter 1 acts as a general introduction to the terminology of extrapolation and how this is incorporated into the tiered approach to risk assessment. It also deals with how extrapolation is used in selecting assessment endpoints and measures of effect and how this is integrated into regulatory frameworks. Extrapolation uncertainty is also discussed.

In keeping with the need to characterize and understand exposures in risk assessment, the second chapter, on matrix and media extrapolation, deals with the very important physical and chemical interactions between the exposure matrix and the biological availability of the substance. This process is key to extrapolation in both the spatial and the temporal contexts, where there are differences in the environments where organisms may be exposed. This chapter reviews the methods of extrapolation that may be used and provides guidance as to the tools to use for this purpose.

Extrapolations of the other significant components of risk assessment, measures of effects, are reviewed in Chapters 3 through 7, which present a hierarchical approach based on biological organization. Extrapolation of effect measures through (quantitative) structure-activity relationships ([Q]SARs) is often necessitated because of the large number of substances for which environmental effects data are not available but that must be regulated under legal instruments that require risk management. Like matrix and media extrapolation, most (Q)SARs make use of the physical and chemical properties of substances and link these with data from toxicity tests with similar substances (training data sets) to predict toxicity to model species based on known or hypothetical sites of action at the molecular level. Chapter 3 reviews the most readily available (Q)SAR approaches, species-to-species extrapolation, and discusses the use of these in a number of regulatory situations.

Chapter 4 deals with extrapolations of effects across levels of biological organization. These range from the biochemical and physiological responses in the target organelles and tissues to the complex interactions that may occur in whole ecosystems and finally to the landscape. The sequence in the chapter follows the concept of tiers in that the responses being characterized become more realistic and closer to the assessment endpoints and ecological entities that are to be protected as one moves to higher tiers.

Although the physical interactions between a toxic substance and a given matrix play an important role in extrapolation (Chapter 2), exposures to mixtures of substances in the environment are a real issue that is receiving increasing attention in the scientific, regulatory, and public communities. Interactions may occur at all levels of biological organization, from the target site, through toxicokinetics and toxicodynamics, to mixture-driven species responses caused by differential toxicities of components of mixtures and their interactions through trophic and functional relationships. Chapter 5 reviews extrapolation tools for mixtures of defined and undefined components and how these may be used to address this complex issue.

Temporal extrapolation is important in terms of the duration of the exposure, the number of exposures, and the nature of the response to these in the organism. Chapter 6 reviews relationships between temporal exposure in relation to acute-to-chronic extrapolation, reversibility, and latency in terms of the interaction between substances and individual organisms. Other temporal extrapolation approaches are needed when considering temporal processes in organisms themselves. These relate to seasonal variability in sensitivity, recovery at the population level, and adaptation to stressors.

Differences in sensitivity between organisms in diverse biogeographic regions are addressed in Chapter 7 on spatial extrapolations. These are key in extrapolating from effect data generated with organisms from temperate regions to other regions, such as the tropics or the polar regions, where, in general, less data are available. Also important is extrapolation from freshwater to saltwater organisms, an issue that is also applicable to matrix extrapolation (Chapter 2). Extrapolation from smaller to larger spatial units is also addressed in Chapter 7, where landscape and watershed extrapolation is reviewed and approaches are assessed.

Areas of uncertainty, future directions, and the resulting research needs are summarized in Chapter 8, and a Glossary of terms is provided in Chapter 9. This is followed by the last chapter, 10, a stand-alone document that offers practical guidance suitable to several levels of users, from those in the laboratory to those making the risk management decisions. This chapter is less detailed in discussing extrapolation but links to a more detailed discussion of relevant questions and choices of methods in the other chapters of the book.

HOW THIS BOOK CAME ABOUT

This book is the end result of a joint submission by a consortium of research groups (nodes) to the American Chemistry Council's (ACC) Long Range Initiative Program. The participating nodes are the Centre for Toxicology at the University of Guelph and Canadian Network of Toxicology Centres (Canada) working in partnership with the Human & Environmental Safety Division, Procter & Gamble Company (United States); Alterra — Department for Water and Climate, Wageningen University and Research Centre (The Netherlands); and the Laboratory for Ecological Risk Assessment — the National Institute of Public Health and the Environment (The Netherlands).

The goals of the Extrapolation Practice for Ecological Effects and Exposure Characterization of Chemicals (EXPECT) project were to collect and review procedures for extrapolation of ecological effects in the context of ecological risk assessment. The process of extrapolation in ERA was the subject of an Organization of Economic Cooperation and Development (OECD) workshop held in 1992, and our efforts built on this previous work (OECD 1992). The focus of this book is on extrapolation in the context of risk assessment of substances as environmental stressors. Extrapolation procedures in the risk assessment of biological and physical stressors were specifically excluded; however, this was not because they were regarded as any less important by the authors. Biological and physical stressors can be major agents of change in ecosystems, but the science of assessing and extrapolating their potential impact and its consequences is, in many ways, very different from that of substances and is beyond the scope of our activities and detailed knowledge.

We reviewed the scientific and technical basis for existing extrapolation methodologies in ERA, and tested several of the extrapolation methods by field responses from existing field studies and studies in controlled static and flowing-water micro- and mesocosm experiments. The findings from these model data were used in assessing the appropriateness of the extrapolation methods. The process that we undertook had a primary focus on criteria-setting and risk assessment methodologies on single compounds, but the implications for mixture toxicity and contaminated site evaluation were considered. The main focus was on extrapolation of effects; extrapolation (modeling) of exposure was not addressed with the exception of the influence of the matrix on exposure and effects. Modeling and extrapolation of exposures are a complex process and deserve a separate text.

Extrapolation methods were tested against critical scientific opinion with several relevant data sets for key model substances and mixtures to stimulate critical assessment of the extrapolation procedures at a stakeholder workshop held in Florida. The stakeholder workshop was held with participation from 31 experts in risk assessment from the United States, Europe, and other countries. Participants were selected from academia, government regulatory bodies, industry, risk managers, and decision makers, and had diverse backgrounds (e.g., ecotoxicologists, ecologists, environmental fate experts, modelers, and regulators). Special efforts were made to ensure participation from countries in climatic regions other than the temperate region.

The workshop followed the familiar "Pellston" formula. A preliminary review was completed and made available to all participants prior to the workshop. At the workshop, workgroups were assigned to consider specific issues identified in plenary sessions, to gather feedback on the utility of reviewed methods, to suggest new or modified ones, and to consider case study substances in relation to the proposed extrapolation methods. These activities were used to prioritize questions and to help formulate a framework (table of contents) for dealing with extrapolation issues in ERA. We used this framework and, with individual contributions from the workshop participants, developed this book. This book incorporates a review and discussion of the tools currently available for extrapolation of ecological effects and the science upon which these are based.

ACKNOWLEDGMENTS

The editors and authors of this book wish to thank the American Chemistry Council for their support of the project. We also thank the participants at the workshop (listed in these acknowledgments and, in some cases, as chapter coauthors) for their important contributions to the writing (acknowledged in the individual chapters) and formulation of the book (through the workshop and in subsequent review). Without them, this book would still be an idea. They are, in alphabetical order, Rolf Altenburger, Donald Baird, Theo Brock, Lars Carlsen, John Chapman, Dick De Zwart, Scott Dyer, Mark Ellersieck, Valery Forbes, Jeff Giddings, Chris Hickey, Mike Hooper, David Johnson, Kannan Krishnan, Thomas La Point, Hans Lökke, Lorraine Maltby, Steve Maund, Vince Nabholz, Willie Peijnenberg, Leo Posthuma, Hans Toni Ratte, Sean Richards, Pete Robinson, Hans Sanderson, Paul Sibley, Keith Solomon, Glenn Suter, Henrik Tyle, Paul J. van den Brink, Amanda Warne, Eva Bay Wedebye, and Maurice Zeeman.

1 Extrapolation in the Context of Criteria Setting and Risk Assessment

Keith R. Solomon, Theo C. M. Brock,
Dick De Zwart, Scott D. Dyer, Leo Posthuma,
Sean M. Richards, Hans Sanderson,
Paul K. Sibley, and Paul J. van den Brink

CONTENTS

1.1 INTRODUCTION AND PROBLEM FORMULATION

Extrapolation is the use of existing information for the prediction or forecasting of events in another situation that is biologically, temporally, or spatially different from that where the existing information was gathered. Extrapolation is an integral part of the process of setting environmental quality criteria (EQC) and for conducting environmental risk assessments (ERAs) for substances. Extrapolation is used routinely in many jurisdictions and for many purposes, ranging from those required in regulatory acts and instruments to those conducted for managerial and economic decision making. Risk assessment and criteria-setting procedures may be ad hoc or formalized, as those used in a number of regulatory agencies. Many of these assessment processes follow prescribed frameworks such as those used in North America, the European Union, and other jurisdictions (US Environmental Protection Agency [USEPA] 1992, 1998; Environment Canada 1997; European Union 1997; Chapman 2001).

1.1.1 CRITERIA SETTING AND RISK ASSESSMENT

As extrapolations are most commonly used in criteria setting and risk assessment, some understanding of the process is necessary to understand the role of extrapolation. Risk can be assessed in a general or a specific sense, and it can be assessed prospectively and retrospectively. General risk assessments are usually region- or country-wide, consider almost all possible combinations of exposure and sensitivity, and are commonly used to set criteria or quality objectives (Figure 1.1). They are usually more conservative as they must consider all eventualities and combinations of sensitivity and exposure. Specific risk assessments are usually confined to a particular site, a particular set of circumstances, or a particular use. They need not consider all possible eventualities and may thus exclude some or many possibilities that would be included in a general risk assessment. Retrospective and prospective risk assessments may be general or specific. Retrospective (diagnostic) risk assessments usually make use of known or measured effects and/or exposure data (Figure 1.1) and would consider the risk from existing or past uses or releases to the environment. Prospective (prognostic)

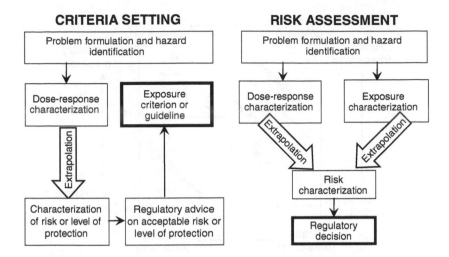

FIGURE 1.1 An illustration of the role of extrapolation in frameworks for ecological risk assessment used either for setting criteria or for assessing risks from existing exposures.

risk assessment is often carried out prior to the marketing, release, or use of the substance in the environment and makes greater use of models to estimate exposures.

Setting of an EQC relies primarily on toxicity and effect data to set a guideline or criterion for exposure concentrations that will be protective of the environment. The ERA utilizes a combination of exposure and effects data as a basis for assessing the likelihood and severity of adverse effects (risks) and feeds this into the decision-making process for managing risks. The process in which these data are used in assessing risk ranges from the simple calculation of hazard ratios to complex utilization of probabilistic methods based on models and/or measured data sets. However, all these processes begin with and utilize an implicit problem formulation or the formal definition of the problem being considered.

1.1.2 PROBLEM FORMULATION FOR RISK ASSESSMENT

Problem formulation is an integral component of all frameworks for criteria setting and risk assessment and is critical to success as it lays down the foundation upon which the process depends (USEPA 1992, 1998; Environment Canada 1997; European Union 1997). This is especially important in ERAs as they are often complex, involving several levels of biological organization, ranging from the organism to the population, the community, and the ecosystem, and possibly include several stressors and/or responses. The process may be informal in its initial stages but will become more formal as one or more iterations are conducted between risk managers, stakeholders, and risk assessors (USEPA 1998).

For logistical reasons, it is frequently necessary to divide complex tasks into smaller components that can be more easily managed. The use of tiers or steps in the process of criteria setting and risk assessment is one method used to reduce complexity and narrow the focus of the process to the key issues, and it has been recommended frequently for use in risk assessments of pesticides (USEPA 1992, 1998;

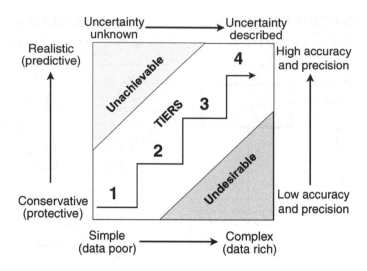

FIGURE 1.2 Tiers in the risk assessment process, showing the refining of the process through the acquisition of additional data.

Suter et al. 1993; Society of Environmental Toxicology and Chemistry [SETAC] 1994; European Union 1997; Ecological Committee on FIFRA Risk Assessment Methods [ECOFRAM] 1999; Campbell et al. 1999). The initial use of conservative assessment criteria (i.e., err on the side of caution) allows substances that do not present a risk to be eliminated from the risk assessment process early, thus allowing the focus of resources and expertise to be shifted to potentially more problematic substances or situations. As one ascends through the tiers, the estimates of exposure and effects become more realistic with the acquisition of more accurate and/or representative data, and uncertainty in the extrapolation of effects is thus reduced or at least better characterized. Likewise, the methods of extrapolation may become more sophisticated as one ascends through the tiers (Figure 1.2).

There are several rules that must be applied to tiers for them to function correctly. First, lower tiers must be more conservative than higher tiers. The function of lower tiers is to exclude substances from further consideration if they truly do not present a problem, and some conservatism is needed to prevent the erroneous release of a potentially harmful substance into the environment. Thus, it is a requirement of the process that any uncertainty factor that is applied in the tiers must respond to the increase in realism of the data and becomes smaller at higher tiers. The second requirement of tiers is that the data available at higher tiers must be more relevant or more realistic than the data at lower tiers. However, this does not mean that the tiers of exposure and effects characterization need to be firmly linked (e.g., Tier-2 exposures can only be compared to Tier-2 effects). Diagonal comparisons can be made as long as they are in the direction of higher tiers, the units of measurement are the same, and the total uncertainty in the data set and the uncertainty factors are reduced by using more realistic information. Similarly, if it is practical and the data are available, tiers can be skipped, as long as the progression is toward higher tiers. An important point is that, in moving through the tiers, the units for exposure remain consistent. Thus, Tier-1 acute estimated or measured exposures of, say, 96 hours must be carried forward to higher tiers with the same time interval.

There are several methods for extrapolation of effects and responses to various levels of organization. These methods may be grouped into certain categories, depending on how they are used or the types of data that are used for input. These are discussed in more detail in the following chapters. It is important to recognize that these methods are essentially tools to be used as dictated by the situation. Thus, extrapolations inherent in a simple hazard quotient (HQ) may be extended and refined though the use of probabilistic methods and further verified or calibrated by other tests such as field observations, micro- and mesocosm (cosm) studies, or whole-effluent testing.

1.2 TERMINOLOGY OF EXTRAPOLATION

Extrapolations always have been a key component of criteria setting and risk assessment, either formally or informally. Basically, extrapolation occurs when data derived in one system or one condition are used to predict effects or responses in another. Extrapolations become important in both the quantitative and qualitative relationships between effect measures and assessment endpoints as well as in estimating exposures in different matrices (Figure 1.1). All risk assessments require some data, and none are able to test all possible combinations of circumstances that may occur. For example, many assessments make use of laboratory-derived effect (toxicity) data from surrogate organisms to postulate effects in organisms in the environment that cannot readily be tested. These extrapolations are made across species within a genus or a family, as well as across a wide range of taxonomic and functional distances. Thus mice, rats, rabbits, and monkeys act as surrogates for humans, and laboratory test species belonging to insects, crustacea, fish, and other vertebrates plus plants and algae act as surrogates for untested organisms in the ecosystem. Additionally, extrapolations may be made between exposures in one system such as water and another such as sediment, where the physicochemical properties of the substance and its environment may alter exposure and, by extension, effects. However, although many extrapolations are formalized or standardized, some are not well based in science, nor are they all harmonized across jurisdictions.

In terms of extrapolation of effects from test organisms to the ecosystem, there are 3 general methods used. The simplest involve the use of generic methods derived from best judgment or traditional experience. Others are based on statistical analysis of data derived from empirical observations in the laboratory and field or on a mechanistic understanding of mechanisms of action. Historically, the earliest and still most common form of extrapolation is the use of mathematical factors in postulating effects in untested situations. Most extrapolations were made in the absence of adequate data, and in order to account for unquantified uncertainty and/or the social value associated with the affected organisms, a factor was commonly used to add a level of protection or conservatism to the extrapolation. These factors go by a number of names: application factor, assessment factor, safety factor, or uncertainty factor (UF), the preferred term that will be used throughout this book.

Uncertainty factors have been commonly used in ERA and criteria setting, and their use in extrapolation takes many forms. The simplest is illustrated in the deterministic calculation of hazard quotients (HQ). Here, the environmental exposure concentration is divided by the concentration that causes a response in a test organism (effect concentration) to give a hazard ratio or an HQ. If the effect concentration is equal to the exposure concentration, the response in the test organisms also would be expected to be observed in the environment where the sample came from. Because the effect concentration is normally a point estimate derived from a concentration-response test, a UF is used to protect against responses at lower concentrations in the same organism as well as the possibility of greater sensitivity in other organisms. These UFs range in value, depending on the response that is measured. For example, the USEPA uses UFs from 1 to 20 for extrapolating the results of laboratory tests on the potential effects of pesticides in the environment (Urban and Cook 1986; Calabrese and Baldwin 1993). Typical UFs applied to hazard quotients are listed in Table 1.1 and illustrated graphically in Figure 1.3. Similarly, UFs are

TABLE 1.1
Comparison of uncertainty factors for assessing risks in aquatic environments by various regulatory bodies

Data	Canada[1]	OECD[2]	OECD[3]	USEPA[4]	EU technical guidance document[5]
Quantitative structure-activity relationships ([Q]SAR)	1000	1000	1000[6]	1000	1000
Acute data (1 or 2 species)	1000	1000	1000	1000	
Acute data (3 taxa)[7]	100	100	100 to 1000	100	100 or 1000
Acute probabilistic (≥5 species)					
Chronic data (1 taxon)			50 to 100		100
Chronic data (2 taxa)			10 to 100		50
Chronic data (3 taxa)	10	10	10	10	10
Chronic probabilistic					1 to 5
Meso- and microcosm data				1	Case by case

[1] Environment Canada (1997), maximum factors.

[2] OECD (1992).

[3] OECD (2002).

[4] Zeeman and Gilford (1993).

[5] This refers to short- and long-term toxicity instead of acute and chronic toxicity. The approach in EU 91/141 is different. According to Annex VI3 of Directive 91/414/EEC, C 2.5.2.2, an uncertainty factor of 100 is required on the LC50 and/or EC50 of the most sensitive of the tested species (Guidance Document on Aquatic Ecotoxicology, Sanco/3268/2001, rev. 3; SANCO 2002). When additional data are available, there is no clear guidance on the lowering of the UF; however, a UF of 10 is generally assumed when large amounts of additional data are available.

[6] Chemicals that can be evaluated using (Q)SARs constrained to ensure that (Q)SARs effectively predict toxicity.

[7] The UF is dependent on whether the substance is a pesticide or not.

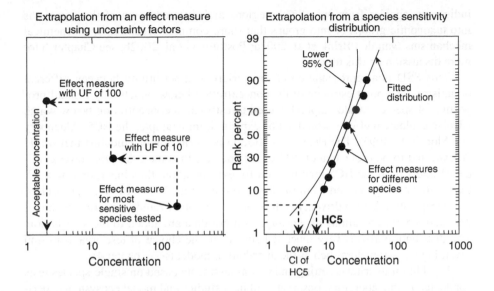

FIGURE 1.3 Graphical illustration of the use of uncertainty factors (UFs) in extrapolation from a single deterministic effect measure and the estimation of UFs from a species-sensitivity distribution. The y-axis scale on the left graphic is intentionally omitted.

used in human health risk assessment. For example, UFs of 10 are used to extrapolate from the average human to a sensitive human, from animal data to humans, from a lowest effect concentration to a no-effect concentration, from subchronic to chronic, and for database inadequacies (World Health Organization International Program on Chemical Safety 1999). Where more than one UF is used, they are multiplied together, and in situations that lack data, the total UF can become very large and impractical (10^5 to 10^6). These UFs are not based on formal analysis of the data and may not reflect the real relationships being extrapolated, leading to errors in interpretation where risk may be underestimated or overestimated. UFs may also be applied to toxicity data, such as in the use of the UF to extrapolate from an effect value to derive a predicted no-effect concentration (PNEC).

When more data have been generated, more formal approaches are available to develop UFs where the data themselves are used to derive the UF by statistical methods. Where enough data have been produced for different species, a species sensitivity distribution (SSD) can be derived (see Figure 1.3 and Chapter 4). SSDs are distributions of toxicity values based on the results of laboratory tests. SSDs offer a level of refinement over single-species hazard quotients that incorporate the range of sensitivity across entire groups of organisms or within specific categories determined from knowledge of the mechanisms of action and the toxicokinetics of the substance. SSDs better facilitate assessment of hazards and risks at the community level, where redundancy and resiliency can play an important role in community homeostasis (ECOFRAM 1999; Solomon and Takacs 2002; Posthuma et al. 2002b). SSD curves may be constructed on the basis of acute or chronic toxicity data, or other effect measures. However, questions have been raised about the need to include indigenous species from specific ecoregions and differentiation of the organisms into taxonomic groups or into groups that share common receptors or biochemical mechanisms (van den Brink et al. 2002a; Posthuma et al. 2002b; see Chapter 6 for more discussion on this topic).

This SSD is a representation of the variation in sensitivity between different organisms. The point estimate of the concentration below which only a small proportion of species would respond is used to estimate a concentration below which adverse ecological effects would not be expected, for example, the HC5 (Aldenberg and Slob 1991, 1993). The HC5 concentration is a criterion or guideline that would be expected to be protective of 95% of the species. In some cases, the lower confidence interval on the HC5 may be used for extrapolation, thus incorporating a UF that is statistically derived from the data (Maltby et al. 2005). In both approaches, the extrapolation is data driven, but the method obviously requires a large amount of data and the extrapolations may change as more data are added to the distribution. The distribution also is very much dependent on the choice of test organisms and several other parameters such as the distribution model (see Chapter 4).

In addition to criteria setting and risk assessments based on single-species tests conducted in the laboratory, observational field studies and model ecosystem experiments are regularly performed to study the effects of contaminants at the community and ecosystem levels. In this ecotoxicological approach, researchers are able to test for possible effects on multiple species and at many levels of biological organization, an option not readily available to human health risk assessors, except where fortuitous exposures in occupational and other settings may be assessed through epidemiological methods. Although not true field studies, these experiments in model systems may be very useful as they provide tools for empirical extrapolations based on observations at the population and community levels. They have the additional advantage of allowing the observation of interactions between species, a process not well addressed in extrapolation techniques except for ecosystem food-web models (see Chapter 4). Thus, ratios of concentrations causing an effect in the laboratory and in the field can be used to derive an extrapolation factor, and where uncertainty can be estimated from multiple studies, a UF can be estimated empirically. An example of this is the modeling approach used in PERPEST (predicts the ecological risks of pesticides; van den Brink et al. 2002c), discussed in more detail in Chapter 7.

1.3 ASSESSMENT ENDPOINTS AND REGULATORY GOALS

One of the key steps in the problem formulation is the statement of the assessment endpoints, or what is to be protected. These are the explicit expressions of the actual valued ecosystem components (organisms, populations, or communities) or ecosystem functions that are to be protected. They are the ultimate focus in risk assessment and act as a link to the risk management process, such as policy goals (Suter et al. 1993; USEPA 1992, 1998). Assessment endpoints will vary depending on the protection goals that are applied to the situation or issue being assessed. Thus, if the protection goal is population structure, the assessment endpoint will relate to populations and may be a quantitative measure of actual population numbers or a change in population numbers that exceeds a threshold. Examples of these are "The population should be 200 organisms/L" or "The change in population numbers must not exceed 10%." By expressing assessment endpoints in quantifiable terms such as these, changes (e.g., impact or recovery) can be characterized and differentiated. Unfortunately, endpoints for assessment and criteria setting are rarely expressed as explicitly as this.

1.3.1 ECOLOGICAL PROTECTION GOALS

Before considering protection goals and their function in extrapolation, it is necessary to consider what constitutes a sustainable ecosystem. Because properties of ecosystems vary in space and time, it is important to have quantifiable and broadly accepted ideas of what constitutes an ecologically important effect, and what constitutes a sustainable ecosystem (Calow 1998a; Brock and Ratte 2002). In ecotoxicology, the concern is rarely for individual organisms but usually for populations and communities in their natural environment. Exceptions are individuals of wildlife populations valued by society or endangered species. Overall, the intention is that populations and communities be sustained in the environment. In this context, to "sustain" is to "hold, keep alive" or, literally, "to be able to last." When considering the concept of sustainability, it is important to recognize that this involves protection from change resulting from a manageable source of risk, not just decreases. Adverse ecosystem responses are usually perceived negatively and are often equated with declines, such as a decrease in population size or a decrease in a function. However, increases in populations, such as algal blooms, or in functional processes may be just as deleterious in the ecosystem (Giesy 2001).

To keep or restore ecosystems in a state resembling more or less pristine conditions may be important for certain environmental uses such as special, highly valued habitats (nature reserves). For ecosystems in areas highly influenced by humans such as by agriculture, industrial activities, and sewage treatment plants, this is neither practical nor feasible; but, in many cases, we recognize the ecological importance of these habitats besides their economic and socioeconomic benefits. The interest of criteria setting and ecological risk assessment is to provide the objectives and criteria for sustainable management of these multifunctional ecological systems. In this context, the question of whether all patches of ecosystem should be protected equally needs to be addressed. In addition, in recent years, ecology has moved from a static view to a dynamic process, in which populations, communities, and ecosystems are considered in their temporal and spatial contexts within the surrounding landscapes and in the context of very long-term changes such as those related to geochemistry and ongoing evolutionary processes.

Within the context of sustainability of communities and ecological functions and services, there are 3 general categories of undesirable effects of substances in the environment. These relate to ecosystem structure, function, and (landscape) aesthetic value to humans (Table 1.2). The structure of an ecosystem is a combination of which organisms are present and how many there are, whereas function relates to what the organisms do in the ecosystem. The choice of protection goals (and, by extension, assessment endpoints) may be based on ecological knowledge or on human value judgments. For example, there is a general tendency to select functional

TABLE 1.2
Examples of the basis for assessment endpoints, measures of response, and measurement techniques

Basis for assessment endpoint	Measure of response	Measurement technique
Structure	Overall species richness and densities	Number of taxa and diversity indices (or scores of multivariate techniques) for the total community or for taxonomic or functional groups within the community
	Population densities of key or important ecological species	Ecological key(stone) species are species that play a major role in ecosystem performance, productivity, stability, and resilience. These species may also play an important role in ecosystem function (see below) and include 1) species that are critical determinants in trophic cascades, such as piscivorous fish and large herbivores; and 2) species that are "ecological engineers," that is, those that have a large influence on the physical properties of habitats such as rooted submerged macrophytes and beavers
	Population densities of indicator species — species with a high "information" level for monitoring purposes	Species protected by law, and regionally rare or endangered species
Ecosystem functioning and functionality	Biogeochemical cycles and energy flow	Environmental quality parameters such as oxygen depletion, changes in primary productivity, and changes in the processing of nutrients such as mineralization of organic matter or fixation of atmospheric nitrogen
		Decreases in the rate of replenishment of harvested resources such as saltwater and freshwater food organisms, and timber
Perceived aesthetic value or appearance of the ecosystem and/ or landscape	Disappearance of species with a popular appeal	Changes in populations of species such as dragonflies, songbirds, and butterflies
		Visual mortality of individual vertebrates or other valued ecosystem components
	Symptoms of eutrophication such as algal blooms in lakes; clarity of water	Taste and odor problems in drinking water

protection goals and assessment endpoints when the populations of the potentially affected organisms may change rapidly for natural reasons, may recover from effects rapidly, or are difficult to characterize. Examples are bacteria and fungi in soil and sediment or algae in aquatic systems. In populations that have lower recovery potential or are easily characterized and/or highly valued, there is a tendency to use structural protection goals such as absolute population numbers. Examples of these are fish, birds, or whales. Choices of protection goals may also be determined on the basis of value judgments; for example, if the risky activity brings great benefits, structural changes may be tolerated if functions are unaffected. These 3 ecological response categories may be further subdivided as shown in Table 1.2 (modified from Brock and Ratte 2002).

With respect to the last category in Table 1.2, it is important to recognize that aesthetic values in society can be ephemeral and are subject to rapid changes. Thus, what society considers important today may be unimportant tomorrow. In addition, societal values are strongly linked to cultural traditions and differ from one culture to another and with the level of economic and social development.

1.3.2 Risk Perception and Ecological Protection Goals

Guidance on how to deal with risks of substances in the environment is provided not only in a regulatory context but also by concepts based on science, ethics, and aesthetics, all of which can be related to the perception of risks by humans. To illustrate this, 4 completely different perceptions of the ecological risks of toxicants in nontarget habitats can be recognized (Brock et al. 2006). These are the 1) pollution prevention principle, 2) ecological threshold principle, 3) recovery principle, and 4) functional redundancy principle. The ecological threshold principle and the recovery principle are, respectively, the conservative and liberal approaches of the ecosystem carrying capacity principle described by Brock (2001).

1.3.2.1 Pollution Prevention Principle

The pollution prevention principle presupposes that all environmental pressure is potentially harmful. Conservative approaches are necessary to protect the environment because multiple stressors due to the presence of low concentrations of more than one substance or unexpected effects of metabolites (e.g., hormone disruption) can never be excluded. This opinion is in line with the community conditioning hypothesis (Matthews et al. 1996), which states that ecological communities tend to preserve information about every event in their history, including stress by substances. It is also in line with the rivet hypothesis (Ehrlich and Ehrlich 1981), which presupposes that each loss of a species (equivalent to a rivet in the analogy) affects ecosystem integrity to a small extent, and, if too many rivets are lost, the system collapses.

The pollution prevention principle considers the "what if" question more important than the "so what" question. Consequently, emission of substances to nontarget sites should be prevented as much as is technologically and socioeconomically feasible. An option in line with the pollution prevention principle is to always set the maximum permissible risk concentration (MPC) in ecosystems on the more or less conservative first-tier approach based on toxicity data from the most sensitive of standard test organisms and application of an uncertainty factor, or on a conservative confidence interval of a small centile, such as the HC5 (Chapter 4) derived from a sufficient number of chronic toxicity data from single-species laboratory tests.

A further conservative approach in line with the pollution prevention principle is to apply an additional uncertainty factor to the MPCs of individual compounds to account for the possible combined effects of the many substances encountered in the environment. For example, in The Netherlands, negligible risk concentrations (NCs) are derived from the MPC by dividing it by 100. These NCs are considered as target values to be reached in the near future (~10 years; Sijm et al. 2002).

The pollution prevention principle should not be confused with the precautionary principle (Graham 1999; Commission of the European Communities 2000; Government of Canada 2001). The pollution prevention principle simply aims to prevent pollution, whereas the precautionary principle is more subtle and is based on precautionary action if the uncertainty of the risk is too great and the intensity is potentially very large. In that case, the measures taken should be proportionate and temporary, accompanied by efforts to reduce the uncertainty, and reviewed again when further information becomes available.

1.3.2.2 Ecological Threshold Principle

The ecological threshold principle presupposes that the environment can absorb and tolerate a certain amount of stress. An approach in line with the ecological threshold principle is to consider a certain concentration of a substance acceptable if the sensitive structural or functional endpoints of the community are not, or are only briefly, impacted. A permissible concentration in line with the ecological threshold principle can be based on the no-observed-effect concentration (NOEC, or effect class 1; see Table 1.3) or the lowest-observed-effect concentration (LOEC, or effect class 2; see Table 1.3) of the most sensitive measurement endpoint as assessed by an adequately performed microcosm or mesocosm experiment. This threshold concentration derived from model aquatic ecosystem experiments may also guarantee that no adverse indirect effects such as algal blooms due to the direct toxic effects on grazing microcrustaceans will occur. Effect classes that can be used to summarize observed effects in aquatic microcosm or mesocosm studies are described in the European Union's "Guidance Document on Aquatic Ecotoxicology" (European Commission 2002; after Brock et al. 2000a, 2000b). In Europe, these effect classes (Table 1.3) are used to evaluate semifield tests submitted for the registration of pesticides.

1.3.2.3 Recovery Principle

The recovery principle also presupposes that the environment can absorb and tolerate a certain amount of stress. The stressor should be limited to an intensity or concentration less than that at which long-term adverse impacts on ecosystem structure and functioning occur. From a scientific point of view, periodically occurring declines in population densities can be considered a normal phenomenon in ecosystems. In the course of evolution, organisms have developed a large variety of strategies to survive and cope with temporally variable unfavorable conditions such as desiccation,

TABLE 1.3
Classes of effects used to assess the effects of substances on communities in microcosms and mesocosms

Class	Effect	Description
1	Effect could not be demonstrated	No (statistically significant) effects observed as a result of the treatment, and observed differences between the treatment and controls show no clear causal relationship. Causality in this context is judged through the use of guidelines similar to those developed for identifying causative agents of disease (Koch 1942; Hill 1965; IPCS 2002).
2	Slight effect	Effects reported in terms of "slight" or "transient" and/or other similar descriptions. Short-term and/or quantitatively restricted response of sensitive endpoints. Effects only observed at individual sample times.
3	Pronounced short-term effect	Clear response of sensitive endpoints, but total recovery within 8 weeks after the last application. Effects reported as "temporary effects on several sensitive species," "temporary elimination of sensitive species," "temporary effects on less sensitive species and/or endpoints," and/or other similar descriptions. Effects observed at some subsequent sampling instances.
4	Pronounced effect in short-term study	Clear effects (such as large reductions in populations of sensitive species) observed, but the study is too short to demonstrate complete recovery within 8 weeks after the (last) application.
5	Pronounced long-term effect	Clear response of sensitive endpoints and recovery time of sensitive endpoints are longer than 8 weeks after the last application. Effects reported as "long-term effects on many sensitive species and/or endpoints," "elimination of sensitive species," "effects on less sensitive species and/or endpoints," and/or other similar descriptions. Effects observed at various subsequent sample times.

flooding, temperature shocks, shading, oxygen depletion, food limitations, toxins in food, as well as anthropogenic stressors (Ellis 1989). In some cases, the stress caused by a substance may more or less resemble that of a natural stress factor. The use of the "normal operating range" of population densities and functional endpoints in specific ecosystems has been suggested as a baseline against which to assess pesticide-induced changes (Domsch et al. 1983). In other words, effects of substances that are restricted in space and time may be regarded, in certain habitats, as ecologically unimportant when they are of a smaller scale than changes caused by other natural or anthropogenic stresses. The recovery principle may be made more complicated because ecosystems may recover to different states that are stable in the long term (Scheffer et al. 2003).

In contrast to single-species laboratory toxicity tests, model ecosystem experiments allow the study of the recovery potential of sensitive populations. An approach in line with the recovery principle is to consider a certain impact of a chemical acceptable when recovery takes place within a short time frame, such as in the effect class 3 (Table 1.3). When environmental stress is restricted in space and time, and noncontaminated habitats are nearby, the recovery processes may ensure the sustainability of sensitive populations. The recovery rate of sensitive populations, however, has been shown to vary considerably between microcosm and/or mesocosm studies (see van Wijngaarden and Brock 1999; Brock et al. 2000a, 2006). In these studies, recovery depended on the degree of isolation of the test systems and the life-cycle characteristics of the populations present. In general, natural populations of species with a relatively short generation time, resistant life stages, and/or well-developed abilities to disperse, such as algae, daphnids, and multivoltine insects, showed rapid recovery from a stressor. The recovery time is dependent on generation time and reproductive strategy, the degree of change induced by the toxicant, and dispersal and/or immigration. For example, recovery of populations of Cladocera in microcosms treated with permethrin was dependent on the decrease in the population after treatment, which was, in turn, dependent on the concentration (Kaushik et al. 1985). In contrast, organisms with a more complex life cycle or with a low ability to recolonize isolated systems, such as the aquatic crustacean *Gammarus* and univoltine insects, may recover only slowly or not at all (van den Brink et al. 1996).

1.3.2.4 Functional Redundancy Principle

The functional redundancy principle presupposes that, for sustainable functioning of an ecosystem, a decrease in biodiversity can be tolerated, as long as key(stone) species and their functions are not impacted (see Table 1.2). This is because of the redundancy in roles and functions provided by the surviving species in the community (Lawton 1994). Functional redundancy is well known in ecology and has been demonstrated experimentally such as in the work of Tilman et al. (Tilman 1996; Tilman et al. 1996).

When adopting the functional redundancy principle, the emphasis is on ecosystem processes; impacts are considered acceptable when functional attributes are not changed, despite possible effects on community structure. Functional endpoints are rarely more sensitive than structural ones (Ellis 1989; Kersting 1994; Klepper et al. 1999); however, an exception is found in the photosynthesis inhibitors such as the triazines and urea classes of herbicides. Effects on functional endpoints indicate the limit of functional redundancy within the stressed community. Once ecosystem processes have changed due to contamination, this is usually an indication of severe effects on structural endpoints.

The functional redundancy principle is most appropriate for intensively used environments such as those impacted by urbanization, industrial, agricultural, and forestry activities. When adopting the functional redundancy principle, however, an important evaluation criterion should be the length of time required for reversibility of the effect, because a completely irreversible change in land (or water) would place the response in another category. The importance of time to recovery has been recognized and is used as a means of prioritizing ecological risks for regulatory purposes (Harwell et al. 1992).

1.4 MEASURES OF EFFECT

Although there was considerable discussion of assessment endpoints above, measures of effect are also key data for risk assessment and a starting point for extrapolation. They are exposure concentrations or doses that cause specific responses, such as mortality, reductions in growth, or reproduction (USEPA 1998). They are usually derived in laboratory studies or toxicity tests and are more easily quantified than assessment endpoints; however, the best effect measures are related quantitatively or qualitatively to the assessment endpoint and thence to the protection goal. Measures of effect combine 2 factors: a response in an organism, population, community, or ecosystem, and an exposure or intensity of the stressor that causes the response. Commonly used effect measures are the concentration that kills a proportion (e.g., 50% or 10%) of the test organisms (LC50 or LC10, respectively), the concentration that causes an effect in a proportion (e.g., 50% or 10%) of the test organisms (EC50 or EC10, respectively), or the NOEC. Variations on the latter include the LOEC, the maximum allowable toxicant concentration (the geometric mean of the NOEC and LOEC), and the benchmark concentration (a concentration causing a response that is not statistically significantly different from the control response; USEPA 1995a). All of these effect measures relate to a specific exposure duration, such as 96 hours.

1.5 TYPES OF EXTRAPOLATION

A number of types of extrapolations are routinely used in criteria setting and risk assessments. These can be categorized on the basis of the starting point and desired outcomes. The process of extrapolation in ERA was addressed at an Organization for Economic Cooperation and Development (OECD) workshop (OECD 1992). Building on the efforts of the OECD workshop, extrapolations can be broadly divided into 4 types: range, matrix and media, spatial–temporal, and ecological data extrapolations (summarized in Table 1.4). These extrapolation categories comprise the basis for the chapters in this book and are summarized here to introduce them in concept.

1.5.1 RANGE EXTRAPOLATION

In range extrapolations, responses for the same endpoint are inferred outside the range of the data from which the model was derived. These are most commonly used to calculate low-effect concentrations such as the LC10, LC25, or "benchmark effect" doses or concentrations from the dose–response line (SETAC 1994), or in the case of human health protection to estimate low-risk exposures such as the 10^{-6} risk of tumor production (USEPA 1995a; see Figure 1.4). In the context of acute responses, the model used for extrapolation (the log dose–probit effect; Finney 1971) is well tested and widely used. However, the possibility of stimulatory or hormetic

TABLE 1.4
Summary of types of extrapolation used in ecological risk assessment

Type	Explanation	Influenced by
Range extrapolation		
Extrapolation beyond the observed data	Responses are inferred or estimated outside the range of the observed data from which the relationship was derived. An example is the estimation of concentrations causing effects below those observed at the lowest response concentration used in a bioassay.	What most influences this is the model chosen to describe the relationship and the model for the extrapolation. Biological thresholds and low-exposure hormesis may affect the outcome of the extrapolation.
Matrix and media extrapolation		
Extrapolation between media	Responses in one medium are inferred from those observed in another, such as between fresh- and saltwater, sediments of different organic matter composition, and waters of different hardness.	Most commonly influenced by chemical and physical interactions between the toxicant and components of the medium. Complexation of metals or binding of organic substances can influence exposure of the organism. Can also be mediated by interactions between the matrix and the physiology and biochemistry of the organism, such as in competition for transport sites.
Extrapolation of exposure regimens		
Extrapolation between exposure regimes	Responses to chronic exposures are inferred from those observed in acute exposures, pulsed from continuous exposures, or from one route of exposure from another, such as oral versus matrix or dermal versus oral.	Time of exposure, pharmacokinetics, and physiological and biochemical processes are the primary determinants of differences in body burden, target tissue concentration, and responses to different exposure regimens. Another important issue is the nature of the endpoint. This is usually different for acute and chronic testing. In this extrapolation, an assumption is made that potency under one condition of exposure is proportional to that under the other condition.
Data extrapolation		
Extrapolation of toxicity from basic physical and chemical properties ([Q]SAR)	Prediction of toxicity from knowledge of the properties of substances such as the octanol-water partition coefficient (K_{OW}) that determine toxicity through basic mechanisms such as narcosis.	The presence of specific receptor-mediated mechanisms such as are common in pesticides. Amount of data available for training sets.

TABLE 1.4 (CONTINUED)
Summary of types of extrapolation used in ecological risk assessment

Type	Explanation	Influenced by
Data extrapolation		
Extrapolation across age and/or developmental stage	Responses in one life stage are inferred from those in another, such as from juvenile to adult, or vice versa, or from one sex to another. This extrapolation is often subsumed by the use of the effect data for the most sensitive life stage.	Cyclical activity such as reproduction and molting may make organisms more sensitive at certain times of the year or life cycle than at others. This variation may be restricted to classes of substances such as endocrine modulators or developmental toxins. The assumption that juvenile stages are more sensitive is not always true because of differences in pharmacokinetics and the staged development of receptors during development.
Extrapolation between species	Response in one species is inferred from that in another. This extrapolation is often subsumed through the use of effect data from the most sensitive surrogate organisms used in laboratory studies to those organisms that occur in the environment. The use of species sensitivity distributions (SSDs) derived from LC50s may be used to extrapolate to LC50s for untested organisms.	Influenced by uncertainty factors that vary according to the measures of effect and the value of the organisms being protected. The use of uncertainty factors is generally believed to be protective because of uncertainty factors and the maximal exposures that occur in most laboratory tests add additional conservation.
Extrapolation between levels of organization	Responses in one system are inferred from those in other systems. Examples include extrapolation from bioindicators at the cellular or physiological level to organisms, from organisms to populations, and from populations to ecosystems.	Requires the use of well-calibrated, specific, and consistent bioindicators when extrapolating from biochemical or physiological responses to the organism. Extrapolation from the organism to the population is usually by way of models or empirical data. These may be affected by uncontrolled or unknown environmental factors or by incorrect parameterization.
Extrapolation from species to communities	Responses in communities are inferred from responses in tests conducted with several to many species, such as in the use of SSDs of effect measures, such as no-observed-effect concentrations (NOECs) to extrapolate from laboratory data to communities.	Size of the data set and types of organisms tested can influence the representativeness of the laboratory data and the model used to characterize the data. Incorrect combinations of species may confound extrapolation of stressors with specificity of action such as pesticides.

(Continued)

TABLE 1.4 (CONTINUED)
Summary of types of extrapolation used in ecological risk assessment

Type	Explanation	Influenced by
Data extrapolation		
Intersubstance extrapolations	Responses to one substance are inferred from responses measured in another, such as extrapolations from one substance to another of the same mechanism of action through the use of laboratory data (potency) and/or field observation data.	Available data from field and/or laboratory studies must be sufficient to reduce uncertainty. Most applicable to well-studied substances with a similar toxic mode of action.
Temporal extrapolations	The temporal variation in response of a population or community is inferred from knowledge of differences in sensitivities and the ability of populations and communities to adapt to or recover from the stress.	Differences in sensitivity resulting from seasonal differences in physiology or biology or in different stages of development of the organism. Resilience of the organisms, populations, or function toward stress and the recovery potential of the organisms or function in relation to reproductive potential and life-cycle characteristics.
Spatial extrapolations	The effects of a particular stressor are inferred to different sizes and types of systems in one region from knowledge of effects in another region.	Spatial variability, temporal variability, biological variability, the presence of gradients, and multiple stressors. Spatial and temporal extrapolations are at the highest tier of extrapolation and build upon the other methods of extrapolation.

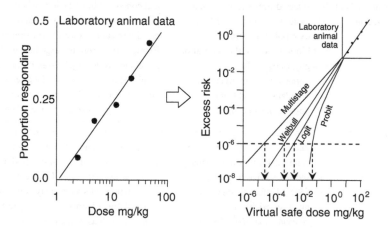

FIGURE 1.4 Illustration of range extrapolation from laboratory animal data (left) to potential responses in humans (right) and the influence of the extrapolation model on the choice of the virtual safe dose.

effects (Calabrese and Baldwin 2001) at small exposures and of nonlinearity or thresholds in the response must be considered.

Thresholds of biological response are also important to consider, especially where the substance occurs naturally and background concentrations are present in the matrix. The choice of the extrapolation model is also relevant, as is illustrated in human health risk assessments (Figure 1.4). Here the most conservative mathematical model is the multistage model, whereas the probit model gives the least conservative estimate of low-risk exposures. None of these models consider threshold responses. Thresholds of effect can be observed in many toxicity studies; however, it could be argued that they are only apparent because of lack of statistical power (e.g., number of observations) to detect subtle effects (Hanson et al. 2003a). Where mechanisms of action are known and these can be shown to act through processes that have demonstrable thresholds, this has been considered in the extrapolation process. An example of this is the carcinogenicity of saccharin in the bladders of male rats. The mechanism is through the formation of tissue-irritating saccharin crystals in the bladder, a phenomenon that only occurs at doses that are great enough to result in the formation of precipitates in the bladder. Hence there is essentially zero risk at exposures below those that result in the formation of precipitates, a thresholded process itself (Cohen et al. 1995).

An approach to estimating thresholds of effect for multiple endpoints within a species has been proposed in which distributions of effect measures from different assay endpoints in a species can be used to extrapolate to a no-effect measure for all possible endpoints (Hanson and Solomon 2002). In practice, a distribution of effect measures is constructed and extrapolated to a low probability (Figure 1.5). This value is used as an estimate of the toxicological benchmark concentration (TBC), below which no effect

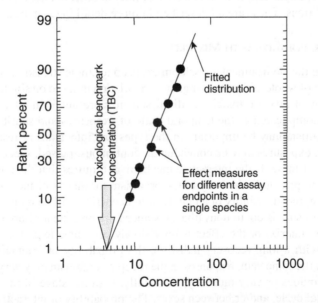

FIGURE 1.5 Illustration of the method for determining a toxicological benchmark concentration (TBC). *Note:* A distribution of endpoints for a species is used to extrapolate to a TBC, below which the likelihood of unmeasured responses being observed is very small.

at the response level used (EC50, EC10, etc.) would be expected, regardless of the endpoint. This approach is similar to that used in developing a threshold of regulation for food additives (Rulis 1996; Rulis and Tarantino 1996; Munro et al. 1999) and has been applied to effect measures for dichloroacetic acid to estimate a practical threshold of response for risk assessment purposes (Hanson et al. 2003b). Like many extrapolation methods, this approach could not predict responses for a novel, unsuspected, and untested effect. Also, many assay endpoints such as length and weight are correlated, and this may affect the characteristics of the distribution.

1.5.2 MATRIX AND MEDIA EXTRAPOLATION

Extrapolations between different media are based on physical and chemical interactions between components of the matrix and the toxic substance and form the basis for Chapter 2 in this book. These interactions may enhance or reduce the biological availability of the substance, thus affecting its apparent toxicity because of the change in exposure (Hamelink et al. 1994). One example of this type of extrapolation is the relationship between the toxicity of substances in fresh- and saltwater environments (Hall and Anderson 1995; Leung et al. 2002; De Zwart 2002; Wheeler et al. 2002a). Euryhaline species are usually more tolerant at isosmotic salinities with increases in toxicity at lower or higher salinity (Hall and Anderson 1995). Whether this response is a double-stressor issue or is due to interactions that change bioavailability is not certain. This is discussed in more detail in Chapter 6. Other examples of matrix and media extrapolation are those related to water hardness, the presence of organic matter, and the formation of insoluble complexes of metals in sediments and soils. Responses may be modified by a number of other abiotic factors relating to the test conditions. These are all described in more detail in Chapter 2.

1.5.3 EXTRAPOLATION WITH MIXTURES

Organisms in the environment are seldom exposed to single stressors; however, with the exception of whole-effluent testing, most regulation is based on single substances. Extrapolation of mixture toxicity is discussed in more detail in Chapter 5 and is made more complicated as the temporal scale of exposures and sensitivities of the receptor organism may be important in the types and intensities of responses seen. For example, exposures may be coincidental, continuous, pulsed, successive, or any combination of these. If the interactive mechanism requires that the components of the mixture be present in the body of the organism (or matrix of the community) at the same time, the probability of an interaction occurring will be reduced. However, where the exposure is continuous (the substances are persistent or are continuously present in the matrix) or the effect is only slowly reversible (e.g., the inhibition of an enzyme with a long recovery half-life), the probability of interactions will be increased. In the same vein, responses in the receptor organism may vary. Sensitivity may be continuous or vary diurnally, seasonally, with the stage in the life cycle or developmental cycle, and/or between sexes. The probability of interactive responses will be smaller when exposures are to nonpersistent substances than when exposure is continuous or the substance is persistent.

Several models are used for extrapolation of mixture toxicity (Chapter 5). These include those of potency (concentration) additivity for substances with the same mode of action (Lloyd 1961; Sprague and Ramsay 1965; Sprague 1970; Sprague and Logan 1979; Bolt and Mumtaz 1996; Könemann and Pieters 1996) that have been applied to dioxins in mammals (Calamari and Marchetti 1973; Giesy and Graney 1989; Ahlborg et al. 1994; Safe 1998, 1990; Birnbaum 1999), dioxins in fish and (Parrott et al. 1995; Tillitt 1999), and polycyclic aromatic hydrocarbons (PAHs) for environmental effects (Schwarz et al. 1995), or that possess carcinogenicity (Collins et al. 1998). For mixtures of substances that have a different mode of action, response or effect addition is an option (Klepper and van de Meent 1997), requiring estimation of the covariation of sensitivities among species for the compounds in the mixture. Because no threshold exists for concentration–addition, the concept can be applied to small environmental exposure levels. This is important, as ecosystems most often experience mixtures of chemicals at low concentrations, excluding spills and other major contamination events (see Chapter 5).

Pharmacologically based pharmacokinetic (PBPK) models have also been incorporated into the assessment of responses to mixtures of some substances where interactions may occur. This was suggested for application in assessing human health risks of contaminants in drinking water (Krishnan et al. 1997) on the basis that most interactions between organic substances occur as a result of induction or inhibition of metabolism. This has been applied in predicting the maximal likely interaction in mixtures of chlorinated and nonchlorinated hydrocarbons (Haddad et al. 2000) and aromatic petroleum hydrocarbons (Haddad et al. 1999). Metabolic and other data are required for PBPK models, and, as these may only be available for a few organisms in the environment, they are not yet widely used for extrapolation.

Responses to simple mixtures can be tested in physical experiments, but this is not possible for complex mixtures with a large number of components. Experiments with complex mixtures are based on factorial designs where responses to each of the substances in a mixture are studied at a range of exposure concentrations. Therefore, factorial designs involve many different combinations and a large number of separate toxicity tests. The feasibility of factorial designs rapidly decreases as the number of chemicals increases, because there is a requirement for more combinations of exposures, thus increasing the assumption that the model fits all the response data. For example, a study involving 3 chemicals tested each at 5 levels results in a 5^3 factorial design requiring 125 total treatment groups. This is clearly impractical if each treatment group were to be evaluated even once, let alone replicated. Box et al. (1978) introduced the fractional factorial design to overcome limitations with the factorial design methods. Fractional factorial designs identify most of the interactions occurring between the compounds and determine which compounds in the mixture cause the effects, yet maintain a manageable number of test groups. Examples of fractional factorial designs can be found in the literature (Box et al. 1978; Groten et al. 1996, 1997). Approaches that use combinations of upper centiles of exposure concentrations in laboratory or field experiments as physical models to characterize mixture extrapolation may be helpful in determining thresholds of response for mixtures.

1.5.4 EXTRAPOLATION OF EXPOSURE REGIMES

Extrapolation used to infer toxicity from one type of exposure regimen to another is often termed "temporal extrapolation." The most common of these extrapolations is that from acute to chronic exposures, but the issue of pulsed versus continuous exposure is also important in assessing possible effects in real-world environmental settings. These extrapolations may involve the use of modified tests with standard species or whole-model ecosystems to simulate realistic exposures such as those of variable duration or those of pulsed exposure for compounds that rapidly dissipate in the environment. In many cases, these involve alterations in exposure route and intensity, both of which can have significant impacts on the toxic responses. Extrapolation from acute responses to NOECs or chronic responses is particularly important as chronic tests are more costly and time-consuming than acute tests. Methods for accurate and precise acute-to-chronic extrapolations have been developed and are available as computer programs such as ACE (Mayer et al. 1999, 2001; De Zwart 2002; Ellersieck et al. 2003) and are discussed in Chapter 6.

1.5.5 ECOLOGICAL DATA EXTRAPOLATIONS

In their simplest form, these are extrapolations between and among different taxa and life stages. The standard single-species (standard protocol) tests are physical models that can be applied to other species with the use of uncertainty factors or assumptions based on the test organism being the most sensitive species. These extrapolations are usually conducted through the use of simple uncertainty factors. Several levels of refinement are possible above this initial level.

1.5.5.1 Quantitative Structure-Activity Relationships ([Q]SARs)

More than 95% of all existing chemicals lack the most basic acute toxicity data for representative ecotoxicological species. This lack of knowledge has drawn attention to the need to develop more pragmatic approaches for estimating toxicity data from chemical and physical parameters of substances. Estimation of toxicity on the basis of (quantitative) structure-activity relationships ([Q]SARs) for a number of representative species has been suggested as an initial alternative to bioassays, especially if the mode or modes of action of the compounds are known (van Leeuwen et al. 1991).

Baseline toxicity (narcosis) is the basis for most (Q)SARs because it is the minimal toxicity of any given substance (Lipnick 1993). The basic cellular structures, functions, and membranes are highly conserved and similar in all biological systems. Therefore, substances that target these systems in the cell are likely to display similar potency in all living systems. Narcosis is believed to be a result of a nonspecific disturbance of membrane integrity and function as a result of the partitioning of xenobiotics into biological membranes. In narcosis, variation in expressed toxicity is controlled primarily by partitioning behavior (i.e., the toxicokinetic processes that control the rate of accumulation and concentration in the membrane). For example, in toxicity tests with fish, this hydrophobic influence can produce LC50 estimates that vary by as much as 6 orders of magnitude, whereas the molar concentration in affected organisms is the same (McCarty et al. 1992).

Most of the ecotoxicological (Q)SAR models are based on the identification of substructures of the molecule, the degree to which they contribute to the physical properties of the molecule, and how they may contribute to the observed activity. Depending on the sophistication, the model will initially attempt to identify possible linear relationships between the octanol–water partition coefficient (K_{OW}) and observed toxicity, when the toxicity of an active compound can be explained by its lipophilicity (narcosis effect). (Q)SAR models for narcosis toxicity based on K_{OW} are available for many endpoints and species, have been applied to preliminary screening of substances for which no data are available, and are discussed in more detail in Chapter 3.

1.5.5.2 Extrapolation across Age and Developmental Stage

Most organisms show differences in response to substances as they pass through varying stages of sensitivity. Commonly, younger organisms are more sensitive than older organisms because they have a greater surface area–volume ratio, allowing the substance to bioconcentrate or equilibrate more rapidly; they may have reduced detoxification potential; or they are in the process of growth and development, where chemical signaling is important. However, this is not always the case. Amphibians are often more sensitive to chemical stressors later in development. Extrapolation across age and stage is discussed in more detail in Chapter 6.

1.5.5.3 Extrapolation between Levels of Biological Organization

Extrapolating between biological scales, such as from physiological and biochemical responses in single organisms (bioindicators; Huggett et al. 1992) to responses at the scale of populations and communities, involves consideration of both temporal and spatial issues. This is a source of uncertainty as well as misinterpretation, such as when a physiological or biochemical adaptation to low exposures is confused with adverse effects. Some bioindicators such as mixed-function oxidase (MFO) activity may be good indicators of exposure but not necessarily good predictors of adverse effects (Munkittrick et al. 1992). A short-term biomarker based on the relationship between oxygen consumption and energy stored in glycogen, protein, and lipid — cellular energy allocation (CEA) — has been shown to correlate well with long-term endpoints such as the 21-day lowest-observed-adverse-effect concentrations (LOAEC) and population-level effects in *Daphnia magna* (De Coen and Janssen 2003). Other biomarkers, such as inhibition of acetylcholinesterase (AChE), can be useful diagnostic indicators of both exposure and effects (death) in, for example, fish, where brain AChE activity is a sensitive biomarker and the correlation between brain AChE activity and mortality is considered to be a valuable diagnostic tool (Murty 1986). AChE has similar usefulness as a biomarker in birds (Mineau 1991). The use of bioindicators in extrapolation is discussed in more detail in Chapter 4.

The extrapolation of laboratory single-species responses to the population level is another ecologically relevant extrapolation that is routinely conducted without extensive confirmation of its appropriateness. Actual population studies in the environment and their interpretation through life tables and other approaches are alternatives, but these are resource- and time-intensive and not feasible for all organisms. However, population models also may be useful in this context. Organisms have evolved several types of life histories and may be "r" (producing large numbers of young with little parental care) or "k" (fewer young and more parental care) reproductive strategists (Stearns 1992). As a result, one species or class of organisms may be more or less able to recover rapidly from population reductions, and these recoveries may be dependent on the life-cycle stage affected. Knowledge of these can be used to identify species- and stage-specific sensitivity to extinction, such as has been done in frogs (Biek et al. 2002). Use of population models in extrapolation is discussed in more detail in Chapter 4.

Current procedures of higher tier risk assessment are often based on the extrapolation of responses observed in relatively simple and short-term (weeks) cosm tests to structurally more complex ecosystems in the field. The predictive value of studies in small cosms (microcosms), however, depends on factors such as fate and exposure of the stressor and the sensitivity and recovery potential of the populations present. The role of cosm studies in extrapolation is discussed in more detail in Chapter 4.

1.5.5.4 Extrapolation from One Substance to Another with a Similar Mode of Action

It is appropriate to assume that chemicals with a similar toxic mode of action have a more or less similar impact on the same type of ecosystem, at least when evaluating similar exposure regimes on the basis of toxic units. Evidence for this is provided by review papers that evaluated the impact of, for example, photosynthesis-inhibiting herbicides and organophosphorus insecticides in semifield tests (Brock et al. 2000a, 2000b). When this assumption is generally true, the existing data sets of higher tier tests with chemicals may be used to assess the potential ecological risks of unassessed chemicals, at least when a basic ecotoxicological data set of the substance is available (e.g., toxicity tests with standard species). In a statistical analysis of single-species aquatic toxicity data, De Zwart (2002) also identified regularities that enable the prediction of ecotoxicological effects for substances with common modes of action. For pesticides, the expert-based model PERPEST has been developed (van den Brink et al. 2002c). PERPEST is an effect model that predicts the effects of a certain pesticide on various community endpoints from empirical data. The model is based on literature review (Brock et al. 2000a, 2000b) of freshwater model ecosystem studies with insecticides and herbicides performed to assess the $NOEC_{ecosystem}$ or $NOEC_{community}$ for individual compounds and to evaluate the ecological consequences of exceeding these standards (see Section 1.4.2.2 and Chapter 4 for more details of the effect categories used to derive these NOECs). Based on the relevant properties of the compound, concentration, and type of ecosystem to be evaluated, PERPEST searches for analogous situations in the database and calculates a prediction using weighted averaging of the effects reported in the most relevant literature references. PERPEST results in a prediction showing the probability of effects on the various groups (van den Brink et al. 2002c).

1.5.6 TEMPORAL AND SPATIAL EXTRAPOLATIONS

Both temporal and spatial extrapolations are important in ecotoxicology when it is necessary to consider variation in sensitivity of individuals during stages of development, or periods of greater sensitivity during specific life-cycle stages, such as reproduction (Tillitt et al. 1998; International Programme on Chemical Safety [IPCS] 2002). Synchronous activities like reproduction or metamorphosis, such as in frogs and the larval and pupal stages of aquatic insects, present particular problems in temporal extrapolation as organisms pass through varying stages of sensitivity. For stressors that act on metamorphosis and reproduction and "are pulsed or present only sporadically" in the environment, extrapolation is difficult without specific data on timing of the sensitive stage and the exposures. Similar needs apply to the seasonal and successional variation in the sensitivity of communities to stressors, particularly in temperate regions where seasonal fluctuations in climate are large and where dormant stages of organisms may be more resistant to stressors occurring in winter.

Adaptation to chemical stress may occur by 2 major mechanisms. Within generations, organisms may adapt to stressors through the induction of repair mechanisms or processes that result in the detoxification of the stressor or its sequestration into less harmful products. This is variously characterized as resiliency or induced tolerance. Examples in animals include the induction of mixed function oxidase (MFO) enzymes responsible for detoxification of organic substances (Munkittrick et al. 1992; Hodson et al. 1997) as well as the induction of metallothionine proteins in plants and animals in response to exposure to metals (Hickey et al. 1995a). Adaptation between generations may result in the selection of more tolerant individuals. This phenomenon has significant implications for the use of pesticides where many species of plants and animals have evolved resistance to these substances to the point that they can no longer be controlled. These forms of adaptation are discussed in more detail in Chapter 6.

Recovery from the effects of chemical stressors is a common occurrence because, over time, substances dissipate from the exposed environment. In cases of slow dissipation, recovery may take considerable time; thus, recovery depends on the nature of the stressor as well as the sources of colonizing organisms and habitat-conditioning species (Ellis 1989). In some cases, recovery may not be to the previous condition but to a different norm or stable state (Scheffer et al. 2003). For biological stressors such as introduced or "alien" species (excluded from this book, but of great ecological significance), recovery to the original state of the ecosystem is less likely to occur than recovery to a new norm (Andersen et al. 2004). The implications of recovery in the extrapolation of ecological effects of stressors are discussed in more detail in Chapter 6.

Spatial extrapolations are particularly important in multiregional and global issues in ecotoxicology. Increasingly, harmonization of national regulatory standards is occurring in the European Union (EU); the North American Free Trade Agreement (NAFTA); the Australian and New Zealand Environment Conservation Council (ANZECC); the Agriculture and Resource Management Council of Australia and New Zealand (ARMCANZ); and international agencies such as the World Health Organization (WHO), the Food and Agricultural Organization (FAO), the OECD, and the European Water Framework Directive. With this globalization, it has become increasingly important to extrapolate toxicity data across large spatial distances, most importantly from one climatic zone to others. As will be discussed in Chapter 7, once the matrix effects are considered, such extrapolations are relatively uncomplicated. Extrapolations from one ecosystem to another across landscapes or from one watershed to another are more complex because of spatial and temporal heterogeneity in the distribution of both the stressor and the more susceptible organisms. These extrapolations are also influenced by the presence of other stressors, be these toxicological (mixtures; see Chapter 5), biological, or the result of physical changes to habitat that result in habitat fragmentation.

1.6 REGULATORY BACKGROUND

The regulatory background to extrapolation follows a hierarchy that is dependent on the amount of data available. In general, extrapolations in low-data situations make use of simple uncertainty factors that are believed to be conservative. As the amount of data increases, uncertainty factors may be reduced or replaced by data-derived statistical measures of uncertainty. This hierarchy is illustrated in the approaches to extrapolation for different substances and situations.

1.6.1 CRITERIA SETTING AND PREDICTIVE RISK ASSESSMENT

1.6.1.1 Generic Chemicals

A large number of organic and inorganic substances are used in commerce in all industrialized countries. In general, we know very little of their ecotoxicology and even their human health significance outside the workplace. This has been recognized in a number of regulatory instruments such as the Toxic Substances Control Act (TSCA) in the United States, the Canadian Environmental Protection Act (CEPA), and the REACH program in Europe (see Chapter 3). There is currently significant effort to assemble basic Tier-1 risk assessment data for many of these substances. In Canada, this activity falls under CEPA, and the screening process to characterize some 23 000 substances for the Priority Substances List (PSL) has recently been completed. Because the effect data for generic chemicals are generally derived from limited data sets or from (Q)SAR models, extrapolation is usually from HQs using UFs (Table 1.1). Substances that trigger concerns in this Tier-1 screening are subjected to a more detailed risk assessment, which may require more data to be supplied.

1.6.1.2 Pesticides

Pesticides were the first substances to be subjected to environmental risk assessment for regulatory purposes. Their use and availability require that they must be registered under regulatory instruments such as the Federal Insecticide, Fungicide, and Rodenticide Control Act (FIFRA) in the United States; the Pest Control Products Act (PCP Act) in Canada; Council Directive 91/414 in Europe; and similar instruments in many other jurisdictions. Registration requires that pesticides are subjected to risk assessment, which requires minimal data sets for environmental properties. These data include toxicity to nontarget aquatic and terrestrial organisms as well as information on fate characteristics and environmental concentrations. Extrapolations are used for both toxicity data and environmental fate data. Risk assessments for pesticides are tiered and different types of extrapolations are used. Tier-1 normally uses hazard quotients for the most sensitive species, with UFs to extrapolate to nontarget species in the environment. More refined levels of risk assessment may make use of smaller UFs where more data are available (Table 1.1) or use probabilistic methods for extrapolation of effects data (ECOFRAM 1999; Hart 2001).

Extrapolations to predict environmental concentrations are also used in pesticide risk assessment. There are several processes where extrapolation models are used, which may involve simple or very complex calculations. These are spray drift and estimations of concentrations in water, soil, and food items. For spray drift, computer models such as AgDrift (Teske and Scott 2000) are used in the United States and other jurisdictions, but tables based on empirical measurements such as the BBA drift tables (Ganzelmeier et al. 1995; Commission of the European Communities 2000) are also used. The drift tables are probabilistically derived and make use of the 95th centile from measured data to extrapolate to a reasonable worst-case deposition value. Concentrations in water may be estimated from simple extrapolations related to deposition and depth of the water, although the assumptions of depth vary from 2 m in the United States (SETAC 1994) through 30 cm in Europe (Riley 1993) to 15 cm in a Canadian forest pool or wetland. More realistic scenarios have recently been developed for extrapolating to various pesticide use patterns for pesticides. For example, in the EU, several scenarios have been developed to estimate exposures to pesticides from drift and runoff (FOCUS 2001; Linders et al. 2002). Models that estimate inputs by integrating drift, runoff, and leaching, such as the GENEEC model, may provide single deterministic concentrations for use in lower tier assessments (Parker 1999), or models such as the MUSCRAT model provide large probabilistic data sets that allow extrapolations based on many years of rainfall, environmental, and physicochemical data (ECOFRAM 1999).

1.6.2 RETROSPECTIVE SITE-SPECIFIC RISK ASSESSMENT

Site-specific risk assessments are directed to a specific question at a specific location and, in general, make less use of extrapolation of effects but may use media extrapolations to address some of the site-specific abiotic factors. In terms of exposures, measurements are usuelly available for these situations and the types of organisms present are often well known, thus reducing the need to extrapolate to responses in unknown species. In addition, biomarkers of exposures and/or effects may be used to reduce the need for extrapolations over great taxonomic distances. Examples of site-specific assessments include studies on large reservoirs (Jones et al. 1999; Cook et al. 1999; Suter et al. 1999b), the use of environmental effects monitoring (EEM) in assessing the effects of pulp mill effluents on fish in receiving waters (Mayer et al. 1988; Munkittrick et al. 1994, 1998; Parrott et al. 1999), and the assessment of contaminated soils (Schouten et al. 2003) and sediments (Chapman 1986, 1996, 2007; Grapentine et al. 2002; Chapman et al. 2002).

1.7 DEFINITIONS AND CHARACTERIZATION OF UNCERTAINTY

Uncertainty is inherent in all sciences and in all measurements (Popper 1979). Without uncertainty, extrapolation would be unnecessary. Differences in types and levels of uncertainty have a significant influence on extrapolation and on all risk decisions that may result from extrapolation. Uncertainty has 3 sources: uncertainty from lack of knowledge, systematic uncertainty, and stochastic uncertainty. These sources of uncertainty apply to all types of scientific models, whether they are empirical, mathematical, or conceptual. The aim of this section is to provide a common framework for understanding the types of uncertainties discussed in the other chapters of this book.

1.7.1 KNOWLEDGE UNCERTAINTY

Uncertainty from imperfect knowledge or ignorance of things that could be known can be important in extrapolation, especially as new substances or processes become widely used and adopted. Examples of this are the lack of prior knowledge that DDT would biomagnify in the food chain, endocrine disruption as a mechanism for ecological responses, and how to consistently extrapolate from an LC50 to a population-level effect. Uncertainty from lack of knowledge can never be addressed to the complete satisfaction of everyone. This is the major reason for strong interest in the precautionary principle. However, the more data that are available, the less likely that errors from lack of knowledge will occur. Knowledge uncertainty is thus the major driver for the collection of more data; however, this data collection is for more diverse knowledge, not that associated with replication to characterize variability. This is acknowledged in the requirement of a minimal data set for many regulatory risk assessments, such as in the derivation of water quality criteria (USEPA 1995b) and the reduction in uncertainty factors applied to a hazard quotient when more toxicity data are provided (Table 1.1).

From a practical point of view, knowledge uncertainty in extrapolation is addressed in several ways. The collection of more data is the most common approach, but this can lead to a never-ending and obsessive search for all of the information, resulting in the waste of resources and the commission of type 3 errors — when the wrong question is asked but is answered very well (Raiffa 1982). Pragmatic approaches to reduce knowledge uncertainty in extrapolation and risk assessment have been used. These usually rely on using deductive reasoning along a number of pathways, such as the guidelines for causality developed for the identification of causes of disease by Robert Koch and Austin Bradford Hill and, more recently, for the identification of endocrine-disrupting substances (Koch 1942; Hill 1965; Bro-Rasmussen and Lokke 1984; Fox 1991; Suter et al. 1993; Ankley et al. 1997; IPCS 2002) and the use of multiple lines of evidence in extrapolating from laboratory and experimental data to the ecosystem in ERA (Hall and Giddings 2000; Grapentine et al. 2002; Schouten et al. 2003). Thus, information on mechanisms of action and observations from other locations or from analogous stressors may be used deductively to support extrapolations for the purposes of risk assessment.

It is commonly believed that knowledge uncertainty leads to underestimation of risks. This may not be the case. For example, lack of knowledge of compensatory mechanisms such as the induction of tolerance in an organism, density dependence of populations of a species, and redundancy of function in ecosystems will all tend to mitigate the severity of impact of a stressor. Similarly, co-occurring stressors may act additively, synergistically, or antagonistically.

1.7.2 SYSTEMATIC UNCERTAINTY

Systematic errors in the risk assessment process are those that may occur through both model and methodological errors (Harremoes 2002). These errors include systematic bias; the use of an incorrect formula or algorithm; and computational mistakes such as incorrect position of the decimal, consistent data entry errors, or error occurring through incorrect instrumental calibration. Provided that systematic errors can be identified, they can be addressed through experimental design, quality control, and quality assurance, or, in the case of analytical errors, by use of a correction factor. Errors of this type result from incorrect sampling where nonrepresentative samples are taken; errors in analysis, such as a lack of correction for recovery of the analytical method; or dependence of errors on concentration, such as in enzyme-linked immunosorbent assay (ELISA) procedures (Scribner et al. 1994). Bias in the selection of species or in assessing endpoints of toxicity tests can cause systematic errors (Solomon et al. 1996). In the calculation of NOEC and LOEC, too widely spaced concentrations may result in incorrect extrapolation to thresholds of effect such as the maximum allowable toxicant concentration (MATC). In the use of species sensitivity distributions (SSDs), a bias in the selection of organisms for inclusion in the toxicity distribution may result in systematic errors. For example, in the distributional analysis of atrazine susceptibility data (Solomon et al. 1996; Giddings et al. 2005), it is apparent that the susceptibility of some groups of organisms is different from that of others. If the toxicity of atrazine to fish were extrapolated in the absence of other data, toxicity to more susceptible organisms such as plants would not be correctly predicted.

1.7.3 STOCHASTIC UNCERTAINTY

Stochastic errors are nonsystematic errors that result from the random nature of the system being assessed and are commonly referred to as "variability." Ecological and environmental data are realizations of stochastic and chaotic processes, as the environment from which data are collected is ever changing. Properties of ecological or environmental systems may subsequently be found that were not detected when the system was first observed or modeled. Changes in the forcing functions of the system, such as those resulting from variations in climate, salinity, oxygen, and so on, may result in new variables having an important influence. These factors combine to ensure that the implicit assumption of stasis necessary for predictive models of extrapolation, whether based on empirical observation or on mathematical models, can never be completely realized for ecological processes (Shelly et al. 2000). These types of errors can be described and quantified but cannot be avoided or corrected for.

1.7.4 PRECISION AND ACCURACY

There are differences in the understanding and interpretation of precision and accuracy between measured values and in models.

1.7.4.1 Precision and Accuracy of Measurements

Precision is the closeness of repeated estimates, observations, or measurements to each other. Accuracy is the combination of small systematic and stochastic errors and reflects the closeness of a measured or observed value to the true value (Jessen 1978). Great precision may imply, misleadingly, that the results are also accurate and, in the absence of knowledge of accuracy, may give a false sense of certainty in an extrapolation and result in an incorrect risk decision. Precision may be increased through a larger number of observations and the use of the mean or median as the estimate of the parameter or observation. However, narrow confidence intervals on the estimate of the mean may incorrectly imply greater accuracy, especially if the uncertainty results from systematic errors of measurement rather than variance in the measured variable. For this reason, protocols for sampling of environmental concentrations include blank samples that are analyzed to assess the potential for contamination and field-spiked samples and internal standards that are used to characterize systematic errors through loss of analyte during transport, storage, sample preparation, and analysis. In some cases, the number of blanks and field spikes may exceed the number of samples for environmental measurement.

1.7.4.2 Precision and Accuracy of Models

Models may be both mathematical and empirical. Mathematical models such as exposure analysis modeling systems (EXAMS; Burns 1997), toxic substances in surface waters (TOXSWA; Adriaanse 1996), or pesticide root zone model (PRZM; Mullins et al. 1993) are well recognized, but all laboratory or field experiments are also models — empirical representations of reality. All models are more or less accurate conceptual portrayals of reality, but some are very useful and necessary to decision making, especially if we understand their limitations and how to interpret them. A model is only as good as its parts or inputs, and it can only describe what we model it to do; if the data used in the model are 15% wrong, the optimal model output will, at best, also be 15% wrong. Usually, the results derived from a model need to be extrapolated to answer the regulatory question that prompted the model (Boesten 2000). This requires an understanding of the uncertainty in the model, which, in turn, increases as the complexity of the model increases.

There is a need to differentiate and characterize types of uncertainties associated with ecological models and extrapolation of results outside the limited domain of the model for regulatory purposes. The first distinction is the difference between model precision and model accuracy of the output. Precision is a quantitative feature we can assess or analyze by various statistical tests and methods. An example is given in probabilistic risk assessment and the use of SSDs to integrate all available information into a risk assessment to provide a more flexible decision-making tool rather than relying on single deterministic numbers in HQs with added arbitrary uncertainty factors. Accuracy in mathematical models is dependent on the correctness of the algorithm and the mathematical processing of the data. For example, use of an incorrect conversion factor can propagate a systematic error throughout the model, and all the output will be consistently wrong. This is particularly a concern when using Monte Carlo models, where the amounts of output data are large and may result in the perception of great accuracy as well as precision. This source of error in models is recognized and is one of the reasons why source codes for models must be available for quality assurance. Modelers often compare the results of models to measurements as a method of assessing accuracy. However, the measurements themselves may be systematically wrong, or the wrong data are selected for the verification or calibration of models (van den Brink et al. 2002a). Worse still, modelers may use calibration factors to adjust output to match one measure of reality without considering all the other unmeasured realities.

1.7.5 ANALYSIS OF UNCERTAINTY

Quantitative model assessment techniques can be broken down into "uncertainty analysis," defined as the process by which parameter uncertainty in measurements or a model is described and quantified, and "sensitivity analysis," by which the consequences of uncertainty are explored. Both accuracy and precision can be increased through the use of quality assurance and quality control (QA/QC; USEPA 2000a, 2003). Depending on the context, the most widely used methods of characterizing uncertainty by decision makers are statistical significance, validated mechanistic causality, and scientific peer review. These criteria include knowledge, systematic, and stochastic uncertainty. Although there is no absolute standard for model assessment, a standard must be set relative to the objectives of a particular regulatory requirement, as has been discussed by scientists involved with modeling aspects of exposure and effects assessments (Shelly et al. 2000). The USEPA uses 5 general categories for data quality assessment to evaluate the quality and relevance of information from external sources for use in regulatory decisions (USEPA 2000a, 2003):

- Soundness: The extent to which the procedures, measures, methods, or models employed to generate the information are reasonable for and consistent with the intended application and are scientifically or technically appropriate
- Applicability and utility: The extent to which the information is applicable and appropriate for the intended USEPA's use
- Clarity and completeness: The degree of clarity and completeness with which the data, assumptions, methods, quality controls, and analyses employed to generate the information are documented
- Uncertainty and variability: The extent to which the variability and uncertainty in the information or in the procedures, measures, methods, or models are evaluated and characterized
- Evaluation and review: The extent of independent application, replication, evaluation, validation, and peer review of the information or of the procedures, measures, methods, or models

1.8 CONCLUSIONS

This chapter has outlined the general principles of the use of extrapolation in criteria setting and risk assessment. Extrapolation is used in the setting of criteria, where only effects are considered, as well as in risk assessments, where both effects and exposures are considered. In formulating approaches to criteria setting and risk assessment, tiers are often used to simplify the process. In keeping with the greater availability of data in higher tiers, methods of extrapolation may become more refined. Thus, different extrapolation methods may be used at different tiers as well as in relation to the protection goals desired. Extrapolation techniques include range extrapolation, matrix and media extrapolation, spatial–temporal extrapolation, and ecological data extrapolation, and each of these areas is discussed in the chapters that follow. The methods used in extrapolations are closely linked to the consideration of uncertainty in criteria setting and risk assessment. Uncertainty may be knowledge driven, systematic, or stochastic and is relevant to both modeled and measured effects and exposures. Uncertainty may be considered in extrapolations through the use of uncertainty factors that are applied to single values in lower tiers or to more complex, higher tiers that make use of species-sensitivity distributions or cosm studies. The uncertainty factors may be based on historical precedent, based on empirical observations, or derived from statistical analysis of the data used in the extrapolation process.

2 Matrix and Media Extrapolation

Dick De Zwart, Amanda Warne-Lorscheider,
Valery Forbes, Leo Posthuma,
Willie Peijnenburg, and Dik van de Meent

CONTENTS

2.1 INTRODUCTION AND PROBLEM FORMULATION

Extrapolations between different media are common in risk assessment. Such extrapolations are based on physical and chemical interactions between components of the matrix and the toxic substance that may enhance or reduce the biological availability of the substance, thus affecting its apparent toxicity because of the change in exposure. In this chapter, the word "medium" is reserved to indicate the major environmental compartments: air, water, sediment, and soil. The word "matrix" is associated with the physicochemical properties of the media. The problems associated with extrapolating between one medium or type of matrix to another are intricate and are generally due to the varying chemical, physical, biological, and spatial characteristics associated with the different media.

The types of media extrapolations routinely required and used in risk assessments include air–water, air–soil, water–sediment, and groundwater–soil. Matrix extrapolations include saltwater–freshwater, hard water–soft water, river–lake–stream–pond, and soil type adjustments. There are, in fact, a large number of different extrapolations possible, each with its own unique problems to be taken into account.

Evaluation of the risk of a particular chemical requires information on fate and exposure and on biological sensitivity (Table 2.1). The extrapolation issues dealing with matrix and media can thus be divided into exposure issues and effects issues. The exposure to and fate of any toxicant are governed by the interaction between the matrix components and the toxic chemical. The influences of chemical and matrix properties are highly interwoven. Therefore it is often difficult to discern between exposure issues due to toxicant properties and exposure issues due to matrix properties. Differences in effects between media and matrices may be related to differences

TABLE 2.1
Key variables that need to be considered when extrapolating among and within media

Category	Variable	Fate	Exposure	Sensitivity
Chemical properties	Fugacity	√	√	
	Solubility	√	√	
	Polarity	√	√	√
	Reactivity	√		√
Medium and matrix properties	Temperature	√		√
	Light	√		√
	Redox	√		√
	pH	√	√	√
	Ions and hardness	√	√	
	Particulates (size/nature)	√	√	
	Organic carbon	√	√	
	Colloids	√	√	
	Exchange capacity	√	√	
	Complexation	√	√	
Species properties	Taxonomic status			√
	Trophic status		√	
	Habitat preference		√	√
	Behavior		√	√
	Body size		√	√
	Life-cycle type			√
	Timing of exposure in relation to life cycle		√	√
	Presence of other stressors			√
	Exposure route		√	√

Note: Columns 3 to 5 indicate major influences of the variable on the fate of the chemical, the exposure of the biota, and biological sensitivity, respectively.

in the inherent sensitivity of the exposed biota originating from communities with a different species composition, and/or to differences in the route of uptake and excretion of the chemical.

Among other determinants, the fate of a chemical is strongly related to its ability to be transported through various media to exert its toxic action elsewhere. Transport is not considered as part of this chapter. For details on transport-related matrix interactions, the reader is referred to Klecka et al. (2000).

2.1.1 TYPES OF MEDIA AND MATRICES

There are 3 basic media types: water, air, and soil. Sediments represent a special type of media. Within these 3 basic categories, there are a number of similar yet complex matrices, and these need to be considered independently.

TABLE 2.2
Water matrix factors that may modify the toxicity of substances to aquatic organisms

Property	Explanation
Suspended particulate matter	Adsorption to particulate matter will make chemicals less biologically active.
Water temperature	Dictates volatilization rate of chemical; affects the chemical activity of contaminants and the physiological processes of organisms.
Water velocity	Dictates transport and dilution of chemical.
Water viscosity	Affects water movement and solute diffusion.
Dissolved oxygen	Reduction of oxygen increases susceptibility to toxicants by a factor of 2 in aquatic organisms (Sprague 1984).
pH	Affects the form, reactivity, solubility, and toxicity of some contaminants.
Salinity	Salts may bind with chemicals, allowing them to become biologically inactive or precipitate out.
Light penetration	Light may speed up the degradation of some chemicals or create harmful metabolites that are more toxic than the original form.

2.1.1.1 Water

Surface water can be defined as any river, lake, stream, pond, marsh, or wetland; as ice and snow; and as transitional, coastal, and marine water naturally open to the atmosphere. Major matrix properties, distinguishing water types from each other, are hard and soft water, and saline and freshwater. Groundwater is typically defined as water that can be found in the saturated zone of the soil. Groundwater slowly moves from places with high elevation and pressure to places with low elevation and pressure, such as rivers and lakes. Partitioning interactions of the groundwater with the solid soil matrix is an important factor influencing the fate of toxicants. Physicochemical properties of water that may affect toxicity of chemicals in all water types are listed in Table 2.2.

2.1.1.2 Atmosphere

The atmosphere consists of 78.09% nitrogen, 20.94% oxygen, 0.93% rare gases, 0.03% carbon dioxide, 0.1% trace elements, as well as dust, water vapor, and anthropogenically emitted substances (Crosby 1998). The vapor pressure of a substance and its fugacity are primary determinants of residence time in the air. Airborne chemicals can travel long distances. During airborne transportation, they may be subject to photodegradation. At low temperatures (polar regions), persistent organic pollutants (POPs) may be transferred to water masses or snow and ice.

2.1.1.3 Soil

There are a number of different types of soils, each varying in the percentage of sand (2.0 to 0.05 mm), silt (0.05 to 0.002 mm), clay (< 0.002 mm), organic matter of different

TABLE 2.3
Soil matrix factors that may modify the toxicity of substances to soil organisms

Property	Explanation
Bulk density	Indicates the pore space available for water and roots; influenced by soil composition (mineral content, mineral type, and organic matter) and soil texture
Mineral type	Affects adsorption of the chemical
Grain size distribution	Affects the surface area where adsorption can take place
Water content	Influences partitioning and availability of chemicals
Permeability	Affects ability of a soil to transmit water or air
Structure	Dictates the porosity of the soil
pH	Affects the form, reactivity, solubility, availability, and toxicity of some contaminants
Metal content	Affects the toxicity of some substances (mainly heavy metals) with binding or antagonistic mechanisms, for example, by alkaline-earth metals and aluminum
Organic matter content, type, and % carbon	Influences soil sorption properties for heavy metals and organic chemicals
Temperature	Affects the chemical activity of contaminants and the physiological processes of organisms
Soil porosity	Influences percolation
Inorganic ions	Can bind to chemicals, rendering them inactive and affecting transportation

types, water content, earth alkaline metals, and other inorganic substances. Soil structure depends strongly on the abiotic characteristics mentioned above. Further, the composition and structure of the soil influence the species composition of the soil biota, and vice versa. Although large particles are generally inactive, smaller particles may be chemically active, complicating the extrapolation among soil types. Soil colloids (finely divided particles of one substance suspended in another) can acquire electrical and surface properties that influence adsorption of gases, ions, and organics. The matrix properties of different soils are the factors that make extrapolation between soils difficult. Soil properties that may affect the toxicity of a chemical are listed in Table 2.3.

2.1.1.4 Sediment

Sediments can be defined as deposits of solid material laid down in water bodies. Minerals usually dominate sediments, but sediments also contain organic substances, including humic substances of different compositions. Sediments may be considered the ultimate sink for hydrophobic chemicals. Although sediments have many of the same properties as soils, the high water content and anaerobic conditions that typically occur within centimeters of the sediment–water interface distinguish this matrix into its own unique category.

2.1.2 Sources of Pollution and Routes of Entry and Exposure

Anthropogenic chemicals may first enter the environment by aquaculture, agriculture, illegal and legal manufacturing processes, human use and disposal, septic or municipal sewage, and animal husbandry. Once in the environment, their fate depends upon the type and physicochemical properties of the media and of the substance itself.

2.1.2.1 Atmosphere

Volatile chemicals reach the atmosphere via direct emission to the air or by volatilization from water, soil, surfaces, and plant and animal respiration. Once in the air, diffusion, advection, and precipitation or deposition are the major sources of movement.

2.1.2.2 Surface Water

Major routes of entry of chemicals into surface waters include precipitation, drift, runoff, industrial and sewage outfalls, groundwater, and human disposal. Once in the surface waters, the chemicals may be transported via advection (bulk movement by currents), molecular diffusion (due to random thermal movement of molecules), turbulent diffusion (mixing), and dispersion. Chemicals may also be transported while adsorbed to suspended particulate matter.

2.1.2.3 Groundwater

Chemicals may enter groundwater as landfill leachates or from deep-well injection of hazardous wastes, leaching from soil and water, or septic tanks. Diffusion and advection are the typical mechanisms of chemical transport in groundwater. Groundwater may be taken up via human use or empty onto the surface waters via a natural spring.

2.1.2.4 Soil and Sediment

Soils may be exposed to toxicants by direct input of chemicals at dumpsites, due to the application of pesticides or accidental spillage, and transfer from elsewhere. Whether originally emitted to air or water, many of the persistent toxicants exhibit their toxicity in the stationary media, sediment, and soil.

2.2 MEDIA- AND MATRIX-RELATED EXPOSURE

The overall aim of this type of extrapolation is to predict exposure and/or effects from one medium or matrix to another. Although physicochemical (fate and exposure) and biological (sensitivity) extrapolations can be addressed separately, it will become clear in the following that there are many links between these elements of risk that can complicate the extrapolation process. The elements of the extrapolation process and the interactions between physicochemical and biological features that are involved are schematically depicted in Figure 2.1.

The development of extrapolation methods for all possible steps strongly relies on both theory and the quality and number of available data. Studies can be conducted

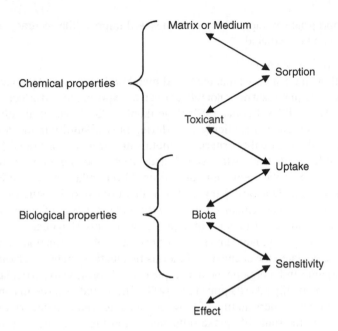

FIGURE 2.1 Schematic overview of the properties and processes that govern matrix and media extrapolations.

with a focus on developing general theories applicable to all situations, but may only be valid for the selected combination of factors studied. Thus, for example, a freshwater to saltwater extrapolation of sensitivity derived for compound A may not be applicable for compound B.

2.2.1 EXTRAPOLATION RELATED TO PHYSICOCHEMICAL PROPERTIES OF MEDIA AND TOXICANTS

2.2.1.1 Toxic Metals

Environmental quality objectives for toxic substances are derived on the basis of risk considerations, where "risk" usually has the meaning of the likelihood and/or extent of an adverse effect. It is the purpose of ecological risk assessment to distinguish between contaminated waters, soils, or sediments in which contaminants may or may not produce effects. In the case of metals, total concentrations in waters, soils, and sediments commonly span several orders of magnitude. Organisms, however, do not respond to total concentrations, and hence quality criteria that are based on total concentrations are unlikely to be predictive of adverse biological effects. The total amount of a substance may not be toxicologically meaningful, as it may partly be unavailable for uptake by organisms. This would not be important if availability were constant in all situations. This, however, is not the case and it is the variation in crucial matrix and medium properties that results in substantially different availability for the uptake of compounds by organisms. Taking this variation into account, applying appropriate extrapolation techniques will improve the accuracy with which (no) effects can be predicted.

2.2.1.1.1 Surface Waters

Various authors have stressed that the "total metal or dissolved metal concentration" approach is inadequate as it does not reflect the true exposure to and effects on aquatic organisms (Morel 1983; Luoma 1983; Pagenkopf 1983; Playle et al. 1992, 1993a, 1993b). The importance of explicitly considering bioavailability in the development of water and sediment quality criteria for metals has been demonstrated (Di Toro et al. 1991; Ankley et al. 1996; Allen and Hansen 1996) and is gaining increased recognition by the scientific community and some (US) regulatory authorities (Renner 1997; Bergman and Dorward King 1997). Laboratory toxicity studies completed during recent years have enhanced the current understanding of the physiological basis of how metals exert toxicity on aquatic organisms (Pagenkopf 1983; Playle et al. 1992, 1993a, 1993b; Janes and Playle 1995). Parallel investigations of metal speciation have elucidated the chemistry of metals in aquatic systems, including the formation of organic and inorganic metal complexes and sorption to particulate organic matter and to biotic ligands (Paquin et al. 2003). These studies have, in combination, led to an improved understanding of how site-specific water chemistry affects bioavailability, how the route of uptake influences exposure, how metals interact with aquatic organisms to exert toxic effects at the organism's site of action, and how toxic effect levels can be predicted. The translation of the above-mentioned studies into toxicity-related bioavailability models and especially the biotic ligand models (BLMs) is gaining interest in the scientific and regulatory community for evaluating metal bioavailability and toxicity in a cost-effective and technically defensible manner (Di Toro et al. 1997; Bodar et al. 2005).

2.2.1.1.2 Soils

In the case of soils, there has been surprisingly little consideration of factors that modulate the bioavailability of metals. In this respect it is necessary to develop methods that contain qualitative and quantitative descriptions of differences in bioavailability between soils typically used for laboratory testing and field soils, between contaminated and noncontaminated (natural background) soils, and among contaminated field soils.

Bioavailability needs to be dealt with as a dynamic process, comprising at least 2 distinct phases: a physico-chemically driven desorption process, and a physiologically driven uptake process requiring the identification of specific biological species as objects of study. Soil organisms potentially have different uptake routes. It is thought that most organisms that live in the soil (e.g., including plants) are primarily exposed via pore water, and that organisms that live on the soil are mainly exposed indirectly via their food. Differences in exposure route are likely to differ between species that possess a soft and permeable skin (sometimes including gills or roots) and those that possess a hard, physically protective, but impermeable integument. This distinction is of quantitative importance, because in the case of uptake via ingestion, gut conditions determine the transfer of chemicals rather than external conditions. There is evidence for predominant pore water uptake of organic substances by soft-bodied animals, but due to their complex physicochemical behavior, such evidence is presently only circumstantial for metals. Free metal ions in pore water are often considered to be the toxic fraction that can actually be taken up by organisms. Clearly, both abiotic (pertaining to soil characteristics) and biotic (species-dependent) aspects determine "bioavailability." Recently, first attempts were made to establish BLM for terrestrial organisms, including earthworms (Steenbergen et al. 2005), plants, invertebrates, and microorganisms (Thakali et al. 2006a, 2006b).

Metal concentrations and metal activities in the pore water are dependent upon both the metal concentration in the solid phase and the composition of both the solid and the liquid phase. In matrix extrapolation, and with emphasis on the pore water exposure route, it is therefore of great practical importance to have a quantitative understanding of the distribution of heavy metals over the solid phase and the pore water. A relatively simple approach for calculating the distribution of heavy metals in soils is the equilibrium-partitioning (EP) concept (Shea 1988; van der Kooij et al. 1991). The EP concept assumes that chemical concentrations among environmental compartments are at equilibrium and that the partitioning of metals among environmental compartments can be predicted based on partition coefficients. The partition coefficient, K_p, used to calculate the distribution of heavy metals over solid phase and pore water is defined as

$$K_p = \frac{[Metal]_{\text{solid phase}}}{[Metal]_{\text{pore water}}} \quad (\text{L/kg}) \qquad (2.1)$$

K_p is not a constant and may vary by several orders of magnitude. It is affected by element properties and both solid phase and pore water characteristics. Knowledge of the relationship between soil characteristics and K_p values enables the calculation of the distribution of heavy metals over the solid phase and pore water for different soils. When coupled to an uptake model for metals by biota that are directly or indirectly exposed via the pore water, the relationships for predicting K_p values may be used to predict metal uptake for these organisms on the basis of the metal concentration in the solid phase and some selected soil properties. The latter should, like the total concentrations, be easily determinable.

2.2.1.1.3 Sediments

An approach similar to that in soils can be applied to metal-contaminated sediments, where sulfides, measured as acid-volatile sulfides (AVS), have been demonstrated as being the predominant factor controlling metal mobility and toxicity in anaerobic sediments. The difference or ratio between SEM (simultaneous extracted metals) and AVS (SEM–AVS) is used to predict toxicity. In cases where SEM does not exceed the AVS, this approach has been shown to consistently predict the absence of toxicity (Allen et al. 1993; Ankley et al. 1996; DiToro, Hansen et al. 2001b). When SEM exceeds the AVS, toxicity is predicted, but the appearance and extent of toxicity may be determined by other binding phases (e.g., organic carbon) in the pore water. Luoma and Fisher (1997) stated that the association of metal bioavailability with AVS in sediments is not, however, straightforward in all cases and should be treated with caution.

2.2.1.2 Organic Compounds

Sorption, or intermedia transport, is of importance in systems that contain more than 1 phase. Chemicals will migrate from one phase to another if the phases are not in thermodynamic equilibrium (i.e., do not have the same fugacity). Octanol is often considered as a surrogate for various condensed lipophilic materials present in natural phases such as nonliving natural organic matter in soils, sediments, or aerosols and certain lipid-like constituents of plants, animals, and microorganisms. Although the fraction of such organic phases on a global scale is quite low, they are major sinks for hydrophobic contaminants (Di Toro et al. 1991).

Experimentally it has been observed that the ratio of concentrations in 2 phases is constant if the concentrations of the chemical in both phases are sufficiently low (thermodynamic equilibrium). In this case, at equilibrium conditions the reversible distribution between phases can be described by a constant, which is known as the distribution coefficient:

$$K_{ab} = C_a/C_b \qquad (2.2)$$

where K_{ab} is the distribution coefficient, and C_a and C_b are the concentrations of the toxicant in phases a and b, respectively.

For air–water systems, this equation is known as Henry's law. For solids–water systems, the equilibrium constant is known as the partition coefficient (K_P) or distribution constant (K_d). Partition coefficients are available for many organic chemicals from laboratory and field measurements. As organic carbon (OC) present in water (dissolved organic carbon, or DOC), sediment, or soil is the main sink for hydrophobic organic contaminants, the partition coefficients for these compounds are often adjusted (normalized) with respect to the organic carbon content of these compartments:

$$K_P = K_{OC} \times f_{OC} = C_S/C_W \qquad (2.3)$$

where K_{OC} is the organic carbon-normalized partition coefficient (L/kg), f_{OC} is the fraction of organic carbon in the sediment or soil, and C_S and C_W are the concentrations of chemical in the solid phase and the water phase, respectively. K_{OC} for neutral organic chemicals is often estimated from the octanol–water partition coefficient (K_{OW}), using simple regression equations:

$$\text{Log } K_P = \log (K_{OC} \times f_{OC}) = a \log K_{OW} + b + \log f_{OC} \qquad (2.4)$$

It may be deduced from $K_P = K_{OC} \times f_{OC}$ that partition coefficients of hydrophobic organic compounds in general are dependent upon the chemical of interest (compound-specific properties affect the value of K_{OC}) and the matrix properties of the medium in which it resides. In addition to the fraction of organic carbon present in the sorption phase, additional environmental factors affect partitioning. These factors include temperature, particle size distribution, the surface area of the sorbent, pH, ionic strength, the presence of suspended material or colloidal material, and the presence of surfactants. In addition, clay minerals may act as additional sorption phases for organic compounds. Nevertheless, organic carbon-normalized partition

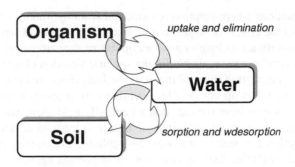

FIGURE 2.2 Schematic overview of the process underlying the equilibrium-partitioning (EP) concept. *Source*: Redrawn from data of Jager (2003).

coefficients for a specific chemical are fairly constant among different soils or sediments, provided that the additional environmental factors impacting partitioning are kept reasonably constant.

Uptake by plants and animals is in general the consequence of a plethora of competing processes, both in the aqueous and in the solid phase, as well as at the interface between the biota and the pore water. As explained by Jager (2003), the leading theory on uptake of chemicals by soil- and sediment-dwelling organisms is the EP theory, formulated and broadly adopted around 1990. Basically, this approach states that organisms do not take up chemicals from soils or sediments directly, but only from the freely dissolved phase in the pore water. A chemical will tend to distribute itself between the soil, water, and organism phases until it is in thermodynamic equilibrium. This implies that the chemical residues in organisms can be predicted when the sorption coefficient of the chemical (partitioning between solids and water) and the bioconcentration factor (partitioning between water and organism) are known. This is schematically depicted in Figure 2.2.

Equilibrium partitioning has become an integral part of chemical risk assessment for soil and sediment to predict toxicity (by extrapolation from aquatic data) as well as body residues (by extrapolation from total concentrations) in soil- or sediment-dwelling organisms. Currently, the term "EP" is often used in a broader sense, relaxing the precondition of equilibrium and denoting the fact that (time-varying) concentrations in organisms can be predicted from the (time-varying) concentrations in pore water.

Despite its popularity in risk assessment practice, limitations of EP have been observed. The most striking deviations are discussed below:

- Sequestration or "aging" is the process by which chemicals tend to become less available with time (for uptake by organisms as well as by "soft" chemical extraction techniques). The most likely mechanism for this behavior is that the chemical is moving deeper into the organic matrix with increasing contact time. Sequestration has been presented as a deviation from EP, but in fact it strongly supports the EP concept. Granted, the use of equations where sorption is estimated from hydrophobicity will fail to predict the effects of sequestration, but EP (in the broad sense) appears to be quite robust as long as good estimates or measurements of pore water concentrations are available for the specific situation of interest.

- Another deviation from EP that is extensively discussed is feeding. Chemicals are taken up not only by organisms from (pore) water through the skin, but also from the gut. It is a generally held view that the existence of multiple routes of entry into an organism leads to deviations from EP predictions, especially for very hydrophobic chemicals. It is predicted (Jager 2003) that feeding becomes an important uptake route for earthworms when log K_{ow} exceeds 5. For sediment organisms, there is evidence that feeding is important for very hydrophobic chemicals and may lead to deviations from EP up to a factor of 5. However, there are few studies that succeed in experimentally separating both uptake routes, and often conclusions on uptake routes are drawn without confirming that equilibrium was established, and without knowing the actual pore water concentrations. Furthermore, it is unlikely that chemicals are transferred directly from a solid phase to an organism without intervention of a solution phase.
- Biotransformation may also lead to deviations from EP, but this process is not well studied. Biotransformation is the uptake and metabolism of toxicants that usually results in more water-soluble metabolites. Especially when the exchange with the pore water is slow, even low levels of transformation may affect the internal exposure to the parent compound.

Despite the limitations, EP is still the reference theory for discussing the accumulation of organic chemicals in soil organisms and for extrapolation between media and matrices. The use of EP requires that body residues and effects observed in biota first have to be related to pore water concentrations.

2.2.2 MEDIA- AND MATRIX-RELATED DIFFERENCES IN DEGRADATION OF CHEMICALS

Time-related spontaneous remediation of toxic risk by degradation or translocation of chemicals is also influenced by medium and matrix properties. However, the reader is referred to Chapters 5, 6, and 7 for extrapolation methods dealing with temporal or spatial topics.

2.3 MEDIA- AND MATRIX-RELATED EFFECTS

2.3.1 EXTRAPOLATION RELATED TO ORGANISM BEHAVIOR

2.3.1.1 Trophic Status

Extrapolating toxicant effects among media may differ according to the trophic status of the species under consideration. Whether a species is a primary producer, primary consumer, or secondary consumer will influence which environmental compartments and media, and hence which routes of uptake, need to be considered.

Thus trophic status should have a relatively large influence on exposure but less of an influence on toxicity per se. Few chemicals act specifically at individual trophic levels, unless the trophic levels are characterized by specifically sensitive taxonomic groups or receptors such as, for example, fish. One of the few definite exceptions to this rule is the action of photosynthesis-inhibiting herbicides that are inherently much more toxic to autotrophic organisms than to heterotrophs. Other possible exceptions are bioaccumulating or biomagnifying chemicals, which tend to have greater effects at higher trophic levels, and plants that are less sensitive to insecticides. However, such effects can be attributed to higher body burdens (i.e., exposure) rather than trophic-level specific differences in sensitivity.

2.3.1.2 Habitat Preferences

Habitat preferences and thus medium and matrix considerations may play an important role in determining the susceptibility of species to toxicants. These can be at the macro scale. For example, certain species may prefer waters or soils of differing mineral, organic, or oxygen content. In aquatic systems, for example, benthic organisms are often more abundant, and communities more diverse, in fine-grained sediments due to higher organic content. For the same reason (higher organic content), fine-grained sediments also often have higher concentrations of organic contaminants. Differences may also occur in microhabitat preference. For example, different species of benthic organisms may be found at different depths in the sediment, and may be oriented with their heads facing down or up. A variety of abiotic factors that are likely to vary among habitats can influence exposure to and toxicity of chemicals. In addition, there may be correlations between habitat preference and taxon susceptibility.

2.3.1.3 Feeding Behavior

Feeding behavior is one of the most important variables that should be taken into account when extrapolating among media. Feeding behavior is known to affect the rates of contaminant uptake from a given medium (e.g., organisms with faster feeding rates are likely to have higher rates of contaminant uptake) and determines which media need to be included in exposure calculations and how. For example, equilibrium-partitioning models used to predict exposure from contaminant concentrations in bulk sediment typically assume that the partitioning of a chemical between sediment organic carbon and pore water is at equilibrium and therefore that ingestion as a route of exposure is not significant (Di Toro et al. 1991). However, it has become clear that EP models underestimate the uptake of sediment-associated contaminants via ingestion, at least for some groups of sediment-dwelling organisms (Landrum et al. 1996; Selck et al. 1998; Selck and Forbes 2003; Timmerman and Andersen 2003). Such animals may accumulate a substantial proportion of their body burden from sediment-bound contaminants during gut passage, which would not be predicted by EP models. Very few studies have tried to relate uptake route (e.g., dissolved contaminants taken up over external body surfaces versus particle-bound contaminants taken up over the gut surface) to toxicity (e.g., Vijver et al. 2003). There is limited evidence to suggest that contaminants taken up over the gut may be less toxic (Selck et al. 1998; Selck and Forbes 2003); however, technical constraints involved in separating effects of starvation from effects of uptake route prevent firm conclusions from being drawn. Clearly the relationship between exposure route and subsequent toxicity is of key importance in the development of media extrapolation models and is one that deserves more thorough study.

2.3.1.4 Avoidance

Perhaps the most difficult aspect of organism behavior to include in media extrapolation procedures is that relating to avoidance. Different species exhibit a diverse array of avoidance behaviors, which they may display in response to natural biotic or abiotic factors in their environment and in response to chemical exposure. Some of the most common include avoidance responses of fish to heavy metals and other contaminants (Sandheinrich 2003), aversion of contaminant-treated diets by birds (Hooper 2003), and drift in stream invertebrates (Sibley et al. 1991; Davies and Cook 1993). Such avoidance behaviors would act to reduce exposure, but are generally ignored in extrapolation approaches. Changes in behavior could potentially have importance in media extrapolation in that they could alter the relative importance of different uptake routes. For example, if a contaminant is partitioned between diet and water, and an exposed species is able to detect and avoid the contaminated food, the relative importance of water as an exposure route could be increased even though total exposure may decrease.

2.3.2 EXTRAPOLATION RELATED TO INTRINSIC SENSITIVITY

2.3.2.1 Freshwater versus Marine Toxicity

Most of the ecotoxicological data available in the open literature are derived from aquatic, freshwater species. For example, the European Centre of Ecotoxicology and Toxicology of Chemicals (ECETOC) Aquatic Toxicity (EAT) database, one of the most extensive ecotoxicological databases, contains a total of 2200 entries covering 368 chemicals and 137 aquatic species (ECETOC 1993). Of the entries, 76% are freshwater species, and the remaining 24% are marine species. Comparisons between freshwater and marine fish and invertebrates (Hutchinson et al. 1998a) using this database indicated that the sensitivities of freshwater and saltwater species were within a factor of 10 for 91% of EC50 values for fish but only 33% for invertebrates. An update of the database (EAT-3) that contains 5400 values (about 24% of which are for marine species) suggests that approximately one-third of the marine fish, invertebrate, and algal species are more sensitive (by a factor of 2 or greater) than their freshwater counterparts (ECETOC 2001). Overall, the data available to date do not indicate systematic or consistent differences in the sensitivity of marine versus freshwater taxa. However, very little is known about the sensitivities of taxa found exclusively or primarily in marine environments (i.e., Ctenophora, Mesozoa, Echinodermata, Nemertina, and Porifera). There is some concern that the comparisons made to date may be biased (ECETOC 2001).

2.3.2.2 Differences in Body Size

Differences in body size within species can have an important influence on effects extrapolation as a result of changes in surface area–volume relationships with size and hence weight (Hendriks et al. 2001; Hendriks and Heikens 2001). Such allometric relationships place constraints on uptake (of chemicals as well as food, oxygen, etc.) and excretion processes, metabolic rates, and most other physiological rate processes (Calow and Townsend 1981). Differences in body size may become an issue in media extrapolation to the extent that media properties influence organism growth rates and body size. For example, the adult body size of marine species may be somewhat lower in estuarine populations than in fully marine populations, and such differences in body size could in principle exacerbate or ameliorate other effects of salt content on chemical exposure or toxicity. Soil infauna species, living in the interstitial space between the soil grains, are generally smaller in size than species living in a less confined habitat.

2.3.3 EXTRAPOLATION RELATED TO MODIFIERS OF TOXICITY

Modifiers of toxicity are generally not specific for particular media and matrices. In applying media and matrix extrapolation techniques, however, these modifiers should be considered as boundary conditions for the validity of the applied models. Examples are discussed below.

2.3.3.1 Temperature as a Modifier of Toxicity

Temperature is one of the most important variables that determines the distribution and abundance of species (Cossins and Bowler 1987) and imposes critical limits on fitness. As a result of increasing metabolic rate, increasing temperature can increase the uptake and toxicity of contaminants by poikilothermic species, but may also increase rates of detoxification and excretion of toxicants (e.g., pyrethroid insecticides; National Research Council of Canada [NRCC] 1987). Temperature extremes in themselves are stressful to organisms, causing induction of various stress proteins, which may be associated with fitness costs (Hoffmann et al. 2003).

2.3.3.2 Low Oxygen as a Modifier of Toxicity

Decreased oxygen concentrations (hypoxia and anoxia) can alter the fate of chemicals and hence influence exposure. However, reduced environmental oxygen levels can act as an additional source of stress, particularly for aquatic species. Some species may decrease their aerobic respiration rates as environmental oxygen concentrations decline, and may or may not increase anaerobic respiration rates in turn (e.g., Linke-Gamenick et al. 2000). From a medium and matrix point of view, oxygen levels in organic-rich, shallow aquatic systems can show substantial diurnal variations as a result of changes in rates of photosynthesis and respiration of primary producers. Organisms living closest to the bottom will likely experience greater temporal variations in oxygen than organisms living closest to the water surface. Because at least some toxicants may be expected to increase oxygen demand (Newman and Unger 2003), without the capacity to move (i.e., diurnal migration), organisms could experience enhanced toxicity at reduced oxygen levels.

2.3.3.3 Resource Limitation as a Modifier of Toxicity

Although populations of organisms employed in toxicity tests (especially chronic tests) are generally maintained under conditions of adequate food and other limiting resources, field populations will often exist under conditions of density-dependent resource limitation. Competition for food, space, light, or other key variables keeps populations from growing exponentially. Although population size may temporarily exceed carrying capacity, over the long term, average population growth rates will not exceed zero. Studies that have compared the combined effects of toxicants and density-dependent factors (largely food availability) on population dynamics are somewhat equivocal. Partly, this appears to be a result of experimental design constraints (Forbes et al. 2001b). At this point, the weight of evidence indicates that toxicant effects on population growth rate are less in density-limited populations than in exponentially growing populations (Sibly et al. 2000; Forbes et al. 2003). However, there are a number of complicating issues related to the form of the density dependence (Grant 1998; Barata et al. 2002a), the relative strengths of the density versus toxicant effects (Linke-Gamenick et al. 1999), and effects on other population-level endpoints such as carrying capacity (Forbes et al. 2003).

2.3.3.4 Presence of Other Toxicants

The presence of other chemicals in the exposure medium (or in other media) can influence the response of organisms to single contaminants in the mixture. Chemicals may interact additively, synergistically, or antagonistically, and there are a number of approaches available to model such interactions (Van Leeuwen and Hermens 1995; and see Chapter 5 on mixture toxicity). Klerks (1999a) has argued that it is less likely for exposed organisms to adapt to complex mixtures than to single chemicals.

2.4 EXTRAPOLATION TOOLS AVAILABLE

Table 2.4 gives an overview of available media and matrix extrapolation tools and models that will be discussed in the following sections. In the layout of the table, a distinction is made between methods that are mainly related to exposure and bioavailability, methods that are mainly related to biological phenomena and properties that modulate effects, and methods that deal with a combination of both. Where possible, it is indicated whether the method is restricted to a certain medium (water or soil) and whether the method is based on knowledge of underlying processes (mechanistic) or on statistical relationships established in dedicated experiments (empirical). Furthermore, the methods have been grouped according to the types of toxicants involved (inorganic or organic or pesticide). The last 3 columns in the table give an indication of the input required for toxicant, media and matrix, and species properties, respectively.

2.4.1 EXTRAPOLATION TOOLS RELATED TO EXPOSURE
AND BIOAVAILABILITY OF METALS

2.4.1.1 Toxic Metal Speciation Models for Water

In aquatic ecosystems, complexation to organic and inorganic ligands and competition between toxic metals and Ca or Mg ions for biological adsorption sites reduce the actual amount of metal available for uptake by organisms. Chemical equilibrium models applicable to natural systems include RANDOM (Murray and Linder 1983;

TABLE 2.4
Tools and models for media–media and matrix extrapolation

Model or tool	Toxicant type	Toxicant properties	Matrix properties	Organism properties
Methods mainly related to exposure and bioavailability				
Metal speciation Mechanistic Retrospective Water	Metal	Identity of metal	pH, ionic strength, ligands, sorption (clay), and interactions (e.g., pH–clay)	NA
Metal speciation Empirical Retrospective Soil or sediment	Metal	Identity of metal	pH, ionic strength, ligands, sorption (clay), and interactions (e.g., pH–clay)	NA
Biotic ligand Mechanistic Retrospective Water Species specific	Metal	Identity of metal	pH, ionic strength, ligands, sorption (clay), and interactions (e.g., pH–clay)	Identity of species and identity of bioligands
BCF-1 EP-organic 1 Mechanistic 1-compartment Retrospective Water	Organic	K_{OW}, K_a	NA	Lipid content
BCF-2 EP-organic 1 Mechanistic Multicompartment Retrospective Water	Organic	K_{OW}, K_a	DOC Suspended OM	Lipid content
BCF-3 EP-organic 3 Mechanistic Multicompartment Retrospective Soil or sediment	Organic	K_{OW}, K_a	OM	Lipid content
McKay, multicompartment models Mechanistic Prospective Diverse media	Organic	$K_{OW}, K_a, K_{wa}, K_{deg}$	Vol, T	NA

(Continued)

TABLE 2.4 (CONTINUED)
Tools and models for media–media and matrix extrapolation

Model or tool	Toxicant type	Toxicant properties	Matrix properties	Organism properties
Methods mainly related to biological properties' modulating effects				
Species sensitivity distributions Mainly water Empirical	All	TMoA EC50 NOEC	NA	Taxon identity Taxonomic level
Mixed methods (both exposure and sensitivity related)				
Ecological models Mixed empirical and mechanistic Very diverse media Site and situation specific	All	K_{OW}, K_a degradation rate (DT50)	Diverse properties	Lipid content, behavior (feeding), body size, and metabolism

Bryan et al. 1997; Woolard and Linder 1999), MINTEQA2 (Allison et al. 1991), PHREEQE (Parkhurst et al. 1990), CHESS (Santore and Driscoll 1995), and WHAM (Tipping and Hurley 1992; Tipping 1994; Tipping 1998). The solution to all chemical equilibrium problems depends on simultaneously satisfying all the mass balance and mass action equations for the defined system. For metal speciation in aqueous systems, this includes inorganic complexation to ligands such as hydroxide or fluoride and mineral surfaces as well as organic complexation to binding sites within natural organic matter (NOM).

As an example, the mechanistic computer program MINTEQA2 (Allison et al. 1991) is an equilibrium speciation model that can be used to calculate the equilibrium composition of dilute aqueous solutions in the laboratory or in natural aqueous systems. The model is useful for calculating the equilibrium mass distribution among dissolved species, adsorbed species, and multiple solid phases under a variety of conditions including a gas phase with constant partial pressures. A comprehensive database is included that is adequate for solving a broad range of problems without the need for additional user-supplied equilibrium constants. The model employs a predefined set of components that includes free ions such as $Na+$ and neutral and charged complexes (e.g., H_4SiO_4 and $Cr(OH)^{2+}$). The database of reactions is written in terms of these components as reactants.

Matrix extrapolation undertaken by this model means that the model calculates the free metal ion concentration as the toxic species, given a total metal concentration and site-specific conditions in terms of water hardness, DOC, salinity, and so on. As an example, according to the MINTEQ model, a type of water with a hardness of 10 mg/L $CaCO_3$, a DOC content of 10 mg/L, a total Zn concentration of 10 mg/L, and a variable pH gives a distribution of Zn species as given in Table 2.5.

TABLE 2.5
Illustration of the distribution of chemical species in an example output of the MINTEQ speciation program

pH	6	7	8	9	10
Species	% of total [Zn]	% of total [Zn]	% of total [Zn]	% of total [Zn]	% of total [Zn]
Zn^{+2}	85.355	82.459	73.196	23.533	0.422
Zn DOC	14.128	15.916	15.312	9.412	0.359
$ZnOH^+$	0	0.766	6.815	22.176	3.994
$Zn(OH)_2$ (aq)	0	0.012	1.059	34.614	62.451
$Zn(OH)_3^-$	0	0	0	1.786	32.168
$Zn(OH)_4^{-2}$	0	0	0	0	0.137
$Zn(CO_3)_2^{-2}$	0	0	0	0.021	0
$ZnCO_3$ (aq)	0	0.384	3.223	8.356	0.465
$ZnHCO_3^+$	0.505	0.463	0.388	0.1	0

Note: Input: hardness = 10 mg/L $CaCO_3$; DOC = 10 mg/L; Total Zn = 10 mg/L; and pH between 6 and 10.

It is obvious that the bioavailable Zn^{2+} fraction is strongly pH related. At all but the highest pH, a considerable part of the total zinc concentration is complexed by DOC and thus rendered biologically inactive. At higher pH, an increasing fraction of total Zn is made less available by hydroxide formation.

2.4.1.2 Toxic Metal Speciation Models for Soil and Sediment

2.4.1.2.1 Soil
There is no mechanistic equilibrium-partitioning model for toxic metals available for the soil and sediment compartments. However, the free metal ion concentration in pore water that is considered relevant for uptake in biota (water exposure route) may experimentally or empirically be related to the total metal content of the soil, according to equation 2.1 (above).

To obtain insight in metal partitioning in field soils, and to enable derivation of (pragmatic) methods for calculating K_p values, experimental data on in situ partitioning of 6 metals (Cd, Cr, Cu, Ni, Pb, and Zn) were collected in Dutch field soils (de Groot et al. 1998). The total metal concentrations in the soil were determined after aqua regia digestion. The free metal ion concentrations in pore water were calculated from the total pore water concentrations measured by applying the MINTEQ model with the input of measured pore water characteristics: pH, DOC, Ca, Mg, Na, K, Fe, Mn, Al, Cl, NO_3, SO_4, and PO_4. In addition, soil properties assumed to affect metal partitioning were quantified: soil pH ($CaCl_2$ extraction), sequential loss-on-ignition percentages of soil (LOI_1 = temperature up to 550 °C, equals percentage of organic matter [OM]; followed by LOI_2 = temperature up to 1000 °C, equals percentage of carbonates), soil cation exchange capacity (CEC), weight percentages of clay particles

TABLE 2.6
Multivariate regression formulas[a]

Metal	Regression equation obtained	Statistics
Cu	$\text{Log } Kp = -1.27 + 0.72 * \text{pH(CaCl}_2) + 0.92 *$ $\log \text{LOI}_1\% - 0.24 * \log \text{Silt}\%$	$R^2_{\text{adj}} = 0.86, n = 46, F = 93.7, P < 0.001$
Cr	$\text{Log } Kp = 1.89 + 0.28 * \text{pH(CaCl}_2) + 0.40 *$ $\log \text{Fe-ox}$	$R^2_{\text{adj}} = 0.72, n = 46, F = 59.9, P < 0.001$
Ni	$\text{Log } Kp = 0.38 + 0.43 * \text{pH(CaCl}_2) + 0.50 *$ $\log \text{Clay}\%$	$R^2_{\text{adj}} = 0.87, n = 44, F = 138.3, P < 0.001$
Cd	$\text{Log } Kp = -1.48 + 0.55 * \text{pH(CaCl}_2) + 0.58 *$ $\log \text{LOI}_1\% + 0.40 * \log \text{Al-ox}$	$R^2_{\text{adj}} = 0.87, n = 45, F = 103.0, P < 0.001$
Pb	$\text{Log } Kp = 0.07 + 0.70 * \text{pH(CaCl}_2) + 0.56 *$ $\log \text{Silt}\%$	$R^2_{\text{adj}} = 0.84, n = 45, F = 113.6, P < 0.001$
Zn	$\text{Log } Kp = -1.04 + 0.55 * \text{pH(CaCl}_2) + 0.60 *$ $\log \text{Clay}\% + 0.21 * \log \text{Al-ox}$	$R^2_{\text{adj}} = 0.87, n = 46, F = 102.9, P < 0.001$

[a] These formulae describe the quantitative relationship between log-transformed partition coefficients of Cu, Cr, Ni, Cd, Pb, and Zn (defined as the ratio of the total metal concentration in the solid phase [aqua regia digestion] and the free metal ion concentration in the pore water) and soil and pore water characteristics.

(lutum fraction < 2 μm), fine silt particles (between 2 and 38 μm), and the amount of aluminum oxide and iron oxide (mmol/kg). Multiple linear regression was applied to the set of soil properties (including DOC [mg/L] in pore water), and K_p values were thus derived. As carbonate acts as an additional sorption phase in soils with pH values > 7.5, a distinction was made between soils with and without detectable amounts of carbonate present.

The extrapolation from total soil concentration to bioavailable pore water concentration of heavy metals makes use of the model illustrated in equation 2.1 (above).

If no information is available on the carbonate content of the soil, or if carbonate is present in detectable quantities, the model is applied with the regression coefficients given in Table 2.6.

If the carbonate content in the soil is not detectable, the model is applied with the regression coefficients given in Table 2.7.

De Groot et al. (1998) also gave regression parameters for calculating the total metal concentration in pore water from the total concentration in the soil. If site-specific physicochemical pore water characteristics are also available, the predicted total metal concentration in pore water can be extrapolated to the bioavailable ion concentration in pore water by site-specific application of the MINTEQ model.

2.4.1.2.2 Sediment
The most widely used approach to model metal bioavailability in sediments is based on the tendency of many toxic metals (Cd, Cu, Pb, Ni, and Zn) to form highly insoluble metal sulfides in the presence of acid-volatile sulfide. Metals are predicted

TABLE 2.7
Multivariate regression formulas[b]

Metal	Regression equation obtained	Statistics
Cu	$\log Kp = -1.03 + 0.64 * pH(CaCl_2) + 0.20 *$ $\log Clay\% + 0.35 * \log Al\text{-}ox$	$R^2_{adj} = 0.92$, $n = 28$, $F = 98.5$, $P < 0.001$
Cr	$\log Kp = 2.46 + 1.03 * \log CEC + 0.26 *$ $pH(CaCl_2) - 0.85 * \log LOI_1\%$	$R^2_{adj} = 0.77$, $n = 28$, $F = 31.3$, $P < 0.001$
Ni	$\log Kp = 0.48 + 0.42 * pH(CaCl_2) + 0.43 *$ $\log Clay\%$	$R^2_{adj} = 0.88$, $n = 25$, $F = 91.8$, $P < 0.001$
Cd	$\log Kp = -0.68 + 0.43 * pH(CaCl_2) + 0.28 *$ $\log Fe\text{-}ox + 0.54 * \log Al\text{-}ox - 0.64 * \log$ DOC	$R^2_{adj} = 0.89$, $n = 26$, $F = 50.5$, $P < 0.001$
Pb	$\log Kp = 0.43 + 0.55 * \log CEC + 0.61 *$ $pH(CaCl_2)$	$R^2_{adj} = 0.83$, $n = 28$, $F = 68.2$, $P < 0.001$
Zn	$\log Kp = 0.94 + 0.36 * pH(CaCl_2) + 1.14 *$ $\log LOI_2\%$	$R^2_{adj} = 0.836$, $n = 28$, $F = 69.7$, $P < 0.001$

[b] These formulae describe the quantitative relationship between log-transformed partition coefficients of Cu, Cr, Ni, Cd, Pb, and Zn (defined as the ratio of the total metal concentration in the solid phase [aqua regia digestion] and the free metal ion concentration in the pore water) and soil and pore water characteristics for soils that do not contain any detectable carbonate.

to be unavailable (and sediments nontoxic) if the molar sum of the concentrations of metals is less than the molar concentration of AVS (Ankley et al. 1996).

2.4.1.3 Biotic Ligand Model

The biotic ligand model is gaining interest in the scientific and regulatory communities for predicting and evaluating metal bioavailability and toxicity of metals, because it takes into account both metal speciation and interactions at receptor and transport sites at the organism–water interface (de Schamphelaere and Janssen 2002). Allen (1999) linked existing water chemistry models such as the CHESS model (Santore and Driscoll 1995) and the WHAM model (Tipping 1994) to ecotoxicological endpoints (e.g., Playle et al. 1993a, 1993b). The resulting BLM (Figure 2.3) incorporates chemical interactions between dissolved organic ligands (humic acids and fulvic acids) and inorganic ligands (Ca^{++} and Mg^{++}), between toxic metal ions and dissolved organic ligands (humic acids and fulvic acids) and inorganic ligands (OH^-, SO_4^{--}, CO_3^-, Cl^-, and HCO_3^-), as well as between cations (Ca^{++}, H^+, Na^+, and Cu^{++}) and biological binding sites (biotic ligands). This BLM is based on a conceptual model similar to the gill site interaction model (GSIM) originally proposed by Pagenkopf (1983) and the free ion activity model (FIAM) as described by Campbell (1995). The model therefore supports the hypothesis not only that toxicity is related to total or dissolved metal concentration, but also that metal complexation and interaction at the site of action need to be considered. The BLM has been calibrated toward acute ecotoxicity endpoints (LC50 and EC50) for fish and invertebrates and is under revision

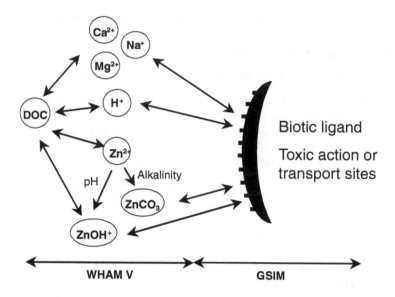

FIGURE 2.3 Schematic representation of a BLM for zinc and fish (de Schamphelaere and Janssen 2002). GSIM is the Gill Surface Interaction Model (Pagenkopf 1983). WHAM V is the fifth version of a chemical equilibrium model for water, sediment, and soil of ion binding by humic substance (Tipping 1994). *Source*: Redrawn from data from de Schamphelaere and Janssen (2002), Pagenkopf (1983), and Tipping (1994).

by the USEPA for integration into the US regulatory framework. To further validate this approach in Europe, an extensive research project has been set up to evaluate the applicability of the BLM not only for acute but also for chronic exposures (de Schamphelaere and Janssen 2002). Acute and chronic models were established and validated 1) in European surface waters of varying physicochemical characteristics (Bossuyt et al. 2004), 2) during a multispecies mesocosm test setup (Schaëfers 2002), and 3) using native organisms collected in European waters (Bossuyt and Janssen 2004).

A detailed description of the biological, chemical, mathematical, and computational aspects of the BLM can be found in Di Toro, Allen et al. (2001a), Santore et al. (2001), De Schamphelaere and Janssen (2002), and Gorsuch et al. (2002).

2.4.1.3.1 BLM Model Application
A BLM for estimating the effects of metal exposure in fish and *Daphnia* is publicly available from the Internet (HydroQual 2005). Thermodynamic information to describe metal accumulation at the biotic ligand is based on the work of Playle and others (Playle et al. 1993a, 1993b; Janes and Playle 1995). Site concentrations and binding constants for gill interactions with metals, hydrogen ions, and cations such as calcium, magnesium, and sodium provide a functional description of metal accumulation, whereas critical biotic ligand concentrations are based on a dose–response relationship of mortality to accumulation at the gill (MacRae 1994). BLM predictions for copper and silver toxicities to freshwater fish have been compared to results from bioassay studies (Janes and Playle 1995; Erickson et al. 1996). For example,

TABLE 2.8
Example input and output of the BLM model for copper toxicity in fish

Input values			Valid range	
Variable	Unit	Value	Min	Max
Temperature	°C	29	0.01	50
pH	unit	8.45	5	9
Dissolved organic carbon	mg/L	5.6	0.01	20
Humic acid %	%	10	0.01	100
Total Ca	mol/L	0.0020	0.01	100
Total Mg	mol/L	0.0007	0.01	100
Total Na	mol/L	0.0033	0.01	100
Total K	mol/L	0.0002	0.01	100
Total SO_4	mol/L	0.0015	0.01	100
Total Cl	mol/L	0.0028	0.01	100
Total CO_3	mol/L	0.0038	0.01	100
Output Values	**Unit**	**Value**	**Unit**	**Value**
LC50 total Cu	mol/L	3.87E-05	µg/L	2456
Free Cu	mol/L	1.10E-07	µg/L	7.00
Active Cu	mol/L	6.28E-08	µg/L	3.99
Organic Cu	mol/L	2.62E-06		
Total organic Cu	mol/L	1.17E-05		
Gill-Cu^{2+}	nmol/g wet weight	4.65		
Gill-$CuOH^+$	nmol/g wet weight	2.68		

Erickson et al. (1996) showed that a wide range in 96-hour LC50 values resulted from adjustments to the pH, DOC, alkalinity, and hardness conditions of the test waters. The BLM was used to predict the effects of copper exposure to fathead minnow. Input data to the BLM included measured water chemistry (pH, DOC, Ca, Mg, Na, K, Cl, SO_4, and CO_3 concentrations). The BLM predicts the total copper LC50 values, based on the amount of copper necessary for accumulating lethal biotic ligand concentrations as presented in Table 2.8. The predicted LC50 can be compared to the measured total concentrations in the field.

2.4.1.4 Equilibrium-Partitioning Models for Organic Chemicals in Water

2.4.1.4.1 One-Compartment Model
According to Bacci (1994), bioconcentration is the accumulation of freely dissolved contaminants in water by aquatic organisms through nondietary routes. In water-only exposures, the primary route of uptake of dissolved contaminants in fish is across the gill epithelium, but depending on the compound, species, and animal body size, a substantial part of the body burden (25% to 40%) may penetrate across the epidermis (Landrum et al. 1996). In many cases, the toxicokinetic behavior of organic contaminants in aquatic organisms can be approximated by a first-order,

1-compartment model. The degree of bioconcentration at steady state, represented by the bioconcentration factor (BCF), depends on both the rate of absorption and the rate of elimination:

$$BCF = k1/k2 = C_b/C_w \qquad (2.4)$$

where k1 is the uptake rate (e.g., ml/h), k2 is the elimination rate constant for the compound (e.g., ml/h), C_b is the concentration in the organism at steady state, and C_w is the concentration in water at steady state. Using the freely dissolved water concentration and assuming no biotransformation, this BCF represents the relative solubility of the compound in water versus the organism's tissue. Biotransformation processes and active elimination can reduce the BCF. The steady-state condition represents the maximal accumulation that can be attained for a given set of exposure conditions (Landrum et al. 1996). However, conditions may change so rapidly that steady state may not be attained except under controlled conditions, a problem that makes actual bioconcentration difficult to predict. The mechanism of bioconcentration from water is comparable to the uptake mechanisms of contaminants from pore water.

Partition coefficients are used to describe the distribution of nonpolar organic compounds between water and organisms. It can be viewed as a partitioning process between the aqueous phase and the bulk organic matter present in biota (Schwarzenbach et al. 1993). The premise behind the use of equilibrium models is that accumulation of compounds is dominated by their relative solubility in water and the solid phases, respectively. Equilibrium models, therefore, rely on the following assumptions (Landrum et al. 1996):

- The compounds are not actively biotransformed or degraded.
- There are no active (energy-requiring) processes dominating the distribution.
- The conditions are sufficiently stable for a quasi-equilibrium to occur.
- Environmental factors, such as temperature, do not change sufficiently to alter the equilibrium conditions.
- Organism and/or organic matter composition is not sufficiently variable to alter the distribution.

A commonly used partition coefficient is the 1-octanol–water partition coefficient, K_{OW}, which is the ratio of a chemical's concentration in 1-octanol to its concentration in water at equilibrium in a closed system composed of octanol and water (Bacci 1994). The 1-octanol is chosen to mimic biological lipids. For organic chemicals, log K_{OW} ranges from –3 to 7. When log K_{OW} exceeds 3, substances are considered hydrophobic (Elzerman and Coates 1987). The K_{OW} partition coefficient has been extensively used as an estimate of the BCF. Under the assumptions of Landrum et al. (1996), together with an estimated lipid content of about 5% in biota and an assumed equal affinity of the compound for both body fat and octanol, the BCF can be calculated by the use of BCF = 0.048 * K_{OW} (Paasivirta 1991). This equation can vary depending on the species used. The relationship between log K_{OW} and BCF can be viewed by scatterplot analysis (Figure 2.4). These plots show a clear relationship for

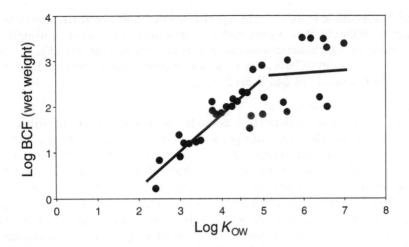

FIGURE 2.4 Plots of log BCF versus log K_{OW} values, showing the lack of linear relationship for very hydrophobic compounds. *Source*: Modified from De Wolf et al. (1992) and Barron (1990).

many compounds and over a broad range of log K_{OW} values, but there are also several discrepancies. In general, there is a good linear relation between log BCF and log K_{OW} for low or moderately hydrophobic compounds (log K_{OW} 3 to 6), but this relation breaks down for strongly hydrophobic compounds (log K_{OW} > 6; Hawker and Connell 1986; Landrum et al. 1996). Hawker and Connell (1986) argued that the lack of linear relationship for these strongly hydrophobic compounds depends on the fact that the time needed to reach equilibrium is generally longer than the exposure time.

The maximal observed value for log BCF is acquired for compounds having a log K_{OW} between 5 and 6 (Landrum et al. 1996; Landrum 1989). Recently, even a negative linear relationship between the biota-sediment accumulation factor (BSAF) and log K_{OW} was demonstrated for very hydrophobic polychlorinated biphenyls (PCBs; log K_{OW} > 6.7; Maruya and Lee 1998). This negative relationship is thought to be due to the difficulty of the relatively large molecules in penetrating membranes because of diffusion and blood flow rate limitations. In addition, it has become evident that 1-octanol is not an ideal solvent for larger molecules (Landrum and Fisher 1998). The equilibration model has been criticized for its fundamental assumptions of negligible metabolism, lack of steric hindrance, and failure to consider blood flow in controlling uptake, distribution, and elimination. The key factor determining uptake of some compounds appears to be molecular size rather than molecular weight, whereas steric hindrance might be of importance for the transfer between water and organism (Barron 1990; Landrum et al. 1996). One of the freely available models to estimate K_{OW} values is KOWWIN™ from the USEPA. KOWWIN has been incorporated into the EPI Suite™, which is available on the Internet (USEPA 2007). See Chapter 3 for more details.

For the matrix extrapolation process for nonpolar organic compounds in most surface waters, the applicability of a first-order, single-compartment equilibrium model means that the total aquatic concentration of the toxicant may be considered as entirely available to the exposed organisms, independent of differences in matrix properties. If the dissolved organic carbon content in the water is low (< 30 mg/L) and the organic toxicant under consideration is moderately hydrophobic (K_{OW} < 3 to 4), the observed toxicity in laboratory toxicity tests can be applied without further correction for bioavailability (see Table 2.9).

2.4.1.4.2 Toxicant Ionization

Ionization presents a complicating factor in the estimation of bioavailability of more polar, dissociating chemicals, like phenols, amines, amides, and a variety of modern pesticides. Dissociation or ionization is strongly pH dependent. Generally, the nondissociated toxicant has a much higher hydrophobicity than the ionized form. Therefore, the overall partitioning is also strongly influenced by pH. The negative log of the acid ionization constant (pK_a) is defined as the ability of an ionizable group of an organic compound to donate a proton (H^+) in an aqueous media. The computed quantity is a measure of its apparent pK_a, or macroscopic dissociation constant, at equilibrium, normally taken at 25 °C. The pK_a value of an organic compound arises from the ionization of either an acid or a base, at a specific center in the structure. The dissociation (apparent) constants for acid and base are commonly expressed as

$$K_a = \frac{[H_3O^+][A^-]}{[HA]} \quad \text{(acid)}$$

$$K_a = \frac{[H_3O^+][B]}{[HB^+]} \quad \text{(base)} \quad (2.5)$$

where the corresponding equilibria are

$$HA + H_2O \rightarrow H_3O^+ + A^- \quad \text{(acid)}$$

$$HB^+ + H_2O \rightarrow H_3O^+ + B \quad \text{(base)}$$

The pK_a of an acid (or protonated base) is simply expressed as

$$pK_a = -\log(K_a)$$

and that for [H^+] as

$$pH = -\log([H^+])$$

Together, in conjunction with the equilibrium expressions above, these result in the useful Henderson–Hasselbach equation:

$$pH - pK_a = \log([A^-]/[HA]) = \log([B^-]/[HB^+]) \quad (2.6)$$

This equation indicates that at a given pH value, the basic form predominates if pK_a < pH, or the reverse if pK_a > pH. This form of the equation is useful when

TABLE 2.9
Bioavailable fraction of organic toxicants as a function of DOC and K_{OC}

		DOC (mg/L)						
		1	5	10	15	20	25	30
	$f_{oc,DOC}$	0.000001	0.000005	0.00001	0.000015	0.00002	0.000025	0.00003
log K_{OC}	K_{OC}	% Available						
−1.21	0.06	100.00%	100.00%	100.00%	100.00%	100.00%	100.00%	100.00%
−0.21	0.62	100.00%	100.00%	100.00%	100.00%	100.00%	100.00%	100.00%
0.79	6.17	100.00%	100.00%	99.99%	99.99%	99.99%	99.98%	99.98%
1.79	61.7	99.99%	99.97%	99.94%	99.91%	99.88%	99.85%	99.82%
2.79	616	99.94%	99.69%	99.39%	99.08%	98.78%	98.48%	98.18%
3.79	6165	99.39%	97.01%	94.19%	91.53%	89.02%	86.64%	84.39%
4.79	61659	94.19%	76.44%	61.86%	51.95%	44.78%	39.35%	35.09%
5.79	616595	61.86%	24.49%	13.95%	9.76%	7.50%	6.09%	5.13%

interpreting the pH–toxicity relationship. The toxicity of a polar compound is given at a particular pH, corresponding to a particular ratio of compound dissociation. If the environmental pH deviates from the pH at which the toxicity is determined, a pH correction is necessary to predict the appropriate partitioning of the chemical between the biota and the matrix. As a rule of thumb, the ionic form of the chemical has a bioaccumulating capacity (K_{OW}) and thus a toxicity that is a factor of 10 lower than the nondissociated parent compound (USEPA 2000b). With the aid of the dissociation formulas, a toxicity correction can be calculated:

- Toxicity (EC50 or NOEC) at any pH is inversely proportional to BCF (a high toxicity corresponds to a low EC50) and thus to $K_{OW\,pH}$, according to BCF = $0.048 \times K_{OW}$ (Paasivirta 1991):

$$K_{OW\,pH} = K_{OW\,nondissociated} * (0.9 * \text{nondissociated fraction} + 0.1) \qquad (2.7)$$

- $pH_{toxicity\,test}$ = x: The nondissociated fraction in the toxicity test ($f_{nd,test}$) is

$$f_{nd,test} = \frac{1}{10^{x-pK_a} + 1} \quad \text{for a monovalent organic acid}$$

and $\qquad\qquad\qquad\qquad\qquad\qquad\qquad\qquad\qquad\qquad\qquad\qquad (2.8)$

$$f_{nd,test} = \frac{1}{10^{pK_a-x} + 1} \quad \text{for a monovalent organic base}$$

- $pH_{environment}$ = y: The nondissociated fraction in the environment ($f_{nd\,env}$) is

$$f_{nd,env} = \frac{1}{10^{y-pK_a} + 1} \text{ and } f_{nd,env} = \frac{1}{10^{pK_a-y} + 1}, \text{respectively} \qquad (2.9)$$

$$\frac{BCF_{test}}{BCF_{env}} = \frac{EC50_{env}}{EC50_{test}} = \frac{K_{OWnd} \times (0.9 \times f_{nd,test} + 0.1)}{K_{OWnd} \times (0.9 \times f_{nd,env} + 0.1)}$$

Thus: $\qquad\qquad\qquad\qquad\qquad\qquad\qquad\qquad\qquad\qquad\qquad\qquad\qquad (2.10)$

$$EC50_{env} = EC50_{test} \times \frac{0.9 \times f_{nd,test} + 0.1}{0.9 \times f_{nd,env} + 0.1}$$

- A model for estimating the dissociation constant pK_a from chemical structure can be found on the Internet (ChemSilico 2005).

As an example, 2,4-dichlorophenol (2,4-DCP) is an acidic compound with a pK_a of 7.68 and a $K_{OW\,nd}$ of 3.06 (Mackay et al. 1997). Consultation of the USEPA Ecotox database reveals a chronic reproduction toxicity NOEC value for *Daphnia*

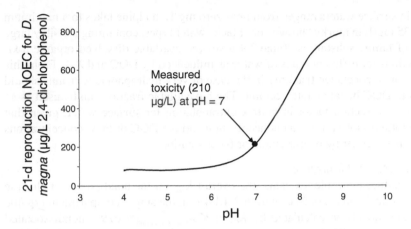

FIGURE 2.5 Extrapolated pH dependence of *Daphnia* sensitivity to 2,4-dichlorophenol.

magna of 210 μg/L at a pH of 7 (Record No. 847). With the aid of the extrapolation technique presented above, the pH dependent toxicity is calculated and presented in Figure 2.5.

2.4.1.4.3 Two-Compartment or Multicompartment Models

To estimate the bioavailability of highly hydrophobic chemicals in waters with a high DOC content or a high content of suspended organic matter, equilibrium-partitioning models with more than only the biological compartment must be used. The reason for this is that the organic matrix of the medium competes with the organisms' lipids for the relative solubility of the contaminant. By the nature of the DOC analysis and the analysis of suspended particulate organic carbon (POC) in water — chemical or thermal oxidation, followed by detection of CO_2 evolution — the partitioning has to be related to K_{OC} instead of K_{OW}. For the type of DOC in surface waters, $K_{OC,DOC} = K_{OW} * 0.62$ (or log K_{OC} = log $K_{OW} - 0.21$; Karickhoff and Brown 1979; Kenaga and Goring 1980). For calculating the sorption to POC, the model PCKOCWIN™ is available from the USEPA. $K_{OC,POC}$ estimations are based on the Sabljic molecular connectivity method with improved correction factors. $K_{OC,POC}$ is the ratio of the amount of chemical adsorbed per unit mass of organic carbon in soils, sediments, or sludge to the concentration of the chemical in solution at equilibrium. PCKOCWIN™ has been incorporated into the EPI Suite™, which is available on the Internet (USEPA 2007).

The fraction of the total dissolved organic toxicant that is available for uptake by biota is considered to be freely dissolved in water. This fraction equals

$$f_{free} = \frac{1}{1 + K_{OC,DOC} \times f_{OC,DOC} + K_{OC,POC} \times f_{OC,POC}} \tag{2.11}$$

where $f_{OC,DOC}$ and $f_{OC,POC}$ are the w/w fractions of the organic carbon contained in dissolved and particulate matter, respectively (5 mg/L = 0.000005). The organic carbon content in surface waters ranges from near zero mg/L in alpine lakes to a maximum of 30 to 35 mg/L in murky "brown" or "black" water types, containing a visibly large amount of humic substances. Table 2.9 gives the available dissolved organic toxicant fractions for different types of water as influenced by DOC and K_{OC}. From this table, it can be concluded that only in the case of very hydrophobic toxicants should sorption to DOC be taken into account. The bioavailable fraction can be used as an availability correction factor in matrix extrapolation for surface water, given that both the total dissolved toxicant concentration and the DOC or POC concentrations are quantified, either by measurement or by estimation.

2.4.1.4.4 Toxicant Ionization

For dissociating ionizable toxicants, the overall K_{OC} (of the nondissociated plus the ionized fraction) can again (see Section 2.4.1.4.1 on the single-compartment organic equilibrium model) be calculated by $K_{OC} = K_{OC\,nondissociated} \times (0.9 \times$ nondissociated fraction $+ 0.1)$ (USEPA 2000b), where the nondissociated fraction is calculated from the pH in the water and the pK_a of the toxicant. If the toxicity of the freely dissolved toxicant is estimated by comparison with experimental toxicity data, the pH shift in toxicity also has to be taken into account.

2.4.1.4.5 AQUATOX Model

The model AQUATOX (Park 1999) is a general ecological risk assessment model that represents the combined environmental fate and effects of conventional pollutants, such as nutrients and sediments, and toxic chemicals in aquatic ecosystems. Unlike the steady-state models mentioned above, AQUATOX is a dynamic model that includes the progress of subprocesses in time, based on process rate information. It may consider several trophic levels, including attached and planktonic algae and submerged aquatic vegetation, invertebrates and forage, bottom-feeding, and game fish; it also represents associated organic toxicants. It can be implemented as a simple model (indeed, it has been used to simulate an abiotic system) or as a truly complex food-web model. The model has been implemented for streams, small rivers, ponds, lakes, and reservoirs (Park 1999). The model is intended to be used to evaluate the likelihood of past, present, and future adverse effects from various stressors including potentially toxic organic chemicals, nutrients, organic wastes, sediments, and temperature. The stressors may be considered individually or together. The fate portion of the model, which is applicable especially to organic toxicants, includes partitioning among organisms, suspended and sedimented detritus, suspended and sedimented inorganic sediments, and water; volatilization; hydrolysis; photolysis; ionization; and microbial degradation. In its simplest use, the AQUATOX model can be applied to calculate matrix-related fate and effects extrapolations in the risk evaluation process. For its more elaborate applications (e.g., toxicity-related aquatic food-web modeling), the reader is referred to the sections on ecological modeling (Sections 4.5.4 and 7.3.2 in Chapters 4 and 7, respectively).

In the following example, the AQUATOX model is run in its simplest form (without biota) for a pond that is 1 m deep, with surface area 400 m^2, a DOC of 10 mg/L, and 1.2 kg/m^2 organic matter in the sediment layer. On day one, a total amount of 50 µg/L of the insecticide chlorpyrifos is added to the water phase of this imaginary water body. The model results are presented in Figures 2.6 and 2.7. From these

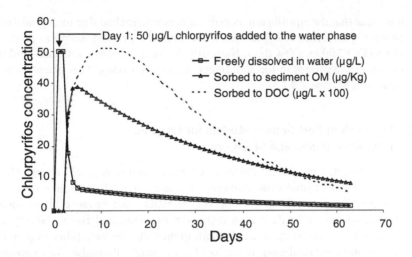

FIGURE 2.6 Results of an AQUATOX model run specifying the fate of chlorpyrifos over time after a onetime addition of 50 µg/L to a small pond.

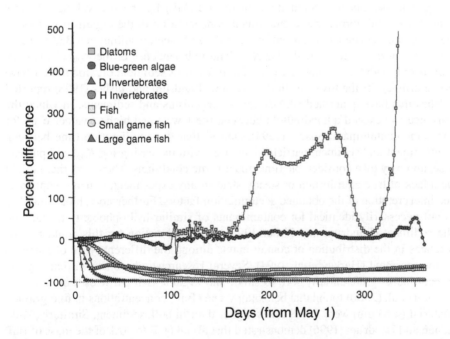

FIGURE 2.7 AQUATOX model run showing the modeled differences in the density of species with and without the normal application pattern of esfenvalerate to a small pond adjacent to a cornfield.

graphs it is clear that the equilibrium condition is never reached due to degradation of the toxicant. Furthermore, it illustrates that even the primary sorption process from water to DOC takes a few days. Naturally, hydrophobic toxicants also tend to adhere to organic substrates in the sediments of the water body. This phenomenon will be treated in Section 2.4.1.5.

2.4.1.5 Equilibrium-Partitioning Models for Organic Chemicals in Soil and Sediment

In soil and sediment, a number of physical, chemical, and biological factors affect the bioavailability of organic contaminants. Because many of these determinants may simultaneously demonstrate site-specific variation, and because all of them influence toxicant uptake in their own right, the exposure and bioavailability of organic contaminants in sediment are difficult to predict. The variability in prediction of contaminant accumulation, using log K_{OW}, is markedly higher for exposure via solid media than for aqueous exposure (Landrum and Fisher 1998). The reason for this is that the organic matrix of the solid medium competes more strongly than water-dissolved organics with the organisms' lipids for the relative solubility of the contaminant. To overcome this uncertainty in prediction, a 2-compartment model using biota–sediment accumulation factors (BSAFs) has been developed. In this model, the lipid-normalized contaminant concentration in the organism is divided by the carbon-normalized concentration in the sediment, resulting in a BSAF that is independent of the compound's log K_{OW}. The independence of the K_{OW} of the toxicant is only valid assuming that body fat and soil or sediment organic matter have equal affinity for the toxicant. In their review, Landrum and Fisher (1998) reported studies that have quantified BSAFs among organisms and sediments, in which the variance can exceed a hundredfold between the lowest and highest values even for the same contaminant. Spacie (1994) showed that the equilibration time between contaminant and sediment particles increases with increasing log K_{OW} and particle size, and may take considerable time under some conditions. Therefore, there might be a lack of true equilibrium or steady state in any experiment, which complicates the interpretation of the obtained accumulation factors. Furthermore, bioavailability is not necessarily identical for contaminants of similar hydrophobicity throughout the sediment (Landrum and Fisher 1998). One explanation for this could be differences in the distribution of contaminants among the different types of particles in the sediment (Harkey et al. 1994). Several researchers have shown that organic contaminants preferentially sorb to small and organic-rich particles. For example, Weston et al. (2000) found that benzo[a]pyrene (BaP) concentrations in fine-grained material (< 63 μm) were 5 to 8 times higher than in bulk sediment. Similarly, Kukkonen and Landrum (1996) demonstrated that about 60% to 70% of the mass of BaP and hexachlorobiphenyl (HCBP) was associated with sediment particles in the 31 to 63 μm range. The results show that the estimation of bioavailability and bioaccumulation of contaminants from sediments is highly complex. The relationship between concentration of contaminants in sediment and bioaccumulation is not linear even when normalized for organic carbon and lipid, and additional factors are probably required to make better predictions.

For calculating the bioavailable fraction of an organic contaminant in soil, essentially the same methods apply as for the water compartment. The difference between the water and the soil or sediment compartments is mainly caused by the possibly of much higher fraction of toxicant-adsorbing refractory organic material in soil or sediment and by the limited pore water and groundwater volume. For peaty soils the organic fraction may be as high as 80% w/w, so that the major part of hydrophobic chemicals will be adhered to the soil particles. Because the transfer of the toxicant, even in soil, is still considered to be water mediated, the toxicity in the soil or sediment matrix will generally be much lower than the toxicity in surface water. By the nature of the type of analysis generally applied to generate information on the organic content of soil or sediment (f_{OM}, fraction weight loss on ignition at a temperature of around 1000°C), the organic sorption coefficient (K_{OM}) has to be related to "organic matter" instead of "organic carbon." Based on the average type of organic soil material, the ratio of K_{OM}/K_{OC} can be fixed to 0.526. For all toxicants, K_{OC} can be calculated using the model PCKOCWIM (USEPA 2007). The bioavailable fraction of the organic toxicant, freely dissolved in pore water, can be calculated using the method already specified for the water compartment:

$$f_{free} = \frac{1}{1 + K_{OM} \times f_{OM}} \tag{2.12}$$

The equation is generalized to include polar compounds by applying the following formula: $K_{OM} = K_{OM\,nondissociated} \times (0.9 \times nondissociated\ fraction + 0.1)$ (USEPA 2000b), where the nondissociated fraction is again calculated from the pH in the pore water and the pKa of the toxicant (see Section 2.4.1.1)

2.4.1.6 Mackay-Type Multicompartment Model: SimpleBox

SimpleBox was created as a research tool in environmental risk assessment. Simple-Box (Brandes et al. 1996) is implemented in the regulatory European Union System for the Evaluation of Substances (EUSES) models (Vermeire et al. 1997) that are used for risk assessment of new and existing chemicals. Dedicated SimpleBox 1.0 applications have been used for integrating environmental quality criteria for air, water, and soil in The Netherlands. Spreadsheet versions of SimpleBox 2.0 are used for multimedia chemical fate modeling by scientists at universities and research institutes in various countries. SimpleBox models exposure concentrations in the environmental media. In addition to exposure concentrations, SimpleBox provides output at the level of toxic pressure on ecosystems by calculating potentially affected fractions (PAF) on the basis of species sensitivity distribution (SSD) calculus (see Chapter 4).

SimpleBox is a multimedia mass balance model of the so-called "Mackay type." It represents the environment as a series of well-mixed boxes of air, water, sediment, soil, and vegetation (compartments). Calculations start with user-specified emission fluxes into the compartments. Intermedia mass transfer fluxes and degradation fluxes are calculated by the model on the basis of user-specified mass transfer coefficients and degradation rate constants. The model performs a simultaneous mass balance calculation for all the compartments, and produces steady-state concentrations in the compartments as output. SimpleBox defines 3 partially nested spatial scales: a regional and continental scale nested in a temperate northern hemisphere scale, a tropical scale, and an arctic northern hemisphere scale. Applicability is limited to environmental situations where spatial differences in concentrations within compartments or boxes are negligible or unimportant. The model is fully described in publicly available RIVM reports (van de Meent 1993; Brandes et al. 1996). Advantages and limitations of multimedia fate models are described and evaluated in a SETAC publication (Cowan et al. 1995). The most recent spreadsheet version (3.0) of SimpleBox is available from the Internet (den Hollander et al. 2004).

2.4.1.6.1 *Example of SimpleBox Use*

Measured concentrations are not always available or may even be nonexistent, as in the case of the planned production of new substances. A producer may consider marketing a new herbicide that is believed to perform better with respect to unwanted side effects (leaching to groundwater, runoff to surface water, volatilization, and subsequent deposition to nearby natural ecosystems). Modeling may then assist decision making.

In the example, SimpleBox 3.0 was used to analyze the environmental performance of a substitute compound for existing pesticides, with atrazine as a reference compound. SimpleBox modeled the environmental fate of the agriculturally used pesticides in a generic river basin (230 000 km^2) where 60% of the land is used for agriculture, 10% is urban and 30% is "natural." In the calculations it was assumed that agricultural soil is loaded at a constant rate, that is, the average annual pesticide use. Pesticides are eliminated from agricultural soil by degradation and transport to other compartments. Steady-state concentrations in air, freshwater, coastal seawater, agricultural soil, and natural soil were calculated for the 47 most intensively used active ingredients in The Netherlands. Thereupon, toxic pressures (PAFs) were derived from the concentrations in water (dissolved) and soil (pore water), using α- and β-parameters of the log-logistic SSDs of each compound. Results are shown in Table 2.10.

Typical use rates run up to 0.16 kg of active ingredient per hectare of agricultural land per year (atrazine and substitute: 0.12 and 0.1 kg·ha·yr^{-1}, respectively). Typically 20% of all of this is degraded in the agricultural soil system: 72% for atrazine, and 98% for the slightly better degradable substitute. Lesser proportions find their way to air, surface water, and groundwater. Logically, the highest pesticide concentrations are predicted in agricultural soil; concentrations in the nontarget compartments are typically much lower. For example, concentrations in seawater are roughly 2 orders of magnitude lower than concentrations in pore water of agricultural soil, for both atrazine and the proposed substitute. Atrazine and the substitute primarily differ with regard to the proportion that volatilizes to air (atrazine 0.4%, substitute $\ll$ 0.1%). As a result, a greater concentration difference between agricultural soil and natural soil is predicted for the substitute than for atrazine. This is also reflected in the calculated toxic pressures on ecosystems. Atrazine and the substitute have approximately equal toxicities, and used in similar quantities. Toxic pressures in agricultural soils are comparable, as are the toxic pressures in aquatic ecosystems.

TABLE 2.10
Modeled fate, concentrations, and toxic pressures (PAF) of pesticides for a generic river basin

	Unit	Typical pesticide[a]	Atrazine	Substitute
Use in agriculture	$kg_{ai} \cdot ha \cdot yr^{-1}$	0.001	0.16	0.12	0.1
Air–water partition coefficient	—	10^{-10}	10^{-4}	10^{-7}	10^{-7}
Solids–water partition coefficient	—	10^{-1}	10^4	8	100
Degradation half-life in water	d	0.2	40	18	8
Degradation half-life in soil	d	0.3	300	50	40
α-value aquatic species	$g \cdot L^{-1}$	−7	−2.7	−3.84	−4.5
β-value aquatic species	—	0.34	0.75	0.62	0.7
Degradation in agricultural soil	%	19	100	72	98
Uptake from agricultural soil by vegetation	%	0	45	16	0.2
Runoff to surface water	%	0	21	6	0.5
Leaching to groundwater	%	0	21	6	0.5
Volatilization to air	%	0	22	0.4	$\ll 0.1$
Bulk concentration in agricultural soil	$g \cdot kg_w^{-1}$	4×10^{-9}	2×10^{-5}	1×10^{-5}	1×10^{-5}
Concentration in pore water agricultural soil	$g \cdot L^{-1}$	4×10^{-11}	3×10^{-5}	4×10^{-6}	3×10^{-7}
Bulk concentration in freshwater	$g \cdot L^{-1}$	6×10^{-12}	3×10^{-6}	6×10^{-7}	2×10^{-8}
Concentration dissolved in freshwater	$g \cdot L^{-1}$	5×10^{-12}	2×10^{-6}	6×10^{-7}	2×10^{-8}
Bulk concentration in coastal seawater	$g \cdot L^{-1}$	2×10^{-13}	4×10^{-7}	8×10^{-8}	2×10^{-9}
Concentration dissolved in coastal seawater	$g \cdot L^{-1}$	2×10^{-13}	4×10^{-7}	8×10^{-8}	2×10^{-9}
Bulk concentration in air	$g \cdot m^{-3}$	1×10^{-16}	1×10^{-9}	2×10^{-12}	3×10^{-14}
Bulk concentration in natural soil	$g \cdot kg_w^{-1}$	8×10^{-14}	2×10^{-7}	8×10^{-8}	3×10^{-9}
Concentration in pore water natural soil	$g \cdot L^{-1}$	4×10^{-16}	1×10^{-7}	3×10^{-8}	8×10^{-11}
Toxic pressure agricultural soil	%[b]	0	24	7	5
Toxic pressure freshwater	%[b]	0	7	2	1
Toxic pressure coastal seawater	%[b]	0	2	0.5	0.3
Toxic pressure natural soil	%[b]	0	1.4	0.2	0.03

[a] 5th and 95th percentiles of 47 active ingredients modeled.

[b] Percentage of species exposed to a concentration higher than their NOEC: potentially affected fraction (PAF).

However, toxic pressures on natural terrestrial ecosystems are predicted to be much lower for the proposed substitute.

2.4.2 EXTRAPOLATION TOOLS SOLELY RELATED TO SPECIES PROPERTIES

There are few specific extrapolation methods that attempt to extrapolate species sensitivities across media and matrices. The types of interspecies (life-cycle models) and intraspecies (allometric models, physiologically based toxicokinetic models, dynamic energy budget models, etc.) extrapolations only marginally deal with the influence of media and matrices. The few extrapolations that have been investigated or applied (e.g., predicting marine species sensitivity from freshwater species sensitivity) are based purely on empirical correlations between sensitivity distributions of selected taxonomic groups in different media and matrices. Pastorok et al. (2002), in their book entitled *Ecological Modeling in Risk Assessment*, dealt with inter- and intraspecies extrapolations. Species extrapolations are dealt with in Chapter 4.

2.4.3 EXTRAPOLATION TOOLS RELATED TO EXPOSURE AND FATE
AND SPECIES SENSITIVITY

A variety of aquatic and terrestrial ecosystem models have been developed to represent biotic and abiotic structures in combination with physical, chemical, biological, and ecological processes. Several of these may be particularly promising for assessing the risks of toxic chemicals to ecological systems. They vary in terms of performance, practical feasibility, the resources required for site-specific parameter estimation, and the extent to which they have been or can be validated. An excellent review of ecosystem models relevant to risk assessment, which assesses each model in terms of its realism, relevance, flexibility, treatment of uncertainty, degree of development and consistency, ease of estimating parameters, regulatory acceptance, credibility, and resource efficiency, is provided by Pastorok et al. (2002, Chaps. 9 and 10). These authors also provided equivalent reviews of food-web models that describe feeding relationships or predator–prey relationships among all or some species in an ecological community (Pastorok et al. 2002, Chap. 8) and landscape models that differ from ecosystem models in being spatially explicit and in potentially including several types of ecosystems (Pastorok et al. 2002, Chap. 11).

The USEPA AQUATOX model (Park 1999) provides a possibility to model toxicant effects in a food web consisting of several trophic levels (see Section 2.4.1.4). The effects portion of the model includes chronic and acute toxicity to the various organisms modeled, and indirect effects such as release of grazing and predation pressure, increase in detritus and recycling of nutrients from killed organisms, dissolved oxygen decline due to increased decomposition, and loss of food base for animals. AQUATOX represents the aquatic ecosystem by simulating the changing concentrations (in mg/L or g/m³) of organisms, nutrients, chemicals, and sediments in a unit volume of water. As such, it differs from population models, which represent the changes in numbers of individuals. As O'Neill et al. (1986) stated, ecosystem models and population models are complementary; one cannot take the place of the other. Population models excel at modeling individual species at risk and modeling fishing pressure and other age or size-specific aspects; but the recycling of nutrients, the combined fate and effects of toxic chemicals, and other interdependencies in the aquatic ecosystem are important aspects that AQUATOX represents and that cannot be addressed by a population model.

2.5 CHOOSING THE EXTRAPOLATION METHODS

An array of extrapolation types for matrix and media extrapolations has been given, with specific approaches being dependent on compound, medium and matrix, and species and ecosystem properties. Guidance to provide a systematic orientation in the array of methods is useful when considering their practical use in a regulatory context. Using a tiered approach (see Chapter 1), the tiers for media and matrix extrapolation listed in Table 2.11 are recommended.

TABLE 2.11
Tiers in media and matrix extrapolation

Tier	Characteristics
exposure	
1	Assume all molecules of the compound to be fully available for uptake by biota (i.e., use total concentrations of a compound, i.e., do not apply any matrix–media extrapolation method to address possible differences in exposure levels between laboratory and field conditions).
2	Assume that the matrix or medium influence the speciation of the compound, which in turn influences exposure (i.e., assume that the matrix or medium determine which proportion of the molecules of the compound is available for uptake [the "supply" side of bioavailability, or physicochemical availability assessment], i.e., apply matrix or media extrapolation methods to address differences in exposure levels between laboratory and field conditions).
3	Assume that the matrix or medium ("supply") influences not only uptake and exposure but also the biological "demand" in a species-specific way, due to an organism's ecological preferences and behavior and its ecophysiological apparatus.
Effects	
1	Assume that the exposed organisms experience a level of stress caused by the matrix or medium similar to the level of stress that they encountered in the laboratory tests (i.e., do not apply extrapolation methods to address possible sensitivity differences between laboratory and field conditions caused by additional matrix or medium-associated stress factors).
2	Assume that the exposed organisms experience a level of stress caused by the matrix or medium dissimilar to the level of stress that they encountered in the laboratory tests, (i.e., do apply extrapolation methods to address possible sensitivity differences between laboratory and field conditions caused by additional matrix or medium-associated stress factors).
3	As in Tier-2, not further specified due to lack of systematic, broadly applicable concepts or approaches.

Note that the number, variety, and type (empirical or mechanistic) of extrapolation approaches are higher for the exposure assessment side of the problem. For the effects assessment, relatively few methods as yet exist, and many of them have a limited applicability (typical compound, typical organism group, etc.). Note further that, over time, when scientific concepts improve and become more operational and when empirical techniques are replaced by mechanistic approaches with broader applicability, the tiers may change accordingly.

Given the tiered system, evidently, any step in the exposure and effects assessment can be considered in a deterministic or probabilistic way. This does affect the outcomes of the risk assessment, but it does not influence the choice of extrapolation methods as guided by the decision tree itself.

With 3 tiers for both exposure and effects assessment, a decision tree can be designed as outlined in Figure 2.8 and described below. In the decision tree, the

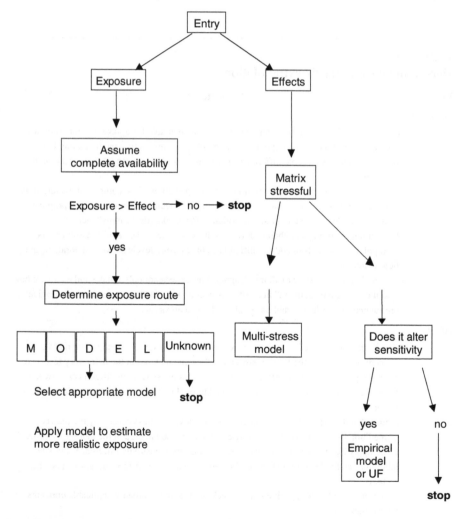

FIGURE 2.8 Decision tree for media and matrix extrapolation.

assessor is asked to answer a series of questions, leading through the tree to the required technique at the appropriate tier in the assessment scheme.

There are 3 levels of questions that can guide users to the available models for matrix and media extrapolation: procedure-, compound-, and matrix-related questions. Procedurally, the user begins the process by responding to a basic question of whether the problem is a typical general problem requiring a tiered and efficient approach that starts from generic approaches and ends in specific, or typical site-specific, assessment problems. In the latter case, one would start with site-specific information in the earliest stages of the assessment. Further, typical procedure questions for matrix and media extrapolation are whether there is a prescribed procedure to be followed (e.g., enforced by the regulatory context), and whether a trigger is defined or to be defined to go to a next tier.

Typical questions related to the compound are whether the toxicity data that are available match with the situation that is being investigated, both regarding the test matrix and the assessed matrix (physicochemical processes determining availability), and regarding the species that were tested and the typical species for the assessed situation (biotic similarity).

Typical questions related to the matrix are as follows:

- What is the medium (water, soil, or sediment)?
- Is there a need for intermedia extrapolation?
- What is the matrix (water type, soil type, sediment type, and their physicochemical characteristics)?
- Is there a need for intramedia extrapolation?
- What properties of the matrix could affect availability (pH, organic matter content, etc.)?
- In which way do these properties act in a compound-dependent way?
- What properties of the matrix could act as stressors?
- Is there reason to believe that there are matrix-specific differences in sensitivity?

Typical questions related to the biota are as follows:

- What is known about the likely routes of exposure?
- Are there behavioral issues that might affect exposure?
- How well do the species for which toxicity data are available relate to the species for which risk is being assessed?

In answering these questions, the risk assessor is guided to the extrapolation methods provided in this chapter.

2.6 UNCERTAINTIES

There are numerous sources of uncertainty and variation in natural systems. These include site characteristics such as water depth and soil type, which may vary from site to site; environmental conditions such as water flow, temperature, and light, which may have a stochastic component; and critical biotic parameters such as maximum photosynthetic and consumption rates, which vary among experiments and representative organisms. In addition, there are sources of uncertainty and variation with regard to pollutants, including pollutant loadings from runoff, point sources, and atmospheric deposition, which may vary stochastically from day to day and year to year; physicochemical characteristics such as octanol–water partition coefficients and Henry law constants that cannot be measured easily; and chemodynamic parameters such as microbial degradation, photolysis, and hydrolysis rates, which may be subject to both measurement errors and indeterminate environmental controls.

Increasingly, environmental analysts and decision makers are requiring probabilistic modeling approaches so that they can consider the implications of uncertainty in the analyses. The user can explore uncertainty by specifying the types of distributions and key statistics for a wide selection of input variables to the risk assessment process. Depending on the specific variable and the amount of available information, any one of several distributions may be most appropriate. A log-normal distribution is the default for environmental and pollutant loadings and concentrations. A sequence of increasingly informative distributions should be considered for most parameters. If only 2 values are known and nothing more can be assumed, the 2 values may be used as minimum and maximum values for a uniform distribution; this is often used for parameters where only 2 values are known. If minimal information is available, but there is reason to accept a particular value as most likely, perhaps based on calibration, then a triangular distribution may be most suitable. Note that the minimum and maximum values for the distribution are constraints that have zero probability of occurrence. If additional data are available indicating both a central tendency and spread of response, such as parameters for well-studied processes, then a normal distribution may be most appropriate.

The extrapolation methods related to differences in media and matrices are mainly governed by sorption processes. Therefore, uncertainty analysis should try to address toxicant properties related to sorption (K_D, K_{OW}, K_a, etc.) and matrix-specific characterization of sorption sites (qualitative and quantitative). Some of the computerized models for exposure and effects assessment explicitly include options for treating key variables as probability distributions (e.g., AQUATOX).

In general, the relationship between the concentration of a contaminant in the pore water and the total concentration in soil is rather uncertain due to geographical differences and lack of knowledge. The following are a few points that can be addressed to improve our ability to make reliable availability predictions:

- Organic matter is an important agent in soils for adsorbing metals as well as organic contaminants. However, part of the organic matter might be in solution (DOC). As a consequence, sorption onto organic matter might, in some cases, increase the total concentration in the pore water and thus the mobility of the toxicant. In other words, the influence of organic matter is not unequivocal.
- In many references that describe the relationship between the concentration of a contaminant in the pore water and the total concentration in soil, the kind of pore water concentration assessed (inclusive or exclusive of organic and inorganic metal complexes) is not mentioned.

- Although many soil organisms are exposed to the freely available metal fraction, other organisms show adverse effects due to intake of contaminated soil particles. In the latter case, the freely available metal fraction is not the right key parameter for assessing ecotoxicological effects.
- In general, the relationship between the concentration of a contaminant in the pore water and the total concentration in soil varies with soil depth due to variations in soil characteristics. Because soil samples in most experiments are taken in the upper soil layer, where at least the organic matter content is relatively high, the calculated available fraction of contaminants generally underestimates the actual average available fraction of contaminants for the whole unsaturated soil profile.
- The best experimental setup to derive the relationship between the concentration of a contaminant in the pore water and the total concentration in soil is under discussion. Field experiments, especially those with in situ contaminated soils, match better with contaminated sites evaluation. Time-dependent influences like pore water content could show a large variation in the pore water concentration with time. Laboratory experiments suffer from lack of reality when spiked soils, sometimes containing high metal concentrations, are used.

2.7 CONCLUSIONS

Media and matrix extrapolation is mainly concerned with bioavailability and fate of toxic substances. To a large extent, the properties of the matrix in combination with the properties of the toxicant determine the uptake of a toxicant by the biota, and thus the consequential effects. A large difference can be observed between the methods available for calculating matrix interference with organic and inorganic toxicants. Organic compounds are considered to follow the rules of equilibrium partitioning between the large-molecule organic constituents of the matrix and the lipid content of the exposed organisms. A number of computer-operated models are available for predicting equilibrium-partitioning exposure. The required input for those models (partitioning coefficients or fugacities and proportions of partitioning compartments) is, in general, easily available. For inorganic toxicants, mainly heavy metals, speciation is considered to govern availability. In the water compartment, metal speciation can be tackled in a mechanistic way. A large number of computer programs are available to calculate the proportion of the metal species capable of entering the exposed organisms. The input to those calculations requires a quantification of a number of water chemical variables (pH, hardness, DOC, etc.). For the soil and sediment compartments, the bioavailable fraction of the metals is, in general, empirically related to a number of soil and sediment characteristics (pH, cation exchange capacity, calcium content, and such).

Different media and matrices may also be characterized by species that have an intrinsic susceptibility that is different for the toxicants they are exposed to. However, knowledge on this aspect is scarce, and it is seldom evaluated on a site-by-site basis.

In media and matrix evaluation, 3 general levels of complexity can be recognized. The simplest approach assumes that all toxicants are completely available to be taken up by the biota. In this case, no extrapolation is required. A slightly more complicated step in the extrapolation process requires the calculation of bioavailable fractions of toxicants. The highest level of complexity additionally includes the action of physiological processes to the expression of the results.

Bioavailability of metals in soil and sediments is a topic that requires strengthening from a scientific point of view. Chemical speciation models for metals in soil and sediment are in the process of being developed. Media- and matrix-related differences in intrinsic sensitivity of species comprise a topic that deserves to be studied.

Bioavailability considerations for both toxic metals and organic chemicals are based either on sound knowledge of chemical processes or on pure empiricism. Both lines are extensively verified.

In the United States as well as in Europe, the use of matrix extrapolation techniques for estimating the bioavailability of toxicants is officially encouraged. An example is the Ohio EPA (1996) recommendation to use hardness-corrected water quality criteria for heavy metals. The EUSES (Jager 2003) computer program of the European Union for predicting the risk of new compounds to be released into the environment corrects for partitioning processes in the calculation for all types of effects.

3 (Q)SAR and Extrapolation

Hans Sanderson, Scott D. Dyer,
and J. Vincent Nabholz

CONTENTS

3.1 INTRODUCTION AND PROBLEM FORMULATION

(Quantitative) structure-activity relationships ([Q]SARs)* are mathematical models, and thus represent an idealized representation of reality based on the theory that a relationship exists between a chemical's structure, its physical or chemical properties, and a measured biological activity. Albert Einstein stated that all models should be evaluated for their usefulness relative to their domain and not to a perceived reality such as experimental data, which themselves are also models. The goal in any (Q)SAR modeling, therefore, is to obtain a mathematical expression that best portrays the relationship between chemistry and biology (Cronin et al. 2003). Between 1950 and 1970, regression analysis and other statistical methods to derive ecotoxicological (Q)SARs increased in use. The driving mechanisms for this were significant reductions in the costs and time consumption for prediction of ecotoxicological and environmental properties of chemicals (Walker 2003). Under the USEPA Toxic Substances Control Act (TSCA) of 1976, the USEPA was required to conduct a risk assessment and show potential risk before it could require test data. Thus the USEPA was forced to use (Q)SARs to predict effect concentrations before they could request toxicity data from the notifier. Three elements comprise the relationship: 1) a descriptor, which could be a physical or chemical entity; 2) an endpoint to be predicted, perhaps another physical or chemical property or a biological activity; and 3) a derived relationship between the descriptors and the endpoint (see Equation 3.1, below). (Q)SARs are used as the first tier in tiered risk assessments and, as such, typically result in the application of the highest uncertainty factors for extrapolation of data, due to the relative remoteness of natural realism in the model. In this chapter, we focus on ecotoxicological (Q)SARs.

3.1.1 REGULATORY POLICIES

In contrast to more basic science, environmental science is often goal driven and should provide the basis for risk assessment, incorporate precautionary measures, and support decision and pragmatic policy making (Escher and Hermans 2002). Additionally, more than 95% of existing chemicals lack publicly available basic acute toxicity data for representative ecotoxicological species. This lack of knowledge emphasizes the need to develop more pragmatic approaches for environmental management to meet the political and public goals of control over our chemical environment, while maintaining a competitive and innovative industry (Zeeman et al. 1995). The European Inventory of Existing Commercial Substances' (EINECS) list of nonassessed commercial chemicals comprises roughly 100 000 different compounds, where chemicals with a production volume of between 1 and 1000 metric tonnes must be registered (~ 30 000 substances). In the EU, risk assessment of chemical substances is driven by the requirements of Directive 93/67/EEC on risk assessment for new notified substances and Commission Regulation (EC) No. 1488/94 on risk assessment for existing substances (a Technical Guidance Document, or TGD). The European Union decided in 2001 to develop a new chemicals policy strategy

---

* (Q)SARs include both QSARs and SARs (structure-activity relationships without quantified predictions).

called REACH (Registration, Evaluation, and Authorization of Chemicals). The policy was adopted in December 2006 and initiated June 2007, and is set to be completely implemented by 2012. With this new proposal for future chemicals policy, the burden of proof is reversed under the slogan "No data — no market." In this procedure, there will be no discrimination between existing and new chemicals. The EU is obliged by the 3 R's (reduce, replace, and refinement) strategy for animal testing in this process and thus places special emphasis on the development and use of (Q)SAR models to reduce the use of animals in toxicity testing — it is estimated that up to 30 animal studies are needed to characterize one substance (Danish Environmental Protection Agency 2001; European Commission 2001). Environment Canada and Health Canada were required by the revised Canadian Environmental Protection Act (CEPA) to categorize 23 000 substances by September 2006 for persistence and/or bioaccumulation potential and inherent toxicity. Japan initiated its assessment of existing chemicals and implemented these into the existing premarket evaluation criteria of new substances on April 1, 2004. The USEPA annually reviews 2300 substances under the TSCA for potential hazards to human health and the environment; 95% of these have no publicly available ecotoxicological data. The substance must be reviewed by the USEPA within 90 days for premanufacture notification (the PMN program; Russom et al. 2003). Within the OECD, (Q)SARs can be applied to a high-production-volume (HPV) chemicals program, under which a screening information data set (SIDS) is produced.

Based on US dollars, cost estimates to produce basic experimental physical property and environmental fate data range from $65,000 to $125,000 for 1 substance. If (Q)SARs were used, it would cost between $17,000 to $25,000 to provide all the basic physical properties and environmental fate parameters, yielding savings of $50,000 to $100 000. In light of time constraints for the registration of chemicals, the number of substances being registered, and the high monetary savings from using (Q)SARs, interest in developing (Q)SARs and their uses in the regulatory arena has increased significantly in recent years (Walker 2003). With this increased use has come recognition of the importance of calibrating (Q)SAR models and of exercising caution during interpolation and extrapolation of data beyond the models domain, both of which can affect the magnitude of the associated uncertainty factors (10–10 000) in risk management. Especially more recently in the European Union, in Canada, and within the OECD countries, this issue is of urgent importance along with approaches to harmonized development and use of (Q)SARs in the regulation of chemicals, and the definition of domain of applicability, use, and magnitude of assessment factors. Therefore, the OECD started a new activity in November 2002 aimed at increasing the regulatory acceptance of (Q)SARs and the identification of validation criteria for (Q)SARs in order to prevent member countries from unilaterally developing and applying national and inconsistent (Q)SARs. General guidance on decision making based on (Q)SARs is not practical, and should be left to the specific regulatory authority (OECD 2003, Section 1.4.4). (Q)SARs are used in the lower tiers of the risk assessment process and have traditionally been used for priority setting rather than actual risk assessment; however, the importance of prioritization may be as great as the risk assessment process itself, because the consequences of not selecting the correct substances in priority setting can be significant in terms of resources and monetary commitment. Because (Q)SARs comprise the first tier in risk assessment, they are intentionally created conservative (to protect against false negatives).

3.1.2 PROBLEM FORMULATION

The aim of this chapter is to introduce ecotoxicological (Q)SARs in general and give examples of current and potential regulatory uses of (Q)SAR models. Because (Q)SARs are typically applied in the absence of toxicological data, we will focus on the uncertainty associated with using (Q)SAR models in extrapolation. To this end, we have conducted a case study with 4 chemicals (2,4-dichloroaniline, pentachlorophenol, nonylphenol, and linear alkylbenzenesulfonate) using the USE-PA's EPIWIN package and the University of Missouri USEPA's ICE (interspecies correlation estimation; Asfaw et al. 2004) program to develop species sensitivity distributions (SSDs) for estimating 5th percentile hazard concentrations (HC5). These results are then compared to those from toxicity tests with and without default assessment factors.

3.2 (Q)SAR DEVELOPMENT

3.2.1 GENERAL

The ideal (Q)SAR should have a well-defined and measurable endpoint based on a diverse data set, and a statistical method that is transparent and appropriate to the end-point data. It should consider an adequate number of chemicals for sufficient statistical representation and include a reasonable distribution of active and inactive chemicals. A wide range of quantified toxic potency (i.e., several orders of magnitude) should be included in the training data set and if possible yield a mode of action or mechanistic interpretation. Data sets used to develop the model must meet the basic requirements underlying the statistical procedure used to develop the (Q)SAR model. The most significant limiting factor in the development of (Q)SARs is the availability of high-quality experimental data. In the European Union, it is policy that if an approved test guideline or protocol is used, then the resulting test data are valid a priori; however, accessible laboratory reports may be too incomplete to provide an adequate validation of the results. The use of test guidelines is not the same as good laboratory practice (GLP). Moreover, the (Q)SAR is typically only applicable to chemicals similar to those used in its development (interpolation; Walker et al. 2003b).

3.2.2 USE OF THE OCTANOL–WATER PARTITION COEFFICIENT (K_{OW})

A significant descriptor for baseline toxicity is the K_{OW}. Baseline toxicity (or narcosis) is believed to be a result of nonspecific disturbance of membrane integrity and function as a result of partitioning of the xenobiotic across biological membranes. Baseline toxicity is the reference case for most (Q)SARs because it is assumed to represent the minimal toxicity of any given chemical. Basal cellular structures, functions, and membranes are highly conserved and similar in most biological entities. Therefore, a large number of toxic effects that target the cell are universal in all organisms and target tissues. On the other hand, there are also more specific modes of action (e.g., uncoupling of oxidative phosphorylation, acetylcholinesterase [AChE] inhibition, etc.) that may be easily modeled, but for which K_{OW} is not a good descriptor of toxicity. The ability of the xenobiotic to interact, pass through, and disturb the cell membrane is the first step for setting up a predictive model across different organisms. The octanol–water partition coefficient is a ratio of the concentration of the chemical in n-octanol and water when the 2 layers are equilibrated with each other. Chemicals with a high log K_{OW} preferentially partition into octanol. It is assumed that octanol may act as a surrogate for tissue, lipids, and membranes, and thus, by plotting the log K_{OW} versus the log LC50 of a compound, it is possible to derive a statistical relationship between the descriptors and the endpoint (see Equation 3.1; Walker et al. 2003b).

Most ecotoxicological (Q)SAR software aims to identify substructures of the compound that appear mostly in active molecules and may therefore be responsible for observed activity. The (Q)SAR software will, depending on the sophistication, generally start by identifying possible linear relationships between K_{OW} and observed toxicity (baseline toxicity or narcosis), when the toxicity of an active compound can be explained by its lipophilicity or when the chemical's mode of action is unknown. Narcosis is a reversible, nonspecific state of arrested activity of proplasmic structures in the cell membrane. The exact narcosis mechanism remains an area of active research with hypotheses centering around lipid membrane perturbations or binding to specific lipid proteins (Bradbury et al. 2003). A survey showed that about 25% of the toxicity of tested chemicals could be explained simply by their narcosis effect; within the remaining 75% of chemicals, the narcosis effect had a strong influence on the overall toxicity (Klopman et al. 2000). Other studies suggest that nearly 70% of all industrial organic chemicals are estimated to act via baseline and polar narcosis modes of action in acute exposures (1 to 14 days; Bradbury et al. 2003). Narcosis related to the lipophilicity of the compound may be modeled using K_{OW} as the only descriptor. Chemicals may have multiple modes of action, which can obscure (Q)SARs that estimate nonspecific narcosis. Niederlehner et al. (1998) found that a (Q)SAR describing the 48-hour EC50 as a function of log K_{OW} accounted for 91% of the variability in response across model chemicals, and a similar (Q)SAR for chronic effects on reproduction accounted for 79%.

Through a careful selection of descriptors and model development, the resulting (Q)SARs may lead to predictions of reasonable accuracy. (Q)SAR models generally work according to

$$P = f(D_{structural}, D_{electronic}, D_{hydrophobic}, D_x) + e \qquad (3.1)$$

where

P: properties (endpoint)

$D_{s,e,h,x}$: descriptors of the molecule

e: noise

The flowchart in Figure 3.1 describes the typical role of (Q)SARs for risk assessment and prioritization of further experimental requirements, especially in the absence or paucity of experimental results, where (Q)SARs will play a significant role.

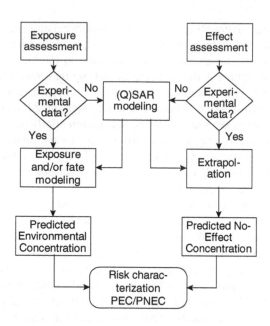

FIGURE 3.1 Flowchart of typical role for (Q)SAR in risk assessment.

3.2.3 TRAINING SETS AND APPLICABILITY DOMAINS

The overall lipophilicity of a compound is typically based on adding all the contributing fragments from the molecule according to the universal quasi-chemical functional group activity coefficient (UNIFAC) method. The UNIFAC model basically calculates differences in energies, which subsequently are used to predict physicochemical endpoints such as solubility, partitioning, and so on (Carlsen 2003). The domain of a (Q)SAR is established by the training set of chemicals. The model domain is defined as the types or classes of compounds that may be studied by the model. Rarely is a training set selected in an ideal manner due to the paucity of high-quality experimental data. It is important in both the training and test set selections to ensure that all substructures that are of interest or are likely to have a significant impact on the results are adequately represented. Extrapolation to chemicals beyond the domain of the model is not warranted. It is critical that the process of (Q)SAR calibration should test the predictive capability of the relationship, examine the restrictions of its applicability, and evaluate its mechanistic hypothesis. Statistically, a (Q)SAR needs to have significant goodness-of-fit statistics that indicate how well the model is able to explain variability in the training data set. The (Q)SAR should also be evaluated for internal goodness-of-prediction with cross-calibration. Finally, it needs to be addressed in terms of its predictive power by using data that were not used in the development of the model (external calibration). External calibration is not a substitute for internal calibration (Walker et al. 2003a).

The USEPA Office of Pollution Prevention and Toxics (OPPT) (Q)SAR Analysis Methods Branch assesses the model domain for ecotoxicity (Q)SAR determined by structural attributes used to predict toxicity. For example, the domain for molecular weight (MW) covers 2 types of toxicants: systemic toxicants and surface-active toxicants. The MW domain for systemic toxicants (i.e., chemicals that have to be absorbed to be toxic) is less than 1000 MW (ECOSAR Suite help files). Chemicals with MWs greater than 1000 are assumed to be too large to be absorbed through membranes. The MW domain for surface-active toxicants has no limits. For example, cationic chemicals, such as polycationic polymers, can show high toxicity with MWs $\geq$ 3 million as long as they are dispersible in water (ECOSAR Suite help files).

The domain for the octanol–water partition coefficient is generally less than log 5.0 with (Q)SARs for neutral chemicals predicting acute toxicity to fish and daphnids and less than 8.0 for (Q)SARs predicting chronic toxicity to fish and daphnids. The USEPA OPPT relies on predicted K_{OW} as the basis of its (Q)SARs because measured K_{OW} is unavailable for more than 96% of chemicals. The OPPT has to predict the toxicity of every new chemical regardless of its position in the chemical universe. Thus, the practical operating range of (Q)SARs for neutral chemicals is –3.0 to 5.0 for acute toxicity and –3.0 to 8.0 for chronic toxicity. The range of log K_{OW} values for chemicals actually contributing to the (Q)SAR may be much smaller, for example, 0.0 to 3.0. In order for the OPPT to do its assessments, toxicity predictions cannot be limited to interpolation within actual test data determining the (Q)SAR, but must include extrapolation most of the time. ECOSAR will provide EC50 values for all substructures of the molecule, and the choice of which value and substructure that are the best descriptors of the molecule's toxicity is left to the user to decide. Often the lowest value is chosen.

The domain of log K_{OW} for chemical classes that can be ionized, such as amines, phenols, and aldehydes, or for organometallics, such as organotins, may range beyond 8.0 and is defined for each chemical class. For example, the log K_{OW} range for fish acute toxicity by aliphatic amines based on predicted log K_{OW} for the free amine ranges up to at least 22 based on measured toxicity data. The USEPA OPPT (Q)SAR for aliphatic amines is based on predicted log K_{OW} for the free amine and toxicity data based on the ionized amine tested as a soluble salt (e.g., chloride at 100% active ingredients [AI] at pH 7, hardness less than 150.0 mg L^{-1} as CaCO$_3$, and total organic carbon [TOC] concentration of dilution water of less than 2.0 mg L^{-1}). The domain for some surfactant (Q)SAR classes is defined by the USEPA OPPT as the number of linear carbons in the hydrophobe of the surfactant and can range from 1 to more than 100 carbon atoms (e.g., the domain of alkyl sulfonate anionic surfactants). The range of surfactants with actual measured toxicity data may be much less, for example, 6 to 20 carbon atoms. In the future, the domain of surfactant classes will be defined as the predicted log K_{OW} of the hydrophobe.

The USEPA OPPT cannot design training sets, nor can it measure the toxicity of industrial chemicals directly. The TSCA prescribes that the chemical industry test chemicals for toxicity; thus, the OPPT is dependent upon what toxicity data are submitted to the USEPA under the TSCA. The OPPT could design a training set for a (Q)SAR such as fish acute toxicity for aromatic diazoniums, but it does not have the ability to get the chemicals in the training set tested. Thus, some (Q)SARs used by the OPPT have training sets composed of two data, one datum, or no data — just assumptions about intercept, slope, and log K_{OW} at which no toxic effects at saturation will occur.

The ecological structure activity relationship (ECOSAR) model does not have any computerized or statistical procedure for removing outliers; the USEPA then relies upon expert judgment. Toxicity data for chemicals not used in a (Q)SAR are listed with that (Q)SAR, and a reason is given for not using the data. If a chemical's measured toxicity is more than 10 times of that predicted by its (Q)SAR, then the chemical is investigated for test conditions that could have enhanced the chemical's toxicity or possible excess toxicity due to a more specific toxic mode of action. If the observed excess toxicity can be explained by theoretical and/or metabolic consider-ations, then a new (Q)SAR class is developed for that chemical, and other chemicals could belong to a homologous series containing the tested chemical. An analogous procedure is followed for a chemical whose measured toxicity is 10 times less than predicted. Test conditions of the toxicity test are investigated for attributes of dilution water, chemical identity, and/or purity, which could have mitigated the bioavailabil-ity and toxicity. If the observed toxicity cannot be explained by testing conditions, then the chemical is investigated for reasons why the chemical was placed in the wrong (Q)SAR chemical class.

3.2.4 Statistical Models

Several different statistical models may be used in the development of a (Q)SAR (Carlsen 2003; Eriksson et al. 2003). These are discussed below.

3.2.4.1 Linear Regression

The simplest model is linear regression. A series of descriptor values together with the corresponding endpoint values are fed into the program, which subsequently calculates the best straight-line fit through the data points, typically requiring an R^2 value of 0.95. The advantage of this method is its simplicity and transparency as well as the general availability of software. On the other hand, the applicability is rather limited because only 1 descriptor can be taken into account in relating endpoints to descriptor values.

3.2.4.2 Multilinear Regression

Multilinear regression can be used where the investigated endpoint is correlated to a linear combination of independent variables (the descriptors). This technique assumes linearity over the whole data set with respect to the descriptors. In addition, normality of the data must be fulfilled, and the descriptors cannot be intercorrelated. Multilinear regression is widely used in (Q)SAR modeling and has the advantage that all numerical information is retained and the predicted endpoint may be better estimated. However, the model may eventually "overfit" the data, after which the addition of further descriptors causes a decrease in accuracy of the model; however, this will typically be disclosed in the calibration step of development.

3.2.4.3 Ordination Methods

Obviously, chemical compounds can be characterized by a wide variety of descrip-tors. Principal component analysis (PCA) typically transforms large numbers of possibly intercorrelated "raw" descriptors into a relatively low number of mutually uncorrelated latent descriptors that are linear combinations of the original descriptors. The developed set of latent descriptors can subsequently be used to summarize the original data set without too much loss of information. However, the problem of finding relationships among latent descriptors, as in multilinear regression, requires a specific functional relationship to be assumed. Even if a reduction to 2 dimensions is possible, the problem of ordering remains. Finally, the interpretation of latent variables is often not easy.

Principal component (PCR) regression combines the PCA and the multilinear regression methodology. Thus, by means of the principal component analysis, the actual number of original descriptors is reduced to a limited number of latent descriptors. These are subsequently applied as descriptors in a conventional multilinear regression model. The advantages of PCR appear to be similar to those of the multilinear regression method; they are relatively simple. The method requires some calculation effort to retrieve the actual endpoint, hampering the transparency of the method.

3.2.4.4 Partial Least Squares

Partial least squares (PLS) regression is a technique related to PCR, although it is performed in a somewhat different way. In contrast to PCR, PLS is a 1-step method, where the latent descriptors are determined and used in 1 operation; PLS does not require a separate regression step as in the case of PCR. PLS is, together with multilinear regression, one of the most widely used techniques for custom-made (Q)SAR model development. A main advantage to this methodology is that PLS, unlike PCR, yields results that are directly related to the endpoints of interest. Further advantages are that it 1) can handle missing data, 2) generally has a robust calibration, and 3) can handle intercorrelated physicochemical descriptors. Disadvantages include reduced transparency and interpretability. In order to obtain accurate models, a large number of samples for calibration are generally required. PLS models can, in principle, be applied for interpolation as well as extrapolation due to the linear nature of the relationships.

3.2.4.5 Partial Order Ranking

Partial order ranking (POR) is based on elementary methods of discrete mathematics (e.g., Hasse diagrams) — if A < B and B < C, then A < C in the ranking procedures. POR does not assume linearity or any assumptions about distribution properties such as normality. The disadvantage is that often a preprocessing of data is needed to avoid the effects of stochastic noise. Combining POR with PCA may improve its usefulness. POR can only be applied for interpolation.

3.2.4.6 Artificial Neural Network

The artificial neural network (ANN) is a relatively new technique and possibly the preferred one for current and future (Q)SAR development. Basically, ANNs can be regarded as multinonlinear regression methods. Thus, the neural network software simply multiplies the input by a set of weights that in a nonlinear way transforms the input to an output value. ANNs are a form of a multiprocessor computer system that aim to mimic the way the human brain works with 1) simple processing elements, 2) a high degree of interconnectedness, 3) simple scalar messages, and 4) adaptive interactions between the elements. Roughly speaking, neural networks operate by assigning individual weights to the single descriptors in such a way that the end-points of the training set are mimicked by the calculation. In this way, the network is trained, which simply means that it uses these examples to establish (learn) the actual relations that exist between the input descriptors and the endpoint by setting these weights. When the relationships between the descriptors and the endpoints are established, the model can subsequently be used for prediction endpoints unknown to the neural network by feeding the network with respective descriptors corresponding to these compounds. The major risk in applying ANN is overtraining, which eventually leads to erroneous predictions, as the model will try to include minor variations in the data of the training set as being significant. ANN can exclusively be applied for interpolation (limited to the model domain). Two of the major advantages of the ANN approach are the generality and applicability of the generated models. Disadvantages include slow data processing, the need for considerable computer power, and a lack of transparency; indeed, ANN models are virtually black box models (Carlsen 2003; Eriksson et al. 2003).

3.2.4.7 Comparative Molecular Field Analysis

Comparative molecular field analysis (CoMFA) enables prediction of the ability of chemical substances to interact with various receptors. The basic principle is that the total energy of a system consisting of the receptor site and the substances under investigation is minimized. The better the substance (often named "ligand") fits into the receptor site, the lower the energy. The technique has been extensively used in relation to the design of pharmaceuticals and recently has been of increasing interest for environmental modeling. The method is not limited to a certain class of chemicals and can be recognized as a noncongeneric method. The method unambiguously relies on a detailed knowledge of the receptor site. However, the number of known structures of receptor sites is rapidly increasing, opening new possibilities in the area of environmental (Q)SARs (Carlsen 2003). The USEPA is developing a (Q)SAR based on CoMFA modeling for estrogen receptor binding that incorporates 3-D chemical structure assessment, identification of a bioactive conformer, and molecular alignment (Schmieder et al. 2003).

3.2.5 Limitations and Outlook of (Q)SAR Models

Accurate prediction of mixtures of chemicals is one of the future challenges for risk assessment and (Q)SAR modeling. Most compounds are present in the environment at concentrations far below their individual median effective concentration (EC50 or LC50) and possibly below their no-observed-adverse-effect concentrations as well, yet they may contribute to substantial effects through combination with other chemicals. It is theoretically possible to construct a (Q)SAR model to predict mixture effects; however, it is generally difficult to validate their predictive power, due to a lack of experimental calibration (Altenburger et al. 2003). Semiempirical quantum chemical and force-field methods are widely used in drug discovery and design, but have not been considered in environmental (Q)SAR contexts, in part due to a lack of efficient computational software and associated hardware to permit real-time desktop 3-D calculations. However, these techniques are becoming more widely applicable, as is shown by Schmieder et al. (2003), and indicate a future challenge for environmental (Q)SAR modeling (Bradbury et al. 2003). For nonassessed chemicals, the main problems are the uncertainty in modes of action of the compound in different organisms, the whole body burden (Escher and Hermans 2002), attributing chemicals to specific classes, and, thus, determining the domain of the model (Bradbury et al. 2003). Octanol is not an optimal surrogate for biological membranes, and, in future work, conventional physiochemical descriptors should be supplemented with other toxicologically relevant parameters (e.g., derived from toxicogenomic analysis). Thus, the internal effect concentration might be over- or underestimated using octanol as the model. A step beyond current environmental (Q)SARs is setting up predictive models that would combine toxicodynamic information with toxicokinetic modeling (physiologically based toxicokinetic, or PB-TK) to analyze and predict concentrations in target tissue and target sites. If combined with dynamic aspects via physiologically based toxicodynamic modeling, a more complete ecotoxicity picture is obtained. This will give insights into rate-limiting steps in, and a theoretically based mathematical model of, the whole chain of events from dose to observable effect. This model could then feed back into the development of (Q)SAR; such models have been developed for human health risk assessments (Yang et al. 1998; Escher and Hermans 2002). However, this process should be done with care, as toxicity is a complicated effect and many steps involved in identifying toxic modes of action are poorly understood and characterized. Ren (2003) compared predicted toxicity obtained using a direct (Q)SAR toxicity prediction based on narcosis and a multiple mechanism identification toxicity prediction (Q)SAR relative to the observed toxicity of 206 phenols. The result indicated that toxicity could be significantly over- or underestimated due to the incorrect identification of the mechanism in the multiple mechanism identification toxicity prediction (Q)SAR approach. The more simple and direct (Q)SAR toxicity based on narcosis outperformed the mechanism-based (Q)SAR. If mechanisms could be correctly predicted in the first place, the mechanism-based (Q)SAR could potentially yield more accurate and precise predictions than the more simple and direct toxicity (Q)SAR (Ren 2003). This cannot be generalized to all bioactivated or receptor-mediated reactions, but it illustrates that the processes in acute toxicity tests can be less specific and therefore less complex than effects in other tests (Hermens 1991). Escher and Hermens (2002) critically reviewed the role of modes of action in ecotoxicology in relation to (Q)SARs, body burdens, and pharmacokinetics and pharmacodynamics, and concluded that these entities are very complex and often oversimplified in ecotoxicology. A better understanding of toxicological modes of action and membrane integrity is required for receptor-mediated toxicity to be discernible in standardized toxicity tests. Currently, standard acute toxicity tests are designed for identification of severe effects, typically as a result of multiple and significant disruptions of membrane integrity, and this is why narcosis-based (Q)SAR models based on the standard toxicity data domain will be predictive of toxicity for most compounds.

Prediction of receptor-mediated toxicity would require a training set based on other than short-term, primarily acute, standardized toxicity tests.

The predictive ability varies with the (Q)SAR model; generally, an acceptable concordance of > 70% to 85% of the chemicals examined is recommended. Of course, a model can never be more precise than the test data on which it was based (the training set). Therefore, it is important to be aware of the precision and reproducibility of the test data used for developing a model. For example, if a biological test gives wrong results 17% of the time, the perfect model based on these results will also be wrong 17% of the time. Furthermore, the compounds screened with (Q)SAR models might possess other hazardous properties that are not taken into account in the model domain or that are receptor driven. (Q)SARs may be developed that either have high sensitivity and low specificity, yielding a high number of false positives, or low sensitivity and high specificity, yielding a high number of false negatives. This is the general optimization dilemma for the development of (Q)SARs (Walker et al. 2003a). Another general dilemma is the risk of overfitting models and decreasing the transparency of the model with increased sophistication, and potential inclusion of additional data uncertainty. ECOSAR output encompasses multiple effect concentrations for a single chemical, which can be seen to represent measurement uncertainty within and among laboratories, which is not insignificant, with coefficients of variation (CVs) of 48.1% for *Ceriodaphnia* 21-day reproduction tests among labs and CVs of 47.6% within labs ($n = 11$) for *Daphnia pulex* acute 48-hour LC50 (Warren-Hicks and Parkhurst 2003).

3.3 (Q)SAR UNDER EPI SUITE VERSION 3.12

3.3.1 ECOSAR

We focus on ECOSAR for the following reasons: 1) in a recent comparative analysis, ECOSAR (Ecological Structure-Activity Relationship v0.99g) under the EPI Suite was ranked highly as an effect prediction program (Moore et al. 2003); 2) the program and training sets have been developed and used extensively by the USEPA for more than 20 years; 3) several countries and organizations use the EPI Suite (Carlsen 2003); and 4) although licenses for many (Q)SAR models can be prohibitively costly for wide application, the EPI Suite package is free and can be downloaded from the Internet from the USEPA website (USEPA 2007). The package contains a series of (Q)SARs ranging from physiochemical to biological and exposure parameters. These include K_{OW}, Henry's Law constant, boiling and melting points, water solubility, atmospheric oxidation, soil absorption coefficients, hydrolysis, bioconcentration factors, fate and exposure predictions, and ecotoxicological endpoints. A predicted K_{OW} from the EPI Suite (KOWWIN v1.66 program based on the structure and components of the compound) is automatically generated as a prerequisite for ecotoxicity predictions with ECOSAR.

The ECOSAR program is used to predict the aquatic toxicity of chemicals based on their similarity of structure to chemicals for which the aquatic toxicity has been previously measured. Since 1981, the USEPA has used (Q)SARs to predict the aquatic toxicity of all new industrial chemicals (Nabholz et al. 1993; Zeeman et al. 1995). The acute toxicity of a chemical to fish (both fresh- and saltwater), water fleas (daphnids), and green algae has been the focus of the development of (Q)SARs. These organisms are group model-organisms and thus not specific species. The type of predictions cover fish 96-hour LC50, daphnids 48-hour LC50, and green algae 96-hour EC50 (data for 72- and 96-hour tests are combined). ECOSAR predicts fish chronic value (ChV) (28 to 120 days), daphnids ChV (14 to 21 days), and green algal ChV (data for chronic 72- and 96-hour tests are combined). Only validated data are used in ECOSAR; ideally, these data are based on the flow-through method or static renewal method with mean measured concentrations; test data based on a static method and nominal concentrations are used in (Q)SARs only when there is a high probability that these data are equivalent to toxicity data based on flow-through or measured methods. ECOSAR will select all the (Q)SAR classes that may apply to a chemical; for example, an ester, phenol, aniline, and epoxide will be selected when all moieties are in a molecule and (Q)SAR predictions are given for each (Q)SAR class selected; it is up to the users to select the effective concentrations (EC) that they want to use for the chemical. In practice, the OPPT will select the prediction that is the most toxic; thus, for the above chemical, the fish 96-hour toxicity value may be based on the phenol (Q)SAR, the daphnids 48-hour EC50 may be based on the aniline (Q)SAR, the algae 96-hour may be based on the ester (Q)SAR, and so on. The test data used are validated by the USEPA OPPT. Mean measured concentrations based on parent material are preferred, but nominal-based data can be used if the data are equivalent to the mean measured data; dilution water hardness = 150 mg L^{-1} as $CaCO_3$ or less, dilution water TOC is less than 2.0 mg L^{-1}, and all data are based on 100% AI of parent material. All chemicals that can be ionized in water (acids, bases, anilines, aliphatic amines, phenols, etc.) are tested as a soluble salt at pH 7, such as aliphatic amine stock solution with HCL added and pH adjusted to pH 7; this is done because effluents in the United States are neutralized to pH near 7 prior to release to the sewer in order to protect the sewage treatment plant — thus, the aquatic environment is only exposed to the ionized chemical near pH 7. Chemicals that are dispersible in water similar to detergents or surfactants are tested as the whole dispersed material. Ionizable chemicals are tested at pH 7 as a soluble salt, generally Cl or Na, and the toxicity test results are expressed in terms of the soluble salt at pH 7 or the ionized state at pH 7; however, the (Q)SARs are in terms of the predicted log K_{OW} for the free amine, the free acid, the free base, the free phenol, and so on. The (Q)SAR therefore expresses the EC50 in terms of the ionized chemical at pH 7 regressed against the predicted log K_{OW} for the free amine. Thus, for an aliphatic amine, use input SMILES (see Section 3.3.2, below) for the free amine, use the predicted log K_{OW} for the free amine, and ECOSAR gives test results for the ionized amine at pH 7. Thus a valid comparison of applicant (industry) data and ECOSAR predictions requires that the industry EC50 be for the ionized amine at pH 7. Industry data for free amines will show either that the low molecular weight amine is more toxic than predicted or that the high molecular weight amine is less toxic than predicted when compared to ECOSAR. The low molecular weight amines change the pH of the dilution water to 8 to 9, and the high molecular weight amines are insoluble in water as the free amine. High molecular weight amines are dispersible in water at pH 7 as a soluble salt just as surfactants or detergents are dispersible in water (i.e., they form micelles; Nabholz et al. 1993; American Society for Testing and Materials [ASTM] 1993).

(Q)SARs are developed for chemical classes based on measured test data that have been submitted by industry, or they are developed by other sources for chemicals with similar structures (e.g., phenols). Using the measured aquatic toxicity values and estimated log K_{OW} values, regression equations (currently more than 150 for more than 50 chemical classes) can be developed for a class of chemicals. A note of caution to users was issued by Kaiser et al. (1999), who questioned the scientific calibration and number of regressions (n) in the ECOSAR program. However, this critique should be considered in view of a study of 8234 discrete organic chemicals on the TSCA inventory; about 95% of the chemicals fell into only 7 chemical classes (neutral organics, esters, acids, amines, phenols, aldehydes, and anilines). The number of regressions for 6 of those key chemical classes were neutral organics (60), esters (26), amines (52), phenols (53), aldehydes (42), and anilines (19). Hence, for the majority (approximately 95%) of TSCA compounds, the size of n is not a domain-limiting problem. The more frequent the chemical structure occurs in tested substances, the more preexisting data (which is the limiting factor for all (Q)SARs), the more regressions, and the more reliable predictions (Zeeman et al. 1995). The ECOSAR class program is a computerized version of the ECOSAR analysis procedure as currently practiced by the USEPA OPPT. It has been developed within the regulatory constraints of the TSCA and is a pragmatic approach to (Q)SAR as opposed to a theoretical approach (Meyland and Howard 1998).

3.3.2 SMILES

SMILES is an acronym for "simplified molecular input line entry system." It is a chemical notation system used to represent a molecular structure by a linear string of symbols in the EPI Suite. A SMILES notation depicts a molecular structure as a 2-dimensional picture as if drawn on a piece of paper. A 2-dimensional drawing of a single chemical structure is possible in many different forms. SMILES notations are comprised of atoms (designated by atomic symbols), bonds, parentheses (used to show branching), and numbers (used to designate ring-opening and -closing positions). With the exception of designating ring positions, numbers are not used in SMILES notation (Howard 1998). The complementary database of Chemical Abstract Service (CAS) registry numbers, with corresponding SMILES notations of more than 103 000 compounds, allows the user to search for and input SMILES notation for a compound via its CAS number. When the ECOSAR identifies a (Q)SAR related to the compound, a "neutral organic (Q)SAR," primarily a log K_{OW}-based output, can be derived. This output is less exact than with excess toxicity predictions due to more specific modes of toxic action of the compound based on the structure and the molecule's entities (aliphatic, phenols, or other properties). With excess toxicity, the toxicity is quantified, and a new (Q)SAR is developed for that class (Nabholz et al. 1993). With the excess toxicity present, the fit to regressions becomes more accurate as more aspects of the compound contribute to the USEPA's continuously updated toxicity database (personal communication, Nabholz 2003 Feb 2002).

3.3.3 CALIBRATION OF ECOSAR

Calibration (external validation) of ecotoxicological (Q)SAR predictions is obviously a major concern for the USEPA because it regulates newly notified chemicals on the basis of (Q)SAR predictions. A joint US and EU validation project (the Minimum Premarket Data–Structure-Activity Relationship [MPD–SAR] study) was performed in the early 1990s (OECD 1994; USEPA 1994) as well as an internal OPPT–USEPA validation study (Nabholz et al. 1993). The MPD–SAR study was conducted with 145 new chemicals from the European Union, and the internal USEPA study was conducted with 462 chemicals, primarily new chemicals (413) but also 49 existing chemicals. In both studies, predicted effective concentrations (ECs) were compared to measured ECs. The ratio for a perfect prediction was equal to 1; thus, ratios of less than 1 indicated overprediction of toxicity and ratios of greater than 1 indicated underprediction of toxicity. Acceptable predictions were determined to be within an order of magnitude of the ideal ratio of 1.0 (i.e., from 0.1 to 1 or from 1 to 10). The internal USEPA validation study was conducted before the MPD–SAR study. The 462 chemicals had 920 individual tests or ECs associated with them and consisted of fish acute, daphnid acute, green algal EC50, fish chronic value (ChV), daphnid ChV, and algal ChV. This allowed for 920 comparisons or ratios to be calculated. Eighty-five percent of the predictions were within the acceptable range. Only 9% of ECs had ratios < 0.1 (overprediction of toxicity by 10 times), and 6% of ECs were underpredicted by a factor of 10. With perfect agreement defined as a (Q)SAR predicted toxicity value to a measured toxicity value equal to 1.0, the average ratio was 0.72, indicating a tendency toward conservative predictions. It was concluded that ECOSAR performed extremely well in predicting the acute toxicity to fish and daphnids (Zeeman et al. 1995).

In the MPD–SAR study, the USEPA and the European Union compared the predicted and measured ECs for the European Union's new chemicals. The measured ECs were those reported in the European Union's MPD set of experimental toxicity summaries. When the USEPA-predicted ECs for fish and daphnid acute toxicity values were compared to the appropriate MPD-measured acute values, there was, respectively, 77% and 59% agreement, 7% and 19% underprediction, and 16% to 23% overprediction by the USEPA. Potential reasons for the under- and overprediction were investigated, and 17 of the underpredictions and 21 of the overpredictions remained unresolved. Therefore, studies that had potential problems were eliminated, and the analysis was repeated. The highest quality subset of the data indicated 87% and 79% agreement between predictions and measured values.

When the USEPA's level of concern for environmental toxicity for a chemical was compared to the level of concern based on the MPD-measured ECs for fish and daphnid acute toxicity tests, there was 54% agreement, 4% to 9% underprediction, and 43% to 38% overprediction by the USEPA, respectively. Potential reasons for under- and overpredictions were investigated; for example, the USEPA based its concern level on both acute and chronic toxicity rather than acute toxicity alone. Once accounted for, only 6 of the underpredictions and 2 of the overpredictions remained unresolved. Therefore, MPD studies that had potential problems were eliminated, and the analysis was repeated. The highest quality subset of MPD-measured ECs indicated 97% and 93% agreement (within one order of magnitude) between predicted and measured concern levels. On average, both the European Union and USEPA experts concluded that the USEPA predictions of environmental toxicity (i.e., all fish, daphnid, and green algal toxicity values regardless of potential quality) were in excellent agreement with the MPD-measured data set. The average validation ratio was 0.71. Eighty-nine

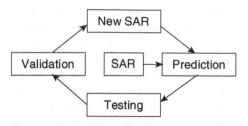

FIGURE 3.2 Evolution of ECOSAR in the USEPA.

percent of all predictions were within a factor of 10, and 74% were within a factor of 5. The MPD-measured ECs caused only 8% of the EPA-predicted concern levels to be changed, but half of these changes (i.e., 4%) might had been found to be false if complete toxicity study test reports had been available for test data validation by the USEPA (Zeeman et al. 1995). The advantage of the USEPA's (Q)SAR analysis over the MPD data set is that the USEPA method assesses all of the potential effects and concerns for a chemical (e.g., both acute and chronic toxicity to fish, invertebrates, and green algae for both freshwater and saltwater environments, including benthic organisms, aquatic insects, and submerged aquatic vegetation). In addition, potential effects to terrestrial organisms (e.g., birds, earthworms, insects, vascular plants, and soil microbes) are evaluated. In the European Union, MPDs were generally restricted to the fish and daphnid acute toxicity tests. This joint EU–US MPD–SAR validation study suggested that the MPD data set could be improved by adding the daphnid chronic toxicity test and the green algal toxicity test, and by requiring a screen for analogy of new chemicals in the EU to known classes of pesticides (OECD 1994; USEPA 1994).

Conservative use of data from ECOSAR overestimated the toxicity 80% of the time relative to experimental-derived toxicity data for pharmaceuticals (Sanderson et al. 2003). However, this comparison only covered some 20 compounds out of the more than 70 pharmaceuticals reported in the environment. Cleuvers (2003) found that (Q)SARs were more conservative than measured data on pharmaceuticals and that narcosis was the main expression of toxicity. ECOSAR has previously been used to develop a framework for prioritizing more than 2141 complex fragrance materials for aquatic risk assessment (Salvito et al. 2002) and has been shown to be applicable to pharmaceuticals for ranking purposes (personal communication, Nabholz 2002; Sanderson et al. 2003; Zeeman 2003). ECOSAR is continuously being developed. The training set in 1995 consisted of 2353 chemicals for 121 structure-activity relationships for 46 chemical classes (Clements et al. 1995). The continuous addition of new data contributes to widening and increasing the robustness of the scope of the training set, which again will result in increasing the applicability of substances within the model domain (see Figure 3.2).

3.4 (Q)SAR TO SSDS: PROBABILISTIC EFFECTS ASSESSMENT

As noted previously, the use of (Q)SAR for effects assessment nearly always uses a static assessment factor to extrapolate to a predicted no-effect concentration (PNEC). Assessment factors may be as high as 10 000 (Environment Canada 2003).

The deterministic hazard quotient (HQ) risk assessment of the most sensitive species approach with application of an uncertainty or safety factor is assumed in a regulatory context to protect a significant proportion of the species (e.g., > 95% of species). This excludes the relative relationships of species sensitivities and their potential impact on effects characterization, rendering the magnitude of conservatism and scientific justification of the deterministic HQ approach questionable. Recently, however, estimating interspecies sensitivity correlations for different test substances has been made easier with the introduction of the Interspecies Correlation Estimates (ICE) software (Asfaw et al. 2004), described in Section 3.4.2 (below). Consisting of derived interspecies correlation estimates for acute toxicity to aquatic organisms, this software allows the user to extrapolate (Q)SAR estimates for fish, invertebrate, and algal sensitivities to a myriad of other aquatic organism sensitivities for a given test substance. To illustrate the functionality of the ICE program, we used ECOSAR estimates for fish, daphnid, and algal acute toxicity data to extrapolate effects values to 27 fish and invertebrate species for 4 model substances: dichloroaniline (DCA), pentachlorophenol (PCP), nonylphenol (NP), and linear alkylbenzenesulfonate (LAS). These substances cover 3 different modes of action (polar narcotic, nonpolar narcotic, and electron transport inhibitor) and a range of uses and physical–chemical properties representing typical industrial chemicals. Perhaps more importantly, each of these chemicals has an established chronic toxicity data set whereby species sensitivity distributions have been created — and to some extent verified via mesocosm tests — and can as such serve as representative non-worst-case model compounds that allow a context-dependent comparison of extrapolated probabilistic predictions with measured deterministic hazard concentrations. The ECOSAR predictions, with excess toxicity, were input into the ICE program without uncertainty of default assessment factors to enable a direct comparison between measured effect concentrations and ICE-extrapolated ECOSAR predictions.

3.4.1 ECOSAR ESTIMATES

Within the USEPA's EPI Suite, descriptions of basic physical–chemical properties for 2,4-dichloroaniline (DCA; CAS No. 554007), pentachlorophenol (PCP; CAS No. 87865), nonylphenol (NP; CAS No. 104405), and linear alkylbenzenesulfonate (C12 LAS; CAS No. 25155300) were obtained using the EPI Suite (Table 3.1), whereas acute and chronic toxicity estimates for the median effective and chronic effect concentrations (the geometric mean between chronic lowest-observed-effect concentration [LOEC] and no-observed-effect concentration [NOEC]) and for fish, daphnids, and algae for each substance were estimated from ECOSAR (Table 3.2). These compounds were chosen based on their widespread use.

3.4.2 ICE ESTIMATES

ICE was developed for estimating acute toxicity of chemicals to species where data are lacking. Interspecies correlations were created for 95 aquatic and terrestrial organisms using least squares regression where both variables are random (i.e., both variables are independent and subject to measurement error; Asfaw et al. 2004). The correlation coefficient (r) is used to describe the linear association amongst the

TABLE 3.1
Physical and chemical characteristics estimated from the USEPA's EPI Suite for 4 materials[a]

Chemical name	Molecular formula	Molecular weight	Log K_{OW}	Water solubility (mg L^{-1})
Dichloroaniline (DCA)	$C_6H_5Cl_2N$	162.0	2.37	302.1
Pentachlorophenol (PCP)	$C_6H_1Cl_5O$	266.3	4.74	1.899
Nonylphenol (NP)	$C_{15}H_{24}O$	220.4	5.99	0.08343
Linear alkylbenzenesulfonate (LAS)	$C_{18}H_{21}O_3SNa$	339.3	3.00	17.44

[a] These materials represent typical industrial chemicals used in the ECOSAR and interspecies correlation evaluation (ICE) extrapolation exercise.

variables. For the users of ICE, the toxicity of 1 species can be input as a surrogate for a predicted species. For example, the user can input an EC50 for the fathead minnow (*Pimephales promelas*) as a surrogate species toxicity value to obtain an estimated EC50 for rainbow trout (*Oncorhynchus mykiss*). The validity of the estimate can be evaluated by observing the correlation coefficient and confidence limits about the regression line. In this example, ECOSAR estimates for fish were limited to fathead minnow responses. Hence, in ICE the user identifies fathead minnow as the surrogate species for a wide array of fish, invertebrate, and algal species. Many of the correlations between fathead minnow and other species were not significant. Only species with at least 10 observations and significant regressions were used in this analysis, leaving 27 predicted species responses (Table 3.3).

TABLE 3.2
ECOSAR estimates of acute and chronic toxicity to dichloroaniline (DCA), pentachlorophenol (PCP), nonylphenol (NP), and linear alkylbenzenesulfonate (LAS)

Organism	Duration	Endpoint	Chemical (mg L^{-1})			
			DCA	PCP	NP	LAS
Fish	96 hours	LC50	26.0	0.803	0.115	2.595
Fish	28 days	ChV	0.16	0.115	0.016	0.399
Daphnids	48 hours	LC50	0.88	1.086	0.277	2.595
Daphnids	21 days	ChV	0.02	0.088	0.012	0.399
Algae	96 hours	LC50	10.0	0.497	0.032	0.007
Algae	96 hours	ChV	2.5	0.242	0.032	0.005

TABLE 3.3
List of predicted species from ICE that have at least 10 observations and
significant correlations ($p < 0.05$) with the surrogate species fathead
minnow (*Pimephales promelas*)

Predicted species	Number of observations	Intercept	Slope	Correlation coefficient
Aquatic sow bug (*Asellus brevicaudus*)	13	−0.02	0.72	0.81
Atlantic salmon (*Salmo salar*)	12	−0.63	1.07	0.94
Black bullhead (*Ameiurus melas*)	11	0.79	0.77	0.89
Bluegill sunfish (*Lepomis macrochirus*)	92	−0.06	0.93	0.91
Brook trout (*Salvelinus fontinalis*)	15	−0.42	0.98	0.91
Brown trout (*Salmo trutta*)	12	−1.20	1.13	0.84
Channel catfish (*Ictalurus punctatus*)	60	0.45	0.83	0.89
Coho salmon (*Oncorhynchus kisutch*)	18	−0.56	0.99	0.95
Common carp (*Cyprinus carpio*)	14	0.20	0.94	0.95
Cutthroat trout (*Oncorhynchus clarki*)	19	−0.54	0.95	0.80
Eastern oyster (*Crassostrea virginica*)	32	1.29	0.43	0.68
Goldfish (*Carassius auratus*)	27	0.33	0.94	0.97
Grass shrimp (*Palaemonetes pugio*)	12	−0.11	0.81	0.85
Green sunfish (*Lepomis cyanellus*)	14	0.24	0.86	0.94
Lake trout (*Salvelinus namaycush*)	10	−0.20	0.83	0.93
Largemouth bass (*Micropterus salmoides*)	20	−0.48	0.97	0.94
Midge (*Chironomous plumosus*)	19	−0.60	1.07	0.77
Mysid (*Americamysis bahia*)	32	−1.62	0.95	0.68
Ostracod (*Cypridopsis vidua*)	10	0.72	0.32	0.65
Rainbow trout (*Oncorhynchus mykiss*)	93	−0.24	0.95	0.92
Scud (*Gammarus pseudolimnaeus*)	16	−1.27	1.07	0.79
Sheepshead minnow (*Cyprinodon variegatus*)	31	0.25	0.87	0.89
Stonefly (*Claassenia sabulosa*)	10	−0.45	0.31	0.69
Stonefly (*Pteronarcella badia*)	11	−0.28	0.29	0.87
Stonefly (*Pteronarcys californica*)	34	−0.06	0.41	0.47
Water flea (*Daphnia magna*)	69	−0.01	0.67	0.52
Yellow perch (*Perca flavescens*)	10	−0.19	0.99	0.98

3.4.3 ICE-BASED SSDs

Two sets of SSDs were created based on ICE extrapolations from ECOSAR fish
values (assuming fish = fathead minnow): 1) all 27 predicted ICE species, excluding ECOSAR values; and 2) the 27 ICE species plus ECOSAR fish, daphnids, and
algae. SSDs were created using a log-normal distribution (SAS Institute 2002).
Figure 3.3 illustrates the SSDs for DCA and NP with and without ECOSAR species,
and Figure 3.4 provides the distributions for PCP and LAS. The predicted hazard

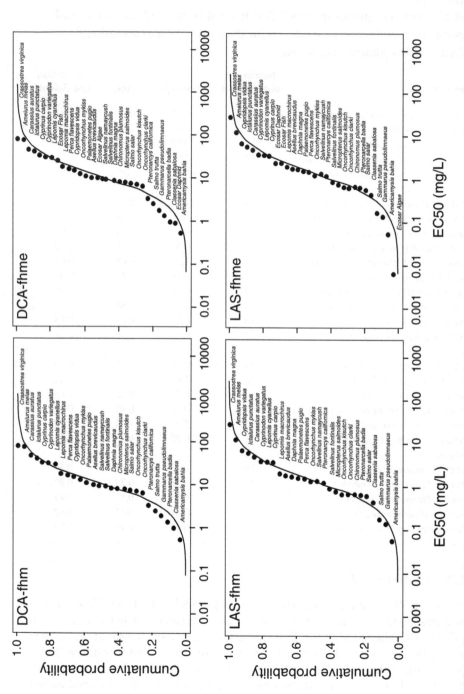

FIGURE 3.3 Species sensitivity distributions (SSDs) for dichloroaniline (DCA) and nonylphenol (NP) using fish estimates from ECOSAR and ICE (Asfaw et al. 2004) using fathead minnow (fhm) as the surrogate species. *Note:* SSDs not including ECOSAR values for fish, daphnids, and algae are noted as "-fhm," whereas SSDs including these values are noted as "-fhme." *Source:* Asfaw et al. (2004).

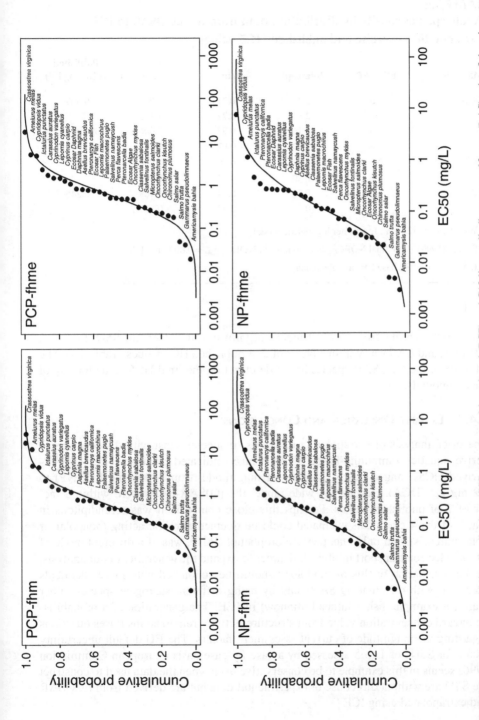

FIGURE 3.4 Species sensitivity distributions (SSDs) for pentachlorophenol (PCP) and C12 linear alkylbenzene sulfonate (LAS) using fish estimates from ECOSAR and ICE (Asfaw et al. 2004) using fathead minnow (fhm) as the surrogate species. *Note:* SSDs not including ECOSAR values for fish, daphnids, and algae are noted as "-fhm," whereas SSDs including these values are noted as "-fhme." *Source:* Asfaw et al. (2004).

TABLE 3.4
Acute species sensitivity distribution data from the ECOSAR to ICE
extrapolation exercise and published HC5 values

Material	ECOSAR	Intercept	Slope	HC5 (mg L⁻¹)	Published HC5 (mg L⁻¹)
DCA	No	2.3	0.71	1.270	0.507[a]
	Yes	2.3	0.72	1.170	
NP	No	−2.16	0.97	0.006	0.006; 0.0039[b]
	Yes	−2.18	0.92	0.007	
PCP	No	−0.54	0.79	0.056	0.028[a]
	Yes	−0.50	0.74	0.068	
LAS	No	0.39	0.74	0.168	0.245[c]
	Yes	0.36	0.82	0.127	

[a] Wheeler et al. (2002a), freshwater log logistic model.

[b] Servos (1999), 4 NP final $NOEC_{fish}$ = 0.006 mg L⁻¹, $NOEC_{daphnid}$ = 0.0039 mg L⁻¹.

[c] Dyer et al. (2003), based on chronic toxicity.

concentration protective of 95% of aquatic organisms (HC5) from the SSDs for DCA, NP, PCP, and LAS was within a factor of 2 of published HC5 values (Table 3.4). The addition of the ECOSAR species had little effect on the final HC5 value for any of the compounds.

3.4.4 EXAMPLE DISCUSSION AND CONCLUSIONS

The brief analysis of extrapolating (Q)SARs to species sensitivity distributions to derive aquatic community-based HC5s clearly indicated promise. The (Q)SAR-derived HC5s ranged nearly 3 orders of magnitude, from 0.006 mg L⁻¹ (NP) to 1.2 mg L⁻¹ (DCA). Despite the wide range, the (Q)SAR-HC5s were within a factor of 2 of published values. Perhaps this close comparison was serendipitous in that 3 of the 4 chemicals evaluated could be deemed narcotic acting (nonpolar or polar narcosis), and only one, pentachlorophenol (PCP), was of a different mode of action (electron transport inhibitor). In order to expand the generality of our analysis, further evaluation of this methodology should be conducted with more chemicals predicted with different (Q)SARs and by using different surrogate species (other than, for example, fish = fathead minnow) in ICE. The generation of a probabilistic hazard concentration value from structure-activity relationships raises questions regarding the magnitude of current assessment factors. The EU default uncertainty factor for SSDs of 1 to 5 assessed by a case-by-case basis (European Commission 1996) seems in this example to be conservative even when the data used to construct the SSD are without direct use of experimental data but are derived from ECOSAR and extrapolated using ICE.

3.5 REGULATORY USES OF ENVIRONMENTAL (Q)SARS

3.5.1 ORGANIZATION FOR ECONOMIC COOPERATION AND DEVELOPMENT

The Organization for Economic Cooperation and Development (OECD) consists of 30 member countries around the world that work together to enhance international trade by minimizing the barriers to trade between countries. One part is by establishing mutually agreed upon common standards (e.g., testing guidelines). As many existing chemicals are international commodities, the OECD SIDS project was jointly undertaken by member countries in the late 1980s to "share the burden" in evaluating the safety of OECD high-production volume (HPV) chemicals (i.e., those manufactured in or imported into any one member country in excess of 10 000 tonnes or in two or more countries in excess of 1000 tonnes). This led to the HPV list of 4843 chemicals with little or no publicly available information, which would then undergo a minimal set of testing, providing an initial screening of their hazards to health and the environment. The required basic information consisted of the following 6 categories: chemical identity, physical–chemical data, sources and levels of exposure, environmental fate and pathways, ecotoxicological data, and toxicological data (OECD n.d.-a). Compounds will be assigned hazard classification in accordance with the United Nations' Globally Harmonized System for classification and labeling of chemicals (GHS) (United Nations Economic Commission for Europe 2004). When no information is available for a given data element, calculations or estimates derived from (Q)SAR models can be provided, but the methods and their quality should be described (Zeeman et al. 1995). The OECD in 2001 suggested that although experimentally derived test data are preferred, when no experimental data are available, validated (Q)SARs for aquatic toxicity and log K_{OW} may be used in the classification process. Such validated (Q)SARs may be used without modification to the agreed criteria if they are restricted to chemicals for which their mode of action and applicability are well characterized (OECD 2001). The OECD is active regarding regulatory uses of (Q)SARs for the registration of chemicals, due to large undertakings in Canada, Japan, the United States, and the European Union (Sanderson and Thomsen 2007). It is proposed that general guidance for decision making based on (Q)SAR results is not practical, but that identification of valid (Q)SARs to avoid unilateral development of national (Q)SARs, and their associated uncertainty, is considered necessary. Further, acceptance of different levels of uncertainty may vary depending on the type of chemical, endpoints, whether other information is available or not, how close one is to making a final decision, and the impacts and consequences (human health, environmental, and economic). Consideration of the applicability domain is of specific importance because it constitutes the major source of uncertainty at all levels. Therefore, it is suggested that a case-by-case uncertainty analysis approach is required. Target values for a maximum acceptable uncertainty cannot be set specifically by the OECD; however, limitations to (Q)SAR estimates should be quantified and provided as to prediction uncertainty, data variability, and model structure uncertainty. The objectives of the OECD program are threefold: 1) to further refine the general and specific development or calibration principles already proposed for the establishment of internationally agreed principles; 2) to provide guidance on the calibration and acceptability of (Q)SARs to different audiences (e.g., developers, calibration bodies, and regulators); and 3) to make validated (Q)SARs available for enhanced and more harmonized regulatory use. The aims of the OECD program include ensuring harmonized and mutual acceptance of predictions, and derive consensus on the role of (Q)SARs in relation to the classification of untested chemicals (OECD 2003). For consideration of a (Q)SAR model for regulatory purposes, the OECD program should apply these agreed upon principles for (Q)SAR model validation:

- Must have a defined endpoint: This is to ensure clarity and identify the experimental system that is being modeled by the (Q)SAR.
- An unambiguous algorithm: This is to ensure transparency in the model algorithm. Without this information, the performance of a model cannot be independently established, which is likely to represent a barrier for regulatory acceptance.
- A defined domain of applicability: It is realized that (Q)SARs are reductionist models that inevitably have limitations in terms of the types of chemical structures that can be predicted; in other words, define the applicability of the model based on the domain.
- Appropriate measures of goodness-of-fit, robustness, and high predictive capacity: The predictive capacity of the model should be determined by external validation. Further work is recommended to determine what constitutes external validation of (Q)SARs.
- A mechanistic interpretation, if possible: Models without mechanistic interpretation can be used in a regulatory context. However, consideration of the mechanistic association between descriptors used in the model and the predicted endpoint should be sought to improve the regulatory applicability of the (Q)SAR. Guides to (Q)SARs currently used in OECD member states has been published (OECD 2005a, 2005b).

The (Q)SAR toolbox was presented in 2007 (OECD 2007) and will serve as an internationally recommended approach to comprehensively assess chemicals and categories of chemicals based on (Q)SARs. It is expected that this set of models will be widely applied in relation to the implementation of REACH.

3.5.2 UNITED STATES

In 1979, almost 62 000 chemical substances were reported to be in commerce in the United States, and these were "grandfathered" into the Toxic Substances Control Act's inventory of existing industrial chemicals. Chemicals not in this list were to be considered new substances, and more than 43 000 new chemical notifications have been submitted by industry for assessment since July 1979. Via inclusion of about 13 000 new industrial chemicals that have been assessed for risk and are now in commerce, the TSCA inventory has now increased to more than 75 000 substances, with a total production and import of 2.7 trillion tonnes per year in 1989. However, the total produced in and/or imported into the United States in 1989 was larger than this estimate. Approximately 25 000 existing chemicals were not reported because they did not reach the 10 000 lbs. per site per year (4.53 tonnes) reporting threshold or because they were inorganic chemicals. In addition, from 1989 to 1995, the production of only the top 50 organic and inorganic chemicals in the United States increased 33% and 15%, respectively. Currently more than 2200 new chemical notices are submitted to the USEPA annually (Zeeman et al. 1999). Section 5 of the Toxic Substances Control Act requires manufacturers and importers of "new" chemicals to submit a Premanufacture Notification (PMN) to the USEPA 90 days before they intend to commence manufacture or import of a new substance (Auer et al. 1993). In the PMN, the only information required is as follows: chemical identity; molecular structure; trade name; production volume, use, and amount for each use; by-products and impurities; human exposure estimates; disposal methods; and any test data that the submitter may have. The manufacturer does not have to initiate any ecological or human health testing before submitting a PMN. Only ~5% of the PMNs contain any chemical fate or ecotoxicological test data (Zeeman et al. 1999). This has resulted in the development and use of general (Q)SARs for estimating a chemical's physiochemical properties; ability to degrade and bioconcentrate; and toxicity to fish, aquatic invertebrates, and algae (Auer et al. 1993), as described in the EPI Suite.

Uncertainty factors are used to compensate for lack of definitive data when comparing effects concentrations with exposure levels and are used as follows for (Q)SARs by the USEPA: 1) 1000 if only 1 acute value (known or predicted) is available; 2) 100 applied to the most sensitive species when the environmental base set of toxicity data is available (acute fish, daphnids, and algae, known or predicted); 3) 10 applied to the lowest ChV (chronic value) when base set data are available (known or predicted); and 4) one applied to the ChV from a field study (e.g., pond) or from a microcosm study (Zeeman et al. 1999). Of the 1500 to 2000 new chemicals the OPPT receives, ~20% are regulated on (Q)SAR predictions alone. In practice, the only assessment factor used by the USEPA OPPT is 10. An environmental toxicity profile (fish 96-hour EC50, daphnid 48-hour EL50, green algal 96-hour E(b)C50 cells mL^{-1}, fish ChV, daphnid ChV, and algal ChV) for all chemicals is generated and integrated into this process using measured ECs, if possible, and then applied with an uncertainty factor of 10 to the lowest chronic value. The USEPA has to assess the risk of all chemicals submitted for notification within 90 days, and unless it is able to do this, the substance will automatically be transferred to chemical inventory without challenge. The requirement for further experimental toxicity data is dependent on the EPI Suite screening, where risk has to be demonstrated. Because the USEPA receives chemicals from every corner of the chemical universe, the (Q)SARs have to have a broad domain (personal communication, Nabholz 2003). Like the OECD, the USEPA has its own HPV program initiated in 1998, where (Q)SAR predictions play a significant role (USEPA n.d.).

3.5.3 European Union

As of 1981, when the European chemicals policy was implemented, the European Union required the submission of base set test data for physiochemical, environmental fate, toxicological properties and health effects, and ecotoxicity parameters for new substances marketed in volumes > 1 tonne. Depending on the exposure–effect ratio, there will be an assessment of whether the substance presents a risk to humans and/or the environment, or further data may be necessary to clarify a concern. When the risk assessor considers the potential need for further (test) data, (Q)SARs may serve as a supporting tool in making this decision. If further testing is needed, (Q)SARs may also be used to optimize the test strategies. Hence, estimates resulting from (Q)SAR models cannot be the only basis for preparing a risk assessment of a substance. (Q)SAR estimates should be seen as a complementary tool, which, when evaluated together with test results, can provide a more complete understanding of the physiochemical and ecotoxicological characteristics of the substance. The result of a (Q)SAR should thus be evaluated for consistency in light of available experimental data and validated estimates from other endpoints. As a general rule, acceptable (Q)SAR estimates should be used only in a conservative manner in the risk assessment process. In the EU particular care is taken to exclude the possibility of reaching conclusions on the risk to humans and environment where those conclusions may have been markedly influenced toward relatively lower risk by use of (Q)SAR estimates (false negatives; European Commission 1996). Typically a maximum default uncertainty factor of 1000 is applied to (Q)SAR predictions in the EU (European Commission 1996). There is no published agreed upon methodology on how the EU will use (Q)SARs in the REACH program under the new chemicals policy that was implemented in June 2007 (European Union n.d.). The most used current models are the European Union System for the Evaluation of Substances (EUSES) models (European Chemicals Bureau 2005), and the expected use of the OECD (Q)SAR toolbox. Implementation methodology of (Q)SARs under REACH is currently being developed. Further details about EUSES and the REACH program's implementation of EUSES and other models are very important but outside the scope of this chapter.

Compounds in existence before September 18, 1981 are listed as existing compounds on the EINECS list of nonassessed commercial chemicals comprising roughly 100 000 different compounds. These compounds are now subjected to risk assessment under the new chemicals policy in the EU, where (Q)SAR models will play a significant role. The exact procedure for using (Q)SARs in the European Union after implementation of the new chemicals policy is still unclear. Currently the uncertainty of (Q)SARs for classification and labeling for human health and environmental effects is limited, on a case-by-case basis, where sufficient experimental data are lacking (European Commission 1996; Carlsen 2003; Walker 2003). One distinct difference from the United States is that (Q)SARs cannot be used as surrogates for data and thus cannot acquit compounds, as the European Union has zero tolerance for predicted false negatives. The subsequent use of (Q)SARs within member states of the European Union is not harmonized; some use them more than others in concordance with EU DIR 1488/94EEC.

Denmark, and the Danish Environmental Protection Agency (DK-EPA), is one of the leading countries in terms of considering the pros and cons to environmental (Q)SARs. It published an advisory list for self-classification of potentially hazardous substances based on predictions from (Q)SAR models (among these the EPI Suite) on hazards to the aquatic environments and human health risks. A total of 46 707 existing substances were examined, identifying 20 624 substances that, according to the models, required 1 or more of the selected endpoints. The list is used as a tool for industry in their assessment of the dangerous properties of chemicals in cases with insufficient or no data for the substance regarding the selected classification endpoints. The models that were used estimated results to an accuracy of approximately 70% to 85% within 1 order of magnitude of measured values (Danish Environmental Protection Agency 2001). The European approach to registration of new substances is more resource intensive than that of the United States, which is also reflected in the number of notifications: on average, 325 new notifications per year versus 2200 in the United States. The single largest current difference between the US and EU approaches to new chemicals is that the European Union requires experimental ecotoxicological data, whereas the United States may accept or acquit a compound based on (Q)SAR predictions alone. The European Commission and several member states are awaiting the outcome of the OECD ad hoc workgroup conclusions and guidelines on the regulatory use of (Q)SARs. The implementation (Q)SARs under REACH is currently being developed as the new EU Chemicals Agency in Helsinki, Finland becomes fully operational during 2008/9.

3.5.4 CANADA

With the promulgation of the Canadian Environmental Protection Act of 1999 (CEPA), the minister of the environment and the minister of health were obligated to categorize approximately 23 000 substances on the Domestic Substances List (DSL) prior to September 2006 (Environment Canada 2005, n.d.). New substances (after January 1, 1987) produced at quantities of more than 10 000 kg per year or that exceed an accumulated total of 50 000 kg require experimental premarket notification data (acute toxicity; LC50 for fish and daphnia). The DSL includes substances that were, in Canadian commerce (January 1984 to December 1986), used for manufacturing purposes, or manufactured in or imported into Canada in a quantity of 100 kg or more in any calendar year. It consists of 50% organics, 10% inorganics, 18% polymers, 2% others, and 20% UVCBs (complex substances of variable composition). Environment Canada is responsible for categorizing the substances on the DSL as being persistent or bioaccumulative and inherently toxic to nonhuman organisms. In order to categorize the substances on the DSL within the mandated time frame and prepare screening assessments as necessary, it is recognized that (Q)SARs will be relied upon to fill data gaps expected for up to 80% of the DSL substances. Given no experimental data, inherent toxicity will be predicted using the following (Q)SARs: ASTER, OASIS-FORCASTING, ECOSAR, TOPKAT, COREPA, and PNN (probabilistic neural network). Exposure assessment will primarily rely upon EPIWIN and level II fugacity modeling. The potential mode of action (MOA; reactive or nonreactive unknown MOA) will initially be determined using ASTER and/or OASIS-FORCASTING. If the MOA is reactive, TOPKAT or PNN will primarily be used to predict toxicity. If the MOA is nonreactive or unknown (narcosis), ECOSAR will also be used. Expert judgment will often guide how model-difficult substances are categorized, and the closet analogue approach may be used in place of unreliable predictions. Environment Canada acknowledges that the process for categorization is not without inherent uncertainty, but given the state of the science, this uncertainty is acceptable. They identify the following primary sources of uncertainty: lack of data, model error and limited training sets, uncertainty associated with measured data in general, extrapolation of biodegradation estimates to multimedia half-life values, use of worst-case media-based half-life rather than overall half-life, bioaccumulation versus bioavailability, unknown MOAs, and toxicity to nonpelagic species (Robinson 2003). To adjust for these uncertainties, assessment factors are used. Narcosis is understood to be the mechanism for the minimal toxicity of a substance. Therefore, if a substance is thought to have an MOA other than narcosis but is not well defined, the uncertainty factor will serve to provide an extrapolation from the less toxic narcosis MOA to a more toxic MOA under worst-case estimates for EC50s. Environment Canada uses assessment factors ranging from 1 to 10000 according to Verhaar et al. (1992), where the DSL substances are placed in 1 of 4 classes depending on the validity of available (Q)SARs and the amount of experimental and predicted data (Environment Canada 2003).

3.6 DISCUSSION AND REGULATORY OUTLOOK

(Q)SARs are highly integrated into effects assessment approaches throughout the world and are likely to be increasingly used for new and existing chemical use and discharge management (OECD 2003). A key message from a stakeholder workshop (held on March 4–6, 2003; hosted by the European Centre for Ecotoxicology and Toxicology of Chemicals; and organized by the International Council of Chemical Associations [ICCA] and the European Chemical Industry Council [CEFIC] as part of their long-range research initiative) was that both industry and regulatory authorities share the same goal, that is, to use (Q)SARs in a much broader scope than is currently practiced for safety evaluation and chemicals management (Jaworska et al. 2003). (Q)SARs statistical models range from simple linear regressions through highly sophisticated neural networks to multivariate ordination techniques (Carlsen 2003; Eriksson et al. 2003). As more structure-activity relationships and complex models are established and make their way into decision making as criteria, there will be a need to provide new and existing users with sufficient education to ensure that (Q)SARs are used properly and not as a "black box." Computer software will increasingly need to make appropriate selection and interpretation criteria transparent to the user (OECD 2003). For example, determination of structure and physical–chemical properties is integral for proper chemical class selection, hence appropriate (Q)SAR selection in the software program is needed. The extent of training sets, their quality, and their level of calibration will need to be explicitly stated to ensure proper use, avoiding common mistakes such as extrapolation beyond the domain of applicability.

Integrating concentration- and effect-addition principles with (Q)SAR opens the door for (Q)SAR-based mixture assessments. As discussed above, linking interspecies correlations (Asfaw et al. 2004) with the USEPA's ECOSAR program allowed for the generation of species sensitivity distributions, hence a probabilistic estimate for aquatic community effects. Estimated HC5s for 4 chemicals were within a factor of 2 of published values, suggesting that current uncertainty factors overestimate NOECs established via data-based SSDs; even SSDs derived from (Q)SARs extrapolated via ICE with assessment factors ranging from 1 to 5 (European Commission 1996) would be appropriate. This challenges the traditional higher confidence in measured data over predicted data. Further studies are needed to evaluate the scientific soundness of this approach with other chemicals, (Q)SAR models, and surrogate species and extrapolation tools. The relationship of predicted chemical properties with predicted chemical fate and transport endpoints can be validated with databases having measured chemical data in various environmental media and exposure regimes (e.g., mixing zones of municipal wastewater treatment plants). Exposure relationships can range from broad partitioning approaches (e.g., fugacity) to site-specific models (wastewater discharge) and watersheds (e.g., the EU GREAT-ER program; GREAT-ER 2005) and the USEPA P2 framework models (USEPA 2005). Tying in effects and exposure can be further evolved via physiologically based pharmacokinetic (PBPK), PB-TK, and/or body-burden-based toxicity models. Although (Q)SARs have been widely used for aquatic effects, exposure, and risk assessment, they are generally lacking for terrestrial or nonaquatic assessments.

Previously, there was some reluctance toward (Q)SAR-based environmental risk assessment (OECD 2003), but (Q)SAR-derived data today receive a Klimisch score of 3 whereas experimental data of low quality receive Klimisch scores of 3 to 4. In such cases (Q)SAR-derived data will and should be used in order to obtain the best available basis for decision making. The significantly different approaches between the United States and the European Union relating to accepting or acquitting a substance on (Q)SAR alone ignores or underestimates the risk of false negatives from experimental data, which, due to the design of the experiment and misuse of hypothesis testing, may be larger than those generated via modeling (Suter 1996). Ecotoxicological data are not without variability. For example, Persoone and Jansson (1994) found that the coefficient of variation (CV = SD/mean) usually exceeded 25% for single-species laboratory tests, and could be as high as 50%. Warren-Hicks and Parkhurst (2003) confirmed these findings in 2003 with CVs of 48% for inter- and intralaboratory tests. Sanderson (2002) found that microcosm data had an overall mean CV of 45%. Thus, ecotoxicological results within 1 order of magnitude are believed to be equal or not different (Carlsen 2003). In order to develop or select the most appropriate model for a given purpose, 2 different types of uncertainties have to be considered: 1) input uncertainty and variability, which arise from missing information about actual values and natural variability due to a heterogeneous environment; and 2) structure (model) uncertainties that arise from the fact that every model is a simplification of reality due to a limited systemic knowledge. The first type relates to descriptor variability as well as endpoints, whereas the second is associated with the actual type of model chosen for the problem. The end choice will aim at reducing the combination (i.e., total uncertainty; Carlsen 2003).

In conclusion, as (Q)SARs are becoming more precise and accurate, and are combined with other environmental- and chemical-relevant software (e.g., ICE) and databases (product registries), and as both regulatory agencies and industries recognize that they will increasingly rely upon (Q)SAR predictions in the future, discrimination between measured and predicted data is being reconsidered in the EU and within the OECD.

ACKNOWLEDGMENTS

This chapter was greatly aided from discussions with Lars Carlsen (University of Roskilde), Maurice Zeeman (USEPA), Eva Bay Wedebye and Henrik Tyle (Danish EPA), Pete Robinson (Environment Canada), Mark Ellersieck (University of Missouri–Columbia), Mike Hooper (Texas Tech University), and Krishnan Kannan (University of Montreal).

IN MEMORIUM

This chapter is dedicated to the memory of "Vince" (J. Vincent) Nabholz, who died suddenly on February 23, 2008. He was a graduate of Christian Brothers University, received a PhD from the University of Georgia, and was senior biologist and ecologist with the US Environmental Protection Agency in Washington, DC, where he served for 29 years, and where he chaired their Structure-Activity Team for the last two decades. He was the principal author of USEPA's ECOSAR computer program, which also is one of the three key elements in Persistence Bioaccumulation Toxicity (PBT), one of the most frequently used (Q)SAR tools of all time.

During his career, he made voluminous, critical, and lasting contributions to our understanding of the relationships between molecular structure and toxicity. He will be sadly missed by literally hundreds of professional scientists in government and industry in the United States and around the World.

DISCLAIMER

This document has been reviewed by the Office of Pollution Prevention and Toxics, USEPA, and approved for publication. Approval does not signify that the contents necessarily reflect the views and policies of the agency, nor does the mention of trade names or commercial products constitute endorsement or recommendation for use.

4 Extrapolation of Effects Measures across Levels of Biological Organization in Ecological Risk Assessment

Paul J. van den Brink, Paul K. Sibley,
Hans Toni Ratte, Donald J. Baird,
J. Vincent Nabholz, and Hans Sanderson

CONTENTS

4.1   INTRODUCTION

An ecosystem, as defined by Odum (1971), includes living organisms interacting with each other and with the abiotic environment. Within the ecosystem, the existence and flow of energy and cycle of materials lead to the development of trophic structures

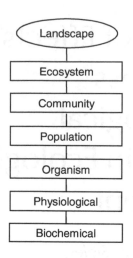

FIGURE 4.1 A hierarchical view of levels of biological organization.

(food webs), biotic diversity, and nutrient cycles. Odum further recognized hierarchical levels of organization within the ecosystem:genes–cells–organs–organisms–populations (species)–communities (Figure 4.1). A key corollary of this concept is that effects at a given level of ecological organization can propagate to higher levels of organization.

Environmental toxicology emphasizes the organism level of ecological organization through its reliance on data generated from single-species toxicity tests. For most industrial chemicals, only toxicity data from single-species tests are available, yet the protection goals of legislation and regulatory authorities include the preservation of populations, communities, and ecosystems (European Union 1997; USEPA 1998). Although the need to evaluate the effects of chemicals at higher levels of biological organization is acknowledged, the use of individual-based endpoints continues to dominate toxicological assessments in ecotoxicology. Establishing methods to bridge this gap has constituted a significant research direction in ecotoxicology in recent years, and considerable effort has been expended in developing quantitative relationships among endpoints across levels of biological organization (Figure 4.1). These kinds of correlations do not prove causation, due to a limited knowledge of biochemical processes and their consequences for individuals. In general, examples of effects propagating from subindividual levels to higher levels of organization are few. It has thus been a common practice to extrapolate effects at higher levels of biological complexity from information derived at lower levels. The most common approaches in this regard are the application of extrapolation factors and models.

Although it is tempting to conclude from Figure 4.1 that the ecological significance of any response becomes more significant in relation to its position in the hierarchy, this is not necessarily the case. "Significance" is determined by the question being posed. This applies to effects at any level of organization, which may or may not propagate to higher levels, depending on the state of the biological system being studied. Conversely, it is not necessarily easier to detect a response at lower levels, nor is it easier to interpret the consequences of a response at higher levels. Experiments and observations at higher levels of biological organization generally require more effort and hence greater cost. This has led to the extensive application of population and higher tier models to extrapolate effects measured at lower levels of biological organization to higher levels of biological complexity. Examples of models that have been applied in this type of extrapolation practice are common, and exhibit a wide range in complexity, reflecting, in essence, the complexity of the level of biological organization to which they are being applied. For practical reviews of the more commonly utilized models in ecotoxicology, see Pastorok et al. (2002) and Bartell et al. (2003).

At the present time, regulatory authorities rely predominantly on the use of simple models that use data from toxic effects on individuals to extrapolate to higher levels of biological organization. Complex models, such as ecosystem simulation models, have not been used extensively because large numbers of explicit and implicit assumptions are needed to parameterize these models. These models require a large amount of data about the fate and effects of a chemical in an ecosystem. When such data are lacking, assumptions have to be made by experts, and the values chosen may be subject to intense debate. Differences between the normal practices of experts and their opinions may lead to differences in simulation model outcomes and ultimately to legal stalemate and inaction (Boesten 2000). Thus, the simplest way to extrapolate from the organism level to higher levels of ecological organization is to use an extrapolation factor. As discussed in Chapter 1, extrapolation factors have been used by the chemical industry and regulatory authorities for more than 30 years and are known by a variety of names: application factors by some research toxicologists, safety factors by the chemical industry, assessment factors by the USEPA, and uncertainty factors (UFs) by the European Union (EU) (see Table 1.1 for an overview of UFs used by different bodies).

The complexity of ecosystems makes it difficult to develop quantitative cause–effect relationships across levels of biological organization. For this reason, there has been a strong tendency among risk assessors to rely on extrapolation procedures, including the application of UF, to assure protection of the ecosystem. In part, this reflects a strong reluctance in ecotoxicology to address ecosystem complexity (Preston 2001). Cairns (1990) suggested that this simplified approach, which is strongly biased toward convenience, reflects the fact that ecotoxicology has evolved from a foundation of methodology and ideology rather than the application of sound scientific principles. Newman (1996) argued that ecotoxicology, if it is to progress as a science, must behave as a science in practice. A key aspect of this must be a willingness to embrace the inherent complexity of ecosystems and a desire to move beyond the use of simplified conceptual models (Preston 2001). It is important to point out that this will only happen if there is much greater dialogue and collaboration between ecologists, who directly address ecosystem complexity, and ecotoxicologists, who, as practitioners of ecological risk assessment (ERA), must manage the effects of stressors on these complex systems.

In this chapter we describe methods that can be used for extrapolating between levels of biological organization, examine their assumptions, indicate their strengths and weaknesses, and look forward to new directions for research in this area. We start the chapter at the biochemical level and move upwards toward the ecosystem and landscape levels. Also, a few special cases of horizontal extrapolation are discussed: direct versus indirect effects, and structure versus function.

4.2   EXTRAPOLATING TO INDIVIDUALS

Measurements on biochemical and physiological responses to environmental stressors have become a common component of site-specific risk assessment, even though they continue to stimulate debate regarding their utility and functionality (e.g., lack of cause–effect between exposure and responses for many biomarkers). As described below, much of this debate revolves around the appropriateness of extrapolating responses across levels of biological organization.

Biomarkers are most commonly applied as indicators of exposure to contaminants at the biochemical level; however, Adams and Rowland (2003) provided examples of biomarkers that are routinely applied at all levels of biological organization. Although there is some inconsistency in the manner in which biomarkers are defined, in general, they fall into 2 categories — those that indicate that an organism has been exposed to a chemical (biomarkers of exposure) and those that indicate that some type of effect has occurred (biomarkers of effect). Biomarkers of effect that can be appropriately correlated or causally linked to effects within and between levels of biological organization have been called "bioindicators" (McCarty and Munkittrick 1996; McCarty et al. 2002). The term "bioindicator" is also used in reference to the application of whole organisms or other biological parameters to assess the degree of contamination or relative status of a particular species, habitat, or ecosystem, though their use in this context could include biochemical measurements (Jamil 2001). Most recently, Melancon (2003) used biomarkers and bioindicators interchangeably in a review of biochemical, physiological, and morphological bioindicators as tools to monitor exposure and effects in aquatic environments.

The number of biomarkers that have been developed to indicate exposure to, or effects from, contaminants is large and continues to grow. It is beyond the scope of this chapter to review all of the available biomarkers, and the reader is directed to several books and reviews on the subject (Huggett et al. 1992; Peakall 1992; McCarthy and Shugart 1992; Di Giulio et al. 1995; Jamil 2001; Melancon 2003). In general, biomarkers can be categorized as follows (Huggett et al. 1992):

- Physiological (e.g., inhibition of acetylcholinesterase or delta-aminolevulinic acid dehydratase)
- Metabolic (e.g., metabolites of xenobiotics and endogenous substrates)
- Genetic (e.g., genotoxic effects and DNA alterations)
- Histopathological (e.g., lesions)
- Immunological (e.g., leukocrit values and blood differential counts)
- Molecular (e.g., activity of cytochrome P450 or concentrations of metallothioneins)

An important characteristic of biomarkers is that they indicate that a contaminant (or contaminants) is (are) present in the environment, that it is biologically available, and that it has reached a target tissue in sufficient quantity and duration as to elicit the observed response (Melancon 2003). In addition, because most organisms are exposed to multiple stressors over time, biomarkers can reflect an integrated exposure of cumulative, synergistic, or antagonistic effects (Adams and Rowland 2003).

4.2.1 BIOMARKERS AS EARLY WARNING INDICATORS

A significant factor leading to the popularity of biomarkers in toxicology is that they reflect initial changes caused by an interaction between a toxicant and biological receptor and thus offer the diagnostic potential to be used as early warning indicators of exposure to stressors. If the specificity of response and duration and intensity of exposure are sufficient, interactions at the biochemical level may lead to a cascade of responses (Carpenter et al. 1985) that eventually are reflected by measurable responses at higher levels of biological organization (Sibley et al. 2000; De Coen and Janssen 2003; Melancon 2003). In this sense, effects that are more commonly measured in laboratory toxicity tests (e.g., survival, growth, and reproduction) can be thought of as representing the culmination of damage that initially occurred at the suborganismal level (De Coen and Janssen 2003), although this may not be true for all effect types (e.g., behavioral effects arising from avoidance). Thus, ideally, biomarkers will be predictive of future harm (Melancon 2003), but one of the criticisms consistently levied against the use of biomarkers in ERA is that contaminant-induced effects that occur at the biochemical level are often very difficult to interpret in terms of responses that occur at the population and higher levels of biological organization. Indeed, there is a significant degree of uncertainty regarding the relevance of extrapolating biomarker-based responses to higher levels of biological organization (McCarty and Munkittrick 1996; McCarty et al. 2002). McCarty and Munkittrick (1996) argued that the foundation of the biomarker approach is the extrapolation of a scalar value, the biomarker itself, to a multidimensional dynamic ecosystem. Attril and Depledge (1997) argued that responses at lower levels of biological organization can be validly related to those at higher levels as long as a clear understanding of their relationship has been established through the use of quantitative, cause–effect investigations rather than from one level to another based solely on statistical correlation, assuming that a single effect is caused by a single substance.

4.2.2 PREDICTIVE VALUE OF BIOMARKER RESPONSES

If responses at higher levels of biological organization do, in fact, represent a cascade of responses initiated by interactions at the biochemical level (De Coen and Janssen 2003), it should be possible to link biochemical responses to contaminant exposure to responses at higher levels of biological organization through experiment, as suggested by Attril and Depledge (1997). Where biomarkers can be linked in a correlation (or, preferably, a cause–effect or dose–response manner) to impairment at higher levels of biological complexity, the diagnostic potential and value of the biomarkers may be greatly enhanced. Indeed, much of the recent research on biomarkers has focused on establishing correlations that will reduce the uncertainty associated with statistical extrapolation. Most of this research has focused on developing quantitative relationships between proximal levels of biological organization (e.g., biochemical–physiological–individual). For example, considerable work has been conducted on linking acetylcholinesterase inhibition, a biomarker of exposure to organophosphorus insecticides, to responses at the individual level (Anderson and Barton 1998; Sibley et al. 2000). Sibley et al. (2000) showed that acetylcholinesterase inhibition in fathead minnows was a consistent and strong predictor of mortality (Figure 4.2). More recently, several biochemical and physiological biomarkers have been developed to assess exposure to endocrine-disrupting compounds (EDCs), and assays have been developed to relate these to effects on reproduction (e.g., egg

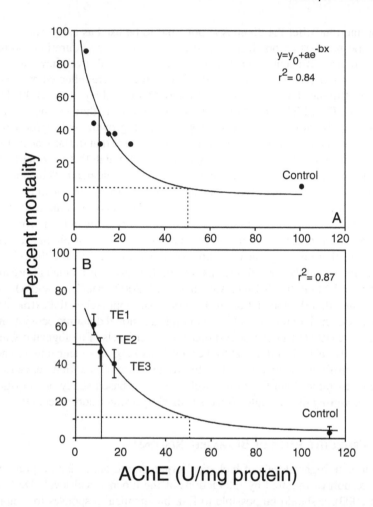

FIGURE 4.2 Mortality of fathead minnows in relation to acetylcholinesterase activity.[a]

[a] As measured in the brain tissue following (A) a 14-day exposure to a ternary mixture of azinphos-methyl, diazinon, and chlorpyrifos in a concentration-response study; and (B) a 7-day exposure to a ternary mixture of azinphos-methyl, diazinon, and chlorpyrifos applied as equipotent (toxic equivalent [TE]) mixtures. *Note:* Dashed lines correspond to 50% reduction in AChE activity; the dashed-dotted line corresponds to 50% mortality. *Source:* Redrawn from Sibley et al. (2000).

production) and development (e.g., somatic responses indicated by gonadosomatic and liver somatic indices). For example, vitellogenin has been used as a biomarker of exposure and effects of estrogenic EDCs, and several assays have been developed for use in various organisms (Sherry et al. 1999). In addition to being highly sensitive to changes that occur at the molecular and biochemical levels, many of the biomarkers developed for EDCs have the potential to be predictive of responses at the population and possibly higher levels of biological organization (Gross et al. 2003), though little research has been conducted in this regard.

Empirical studies that quantitatively link the responses to stressors at the biochemical level with those at the population, community, and ecosystem levels of biological organization are few. One approach that has been used is the development of dynamic energy budgets (DEB), which simulate energy flow as a way of linking molecular processes to population dynamics and higher levels of biological organization (Nisbet et al. 2000). For some organisms (e.g., *Daphnia magna*), elaborate DEB models have been constructed to relate observed patterns of growth, development, reproduction, and mortality to energetic information such as feeding rates and maintenance requirements (Gurney et al. 1990; McCauley et al. 1990; Ross and Nisbet 1990; Kooijman and Bedaux 1996). Although these studies tended to focus on energy flow between proximal levels of biological organization, it has been clearly recognized that DEB models hold strong potential to relate stressor-induced changes in the energetic profile of organisms to the population level of biological organization. Indeed, Nisbet et al. (2000) provided concrete examples of studies that have attempted to do so. More recently, De Coen and Janssen (2003), using a suite of model toxicants, were able to develop relatively strong correlations ($R^2 = 0.66$ to 0.77) between several energy-related biomarkers, combined with parameters related to oxidative stress and DNA damage, and long-term effects at the population level for *D. magna*.

Although significant progress has been made with respect to the development and validation of specific biomarkers of exposure and effects as a basis for understanding interactions of contaminants with aquatic organisms, there remains considerable uncertainty regarding the validity and ecological relevance of extrapolating these responses across levels of biological organization. The role of biomarkers in ecological risk assessment is perhaps best viewed as providing supplemental lines of evidence that can be used to demonstrate links between sublethal biochemical responses and adverse effects in natural populations in field studies (Hyne and Maher 2003).

4.3 EXTRAPOLATING TO POPULATIONS

Population simulation models typically deal with the dynamics of the abundance or distribution of a single species (Pastorok et al. 2002) and are increasingly being used to predict field-level responses from laboratory toxicity data and life cycle characteristics of a species. Population models (and ecosystem models; see Section 4.5.2) serve an important role in ecological risk assessment because they provide information that supports the goal of ERA, which is often the protection of populations, biodiversity, and system function, rather than individuals. The main focus of population models is to estimate the response dynamics, particularly recovery, of populations following short-term exposure to a chemical and to estimate the long-term consequences of sublethal effects. Depending upon the class of organisms, recovery and sublethal effects may be difficult to evaluate using current test methods, even in semifield experiments (see Section 4.5.1), because the limited time span of most studies (typically not longer than a year and often much shorter) may not adequately capture life history dynamics. Furthermore, the isolated nature of microcosms and mesocosms may not present the same opportunity for recovery through immigration as exists in natural habitats (van den Brink et al. 1996). For some fast-reproducing species that do not have high habitat or food requirements, like algae and *Daphnia*, population-level effects can be evaluated experimentally (e.g., van der Hoeven and Gerritsen 1997). For species that have a more complex life cycle, performing population-level experiments is difficult, and, as stated above, it is even difficult in mesocosms.

4.3.1 GENERIC POPULATION MODELS

In ecotoxicology, the most commonly used population simulation models are generic models that take into account organism life history characteristics and involve scalar abundance, such as the logistic growth model (Gotelli 2001) or the estimation of a direct inhibition of population growth rates by a chemical via measurements of reproduction or mortality rates (Newman 1995; Tanaka and Nakanishi 2001). Outcomes in these models include density and growth rate estimations, the probability of extinction, and time to recovery. One of the most common descriptors in these types of models is the instantaneous rate of increase (r_i) or the intrinsic rate of increase (r_m). The intrinsic rate of increase measures the ability of a population to increase exponentially in an unlimited environment (Walthall and Stark 1997b) and has been widely applied in toxicological studies to extrapolate from survivorship and reproductive data (individual-level information) to population-level effects (Walthall and Stark 1997a, 1997b; Forbes and Calow 1999). The instantaneous rate of increase provides ostensibly similar information compared to that provided by r_m but requires fewer resources to measure as it is calculated from the number of individuals in a starting and ending population and does not require detailed schedules of survivorship or fecundity (Walthall and Stark 1997b). Forbes and Calow (1999) concluded that r is a better measure of responses to toxicants compared to individual-level effects because it integrates complex interactions among life history traits and provides a more realistic measure of ecological impact. The concept, however, suffers from the fact that any effect on r is context dependent, and thus difficult to interpret ecologically (e.g., see Barata et al. 2002a). Moreover, it is often estimated using data from laboratory-reared individuals, and thus ignores environmental externalities such as habitat loss and extrinsic mortality.

More recently, studies have applied the probability of extinction as an endpoint to extrapolate short-term effects on long-term population consequences. Based on population viability analysis (Boyce 1992; Groom and Pascual 1997), population size is projected into the future using demographic rates and models that incorporate stochastic effects (Snell and Serra 2000). In practice, it would be difficult to determine extinction rates experimentally due to the need to conduct experiments over multiple generations. Thus, the probability of extinction is typically modeled using the instantaneous rate of increase (Snell and Serra 2000).

Another simple approach to predict effects at higher levels of biological organization through extrapolation is the HERBEST model (van den Brink and Kuyper 2001). HERBEST estimates direct effects and potential recovery of field populations of aquatic organisms from predicted or measured pesticide concentrations, laboratory toxicity data, and life history characteristics. It is important to realize that simple population models do not incorporate potentially important aspects of population dynamics such as age structure (stage specificity) and therefore often lack realism (Caswell 1996; Sauer and Pendleton 2003; Barnthouse 2004). However, one of the main uses of simple models is for screening assessments of the ecological risks of chemicals, and, because such assessments are often conducted in the absence of high-quality data, scalar models may be highly appropriate (Ferson 2002).

4.3.2 COMPLEX POPULATION MODELS

A reduction in uncertainty associated with extrapolation between individuals and populations can be achieved by using more sophisticated (and often more complex) population models. Population models based on life history characteristics defined in terms of age or stage have become increasingly popular in recent years (Klok and De Roos 1996; Caswell 2001; Kuhn et al. 2001; Newman 2001; Carroll 2002a). Because these size-structured matrix models track characteristics of organisms as a function of age or stage, unlike the simple scalar models, they provide enhanced biological realism. These models, which may be deterministic or stochastic, incorporate information such as survivorship and fecundity defined by size or morphological state (e.g., stage) or age (Carroll 2002a).

Individual-based population models have also become increasingly prevalent in recent years. Examples of this class of model are dynamic energy budget models such as those developed by Gurney et al. (1990), McCauley et al. (1990), and Kooijman and Bedaux (1996). This class of models incorporates the variation that typically exists among individuals of a population by dividing the entire population into subsets of individuals with similar attributes (Maltby et al. 2001; Regan 2002). Such models can also include migration and dispersal patterns and therefore be highly complex. Examples of the latter are the terrestrial ALMaSS model (Topping 1997; Topping et al. 2003) and the aquatic MASTEP model (van den Brink et al. 2007).

A metapopulation is a group of several local populations in the same geographical area that are linked by immigration and emigration (Gotelli 2001; Akcakaya and Regan 2002). Metapopulation models can model both population persistence and population abundance. In the context of persistence, a population either is present (locally persistent) in an area or is not (extinct). Although metapopulation models have been used widely in ecological circles for almost 2 decades, their application in ecotoxicology has been far less common (Sherratt and Jepson 1993; Mauer and Holt 1996; Spromberg et al. 1998).

Each of the classes of models discussed above can be used for different kinds of extrapolations, including extrapolation of biochemical or mortality effects observed in individuals to effects at the population level, extrapolation of results obtained from semifield experiments to the real field, and integration of population and landscape characteristics to landscape-level population consequences of chemical stressors. For additional information on population models and their use in ecological risk assessment, the reader is directed to the reviews found in Sauer (1995), Caswell (1996), ECOFRAM (1999), Newman (2001), Pastorok et al. (2002), Regan (2002), Bartell et al. (2003), and Sauer and Pendleton (1995; 2003).

4.4 EXTRAPOLATION TO COMMUNITIES

In ecological risk assessment, the target for protection may often be the community. It is recognized that not all populations can be tested and protected, and by maintaining biodiversity at the community level it is assumed that ecosystem structure and function are maintained. A community is, however, more than an agglomeration of species, but is defined in terms of the species present and their interactions. Abercrombie et al. (1973, 70p) defined "community" as an "ecological term for any naturally occurring group of different organisms inhabiting a common environment, interacting with each other especially through food relationships, and relatively independent of other groups."

4.4.1 EXPERIMENTAL APPROACHES

Aquatic microcosms and mesocosms offer the ideal situation to investigate populations of species interacting in their natural environment (i.e., to study communities stressed in structured systems; see Section 4.5.1). It is, however, only recently that these experiments have been analyzed at the community level (Van Wijngaarden et al. 1995; Sparks et al. 1999; van den Brink and Ter Braak 1999). Until 10 years ago, experiments were evaluated at the population level, largely ignoring species interactions and energy flows in the systems. The development of community-level endpoints offered the possibility to evaluate the experiments on a community level (i.e., they offered the opportunity to scale up the level of evaluation; Kedwards et al. 1999). Summary community-level endpoints calculated from the results of these experiments are mostly structural ones: measures of diversity (e.g. numbers of species, and the Shannon-Weaver diversity index) and similarity of the treated systems compared to the untreated controls (e.g., the principal response curves method, Bray-Curtis dissimilarity, or Stander's index; see van den Brink and Ter Braak 1998 for a comparison).

For example, Sibley et al. (2001a) summarized the effects of creosote on the zooplankton community in 1 diagram using the principal response curves (PRC) method (Figure 4.3). Instead of showing the response of the 86 individual zooplankton taxa that were identified, their response is captured in a single diagram. It must be noted that other measures of diversity or similarity can be used, but these are generally less sensitive and do not offer the possibility to go directly back to the species level (van den Brink and Ter Braak 1998; Smit et al. 2002; van den Brink et al. 2003). Measures of diversity and similarity, like principal response curves, offer the possibility to evaluate changes in the structure of the community, although they also can be applied to functional parameters (van den Brink and Ter Braak 1999). The main disadvantage of community-level endpoints is that they do not explicitly capture species interactions and interdependence among species. PRC diagrams, however, do offer the possibility to develop hypotheses concerning such interactions, and can be used as aids to investigate hypothesized cause–effect relationships.

Microcosms and mesocosms are relatively expensive and time-consuming approaches, and thus may not always be appropriate to evaluate effects for many chemicals. Often, the only effects data that exist for a chemical are results from standardized laboratory single-species toxicity tests. In criteria setting there are 2

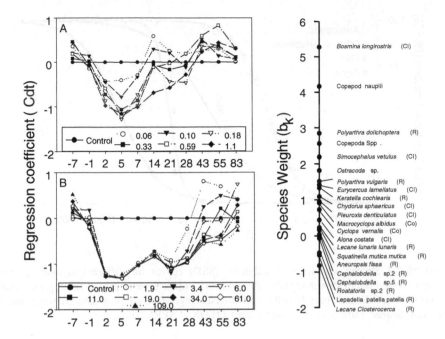

FIGURE 4.3 Principal response curves and species weights (b_k) of zooplankton exposed to liquid creosote in aquatic microcosms. *Note*: Of all the variances, 32% could be attributed to the sampling date; this is displayed on the horizontal axis. Sixty-five percent of all variance could be attributed to treatment level. Of this variance, 22% are displayed on the vertical axis. The lines represent the course of the treatment levels in time. The species weight (b_k) can be interpreted as the affinity of the taxon with the principal response curves (c_{dt}). Only species comprising greater than 5% of the zooplankton community at any point in time, or that occurred in more than 75% of samples, are presented in the species weight line graph. For clarity, the graph is divided into low (A) and high (B) concentrations. Cl = Cladocera; Co = Copepoda; and R = Rotifera. *Source*: Redrawn from data from Sibley et al. (2001a).

methods to extrapolate these values to the community level, the use of uncertainty factors and the use of probabilistic methods. There is an extensive literature on comparing the 2 approaches, and the reader is referred to Scheringer et al. (2002), Volosin and Cardwell (2002), Crane et al. (2003), and Brock et al. (2004) for examples. Microcosms and mesocosms are designed to mimic the field but are not representative for all ecosystems present in nature. Therefore, extrapolation from one ecosystem type (e.g., a macrophyte-dominated aquatic ecosystem) to another ecosystem type (e.g., a plankton-dominated aquatic ecosystem) might still be needed. Recent reviews, however, indicated that for compounds for which multiple mesocosm tests were available, variation of the safe threshold value was relatively low, but effects observed at concentrations higher than the threshold value may differ substantially (van Wijngaarden et al. 2005b; Brock et al. 2006; and see Chapter 7). Section 4.5.1 provides more information on how microcosms and mesocosms can be used to assess the ecosystem level.

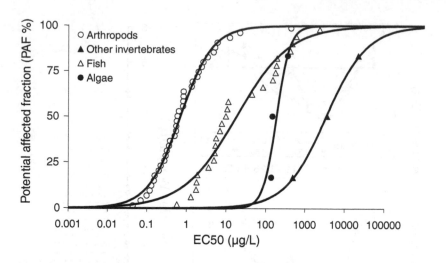

FIGURE 4.4 Species sensitivity distribution (SSD) curves for different aquatic species groups for the insecticide chlorpyrifos. *Note*: Shown are the toxicity data of arthropods, other invertebrates, fish, and algae. Results of the logistic regression on these data are represented by lines. *Source*: Data were obtained from ECOTOX (USEPA 2001).

4.4.2 Species Sensitivity Distribution Concept

On the probabilistic side, the species sensitivity distribution (SSD) concept offers the possibility to predict effects at the community level using all available data (Posthuma et al. 2002b). SSD is defined as a probability density function of the cumulative distribution function of the toxicity of a certain compound or mixture to a set of species that may be defined as a taxon assemblage or community (Figure 4.4). In the United States and European Union, the SSD concept has been used during the last 10 years to set water quality criteria and estimate risks based on results of water quality-monitoring programs (Knoben et al. 1998; Stephan 2002; van Straalen and van Leeuwen 2002; Preston and Shackelford 2002). To estimate a concentration below which the fraction of species exposed above their no-effect level is considered acceptably small, a cutoff value of the distribution must be chosen. Usually a cutoff value of 5% or 10% is chosen, and their corresponding concentrations are named HC5 and HC10 (hazardous concentration of 5% or 10%). The use of the SSD concept in ecological risk assessment is based on some assumptions (Versteeg et al. 1999; Forbes and Calow 2002), some of which conflict with the definition of a community stated above. These assumptions, discussed in detail below, are as follows:

1) The sample of the species on which the SSD is based is a random selection of the community of concern, and is herewith representative for this community.
2) Interactions among species do not influence the sensitivity distribution.
3) Because functional endpoints are normally not incorporated in the SSD, community structure is the target of concern.

4) The laboratory sensitivity of a species approximates its field sensitivity.
5) The endpoints measured in the toxicity tests on which the SSD is based are ecologically relevant.
6) Because in SSD all species have equal weight, it is assumed that all species are equally important for the structure and functioning of the ecosystem of concern.
7) The real distribution of the sensitivity of the community is well modeled by the selected statistical distribution.
8) The number of species data used to fit the distribution is adequate from a statistical, ecological, and animal welfare point of view to describe the real distribution of the sensitivity of the community.
9) The protection of the prescribed fraction of species ensures an "appropriate" protection of field ecosystems.

Unlike the definition of a community that is provided above, the toxicity data used to construct an SSD normally are not derived from species in the community of concern (Assumption 1). So the SSD does not represent a known community, but is often interpreted as if it does (Forbes and Calow 2002). When used for criteria setting, this is less problematic because one does not have a particular community in mind but wishes to protect a generic community, like "the arthropods" (Suter et al. 1993). The most simple and inexpensive way of constructing an SSD for this purpose is to use (Q)SAR to predict toxicity (e.g., Chapter 2) and construct an SSD using these predicted toxicity data. If needed, information on slope and placement of species within the SSD can be derived from other chemicals (De Zwart 2002).

A second assumption inherent in applying SSDs to predict community-level effects is that ecological interactions among species — so-called "indirect effects" — are absent or insignificant (see Section 4.5.2, Assumption 2, below; Wagner and Løkke 1991). Indirect effects are not necessarily linked to a certain centile of the SSD curve (for example, they could occur at the HC10 or HC90 level); therefore, values higher than HC5 may not correspond to a certain overall effect (direct and indirect effects together; van den Brink et al. 2002a). Food-web models that are discussed later in this chapter (Section 4.5.4) can be used to extrapolate from populations to communities taking indirect effects into account.

It is assumed that by protecting the structure of a community, its functions, including energy flows within food webs, are also protected (Assumption 3). However, the validity of this assumption is questionable; for example, Brock et al. (2000a) showed that photosynthesis-inhibiting herbicides affect ecosystem functions such as community metabolism at lower concentrations than structure. It should be noted, however, that this could also be a result of the higher accuracy of measurement and the lower natural variability of functional parameters compared to structural ones. Also, the extent of the difference is rather small; slight changes in structural parameters were generally observed at concentrations 1.6 times greater than those at which slight effects on functional parameters were found (Brock et al. 2000a). For an overview of how to incorporate species interactions and functional parameters into the risk assessment, the reader is referred to Kersting (1994) and to Sections 4.5.2 and 4.5.3. In principle, by combining ecological (food-web) simulation models to define the ecological interactions with the SSD concept to define the toxicology component, it is possible to address this problem (Brix et al. 2001; Solomon and Takacs 2002). This can be done by assigning a distribution of values to functional groups rather than a single sensitivity value. By running multiple simulations, a distribution of outcomes is obtained that also reflects uncertainty regarding the sensitivity of functional groups.

There is an extensive literature on the comparison between laboratory and field sensitivity of species (Assumption 4). The general conclusion of these comparisons is that, when exposure is comparable, the laboratory sensitivity of a species to certain chemicals is representative of its field sensitivity. On an SSD level (i.e., comparing the laboratory and field sensitivities of agglomerations of species), only a few examples are available. van den Brink et al. (2002a), Schroer et al. (2004b), and Hose and van den Brink (2004) used microcosm and mesocosm experiments (see Section 4.5.1) performed with the insecticides chlorpyrifos, lambda-cyhalothrin, and endosulfan to calculate acute field EC50s for species inhabiting the systems. They compared the SSD based on these values (field-SSD) with an SSD based on results from standard laboratory experiments (lab-SSD). van den Brink et al. (2002a) and Schroer et al. (2004b) found an exact match between the two when both SSDs were based on sensitive groups (in these cases, arthropods) and the lab-SSD was based on the same concentration range used in the field experiment. Hose and van den Brink (2004) found a difference of a factor of three between lab-SSD- and field-SSD-based HC5 values, although the confidence intervals overlapped. The HC5 value derived from laboratory data was less than that derived from the mesocosm data and was thus protective of those populations. The findings of these studies suggest that, at least for these insecticides, the exclusion of natural environmental factors (such as light, temperature, habitat suitability, and shelter) in laboratory tests did not significantly change the results of the sensitivity distribution of the arthropods.

Assumption 5 states that the endpoints measured in the toxicity tests on which the SSD is based must be ecologically relevant. Mortality is the most frequently studied endpoint in laboratory tests. In chronic tests, endpoints such as reproduction and inhibition of growth are also studied. Forbes et al. (2001a) argued that individual-level endpoints like survival, fecundity, and growth may not reflect effects at the population level (Assumption 5). They recommended that additional consideration be given to the relative frequency of different life cycle types, to the proportion of sensitive and insensitive taxonomic groups in communities, and to the role of density-dependent influences on population dynamics (see also Forbes and Calow 2002).

For the SSD concept, all species are considered to be equal (Assumption 6), although we know that some species are more important for the functioning of ecosystems than others, for example, the role of earthworms as allogenic engineers in soil ecosystems (Lavelle et al. 1997) and macrophytes as autogenic engineers in the aquatic environment (Jones et al. 1994). Also, the concept of keystone species (i.e., species that have effects on ecosystems that are disproportionate to their relative abundance) acknowledges that not all species are equal. It is possible to weight species in an SSD based on their importance for the ecosystem. There is, however, no consensus in the field of ecology on what to measure as an indicator of the importance of a species to an ecosystem, nor does such a consensus seem likely (Bond 2001).

Most ecologists, however, now agree that we need to focus on 2 categories: dominants (those abundant species that play a key role in ecosystem function, e.g., the crustacean detritivore *Gammarus pulex*; Maltby et al. 2002) and keystone species. What is rare now may become important or dominant later due to phenomena such as climate change. Hence, there is merit in protecting and studying less common species also, a point that has been recognized by ecologists in the form of the biodiversity insurance hypothesis (Naeem 1998; Yachi and Loreau 1999).

When choosing a model to fit the distribution (Assumption 7), one can choose between distribution-free and distribution-based methods (Jagoe and Newman 1997; Grist et al. 2002), and between methods based on classical or Bayesian statistics (Aldenberg and Jaworska 2000). The most popular are those based on the logistic (Aldenberg and Slob 1993) and log-normal (Aldenberg et al. 2002) distributions, because they require less data than distribution-free methods and are relatively easy to fit with standard statistical software. A few authors compared the outcomes of logistic and log-normal distribution fitting. Forbes and Calow (2002) found that logistic HC5s were always lower than log-normal HC5s. Grist et al. (2002) found the opposite: log-normal HC5s were always lower than logistic HC5s. Wheeler et al. (2002a) came to the same conclusion as Grist et al. (2002), but their assessment was partly based on the same data. Newman et al. (2000) compared log-normal models with a distribution-free bootstrapping method using 30 data sets of metals and pesticides. They reported that half of the data sets failed the test on log-normality, which could be a result of the multimodality of the data due to the inclusion of sensitive and insensitive species. They concluded that the distribution-free method performed better but needed much more data. Verdonck et al. (2001) compared bootstrap, maximum likelihood estimation, and Bayesian approaches and concluded that the Bayesian and maximum likelihood estimation methods are superior to parametric bootstrapping because they are easier to use and not so computationally intensive. Jagoe and Newman (1997) also concluded that distribution-based parametric methods appeared to have higher precision compared to distribution-free methods, but the estimate does not reflect the actual value with 95% confidence if the real distribution of the data deviates from log-normal. In order to meet the assumption of normality, van den Brink et al. (2002a) and Maltby et al. (2005) noted the importance of considering the SSDs for taxonomic groups separately when a specific mode of action for the toxicant is known. Figure 4.4, for example, shows that for insecticides, separate SSDs for vertebrates, arthropods, nonarthropod invertebrates, and primary producers should be constructed. Median HC5 values derived from arthropod SSDs were significantly lower than those derived from vertebrates and nonarthropods (Maltby et al. 2005). van Straalen (2002) evaluated 4 distributions — uniform, triangular, exponential, and Weibull — to determine a finite lower threshold (an HC0) to overcome the problem of choosing arbitrary cutoff values (e.g., the 5% centile being the HC5) and the associated problem of communicating the chosen level of protection to the general public.

The adequate number of data points needed to construct an SSD depends on the method used (Assumption 8). Generally distribution-free methods need more data points (30 or more) than distribution-based methods (Newman et al. 2000). Crane et al. (2003) stated that, for chlorpyrifos, species sensitivity followed a log-normal distribution when fitted to all available aquatic (in)vertebrate data and to arthropod data alone. The quantity of toxicity data had little influence on species sensitivity distribution when fitted to all data, when *n* was greater than 10 species. Unfortunately this exercise was not repeated for arthropod taxa alone. It might be expected that arthropod data alone will fit a log-normal distribution better because nonarthropods have a low susceptibility to chlorpyrifos, and inclusion of these data can lead to the combination of 2 different distributions. Suter et al. (2002) reviewed the literature on this matter and found numbers that were considered adequate between 3 and 30. Eight species from different families are required by the USEPA (1995b). The Higher Tier Aquatic Risk Assessment for Pesticides (HARAP; Campbell et al. 1999) workshop recommended the inclusion of 8 relevant species and 5 vertebrate fish species when SSDs are used in the admission procedure of pesticides. For example, for insecticides, arthropods are considered relevant, whereas for herbicides, relevant organisms for SSD construction belong to primary producers. This number is based not only on statistical examination but also on practical, ethical, and economic arguments and expert knowledge.

More and more studies evaluating whether the HC5 values are protective for ecosystem structure and function are becoming available (Assumption 9; Emans et al. 1993; Solomon et al. 1996; Versteeg et al. 1999; Smit et al. 2002; Selck et al. 2002; van den Brink et al. 2002a; Brock et al. 2004; Hose and Van den Brink 2004; Maltby et al. 2005). In general, all authors concluded that the SSD concept (in the form of an HC5) can provide a cost-effective risk evaluation to establish "acceptable concentrations" to set targets for pesticides in the aquatic environment (see Figure 4.5

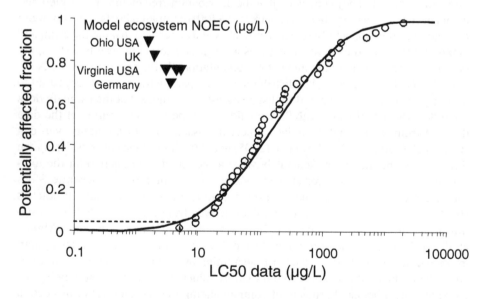

FIGURE 4.5 Single-species distribution for copper based on acute invertebrate toxicity data. *Note*: Arrows denote model ecosystem NOECs for studies performed in Virginia and Ohio, United States; the United Kingdom; and Germany. Single-species data are taken from Brix et al. (2001), and model ecosystem data from Versteeg et al. (1999). All data have been adjusted to a hardness of 50 mg/L CaCO$_3$. Median HC5 is 5.8 μg/L copper (indicated by broken line), and mean ecosystem NOEC is 4.1 μg/L.

as an example for copper). Although the literature seems extensive, large data and theory gaps on the verification of the concept exist for the terrestrial environment and for compounds other than pesticides. For the terrestrial compartment, achieving this verification is more complicated; that is, more well-designed field experiments are needed, including an estimation of the bioavailable fraction of the compound and the spatial heterogeneity of this fraction (van den Brink et al. 2002a). Solomon and Takacs (2002) argued that the HC5 could be protective for ecosystems. They stated that laboratory-based risk assessment overestimates risks at low concentrations because of the presence of community resilience, species redundancy, and adaptation in the field and underestimates risks at high concentrations because of the presence of interaction between species that leads to indirect effects. These arguments, however, are at variance with the precautionary principle, so more validation of a protective percentage and insight into effects of low concentrations of chemicals are needed.

Besides meeting its assumptions, other problems in the application of SSD in risk assessment to extrapolate from the population level to the community level also exist. First, when use is made of databases (such as ECOTOX; USEPA 2001) from which it is difficult to check the validity of the data, one does not know what is modeled. In practice, a combination of differences between laboratories, between endpoints, between test durations, between test conditions, between genotypes, between phenotypes, and eventually between species is modeled. Another issue is the ambiguous integration of SSD with exposure distribution to calculate risk (Verdonck et al. 2003). They showed that, in order to be able to set threshold levels using probabilistic risk assessment and interpret the risk associated with a given exposure concentration distribution and SSD, the spatial and temporal interpretations of the exposure concentration distribution must be known.

4.5 EXTRAPOLATING TO LANDSCAPES AND ECOSYSTEMS

4.5.1 SEMIFIELD TESTS FOR EXTRAPOLATION

Semifield tests, sensu stricto, are experimental ecosystems that run under fieldlike outdoor conditions. Aquatic micro- and mesocosms have been widely used in assessing effects of substances under semifield conditions (Stephenson et al. 1986; Arnold et al. 1991; Hill et al. 1994; van den Brink et al. 1996; Kennedy et al. 1999; van den Brink and Ter Braak 1999; Culp et al. 2000a, 2000b; Sibley et al. 2001a, 2001b, and many others). Also, terrestrial model ecosystems (TMEs) have been developed that are, in contrast to the use of field plots, separated from their source site and run under laboratory conditions (Knacker et al. 2004). Because experience is limited, they will be considered here only marginally.

Because semifield tests are used for extrapolation between laboratory and field (i.e., to extrapolate from individuals and populations to communities and ecosystems), it is important that semifield tests are representative of natural ecosystems. Besides field experiments, aquatic semifield tests are seen as the closest approximation to natural aquatic communities of the real world. As a rule, soon after installation a semifield test exhibits many features of a natural ecosystem, such as a dynamic assemblage of organisms (primary producers, consumers, and decomposers), material cycling and energy flow, and assimilation and decomposition of biomass. It becomes a sustainable system (see Section 1.4.1 in Chapter 1) for a limited period; only occasionally some modifications are necessary such as adjusting the water level to ensure definite toxicant-concentration relations, or some watering to prevent a terrestrial system from damage by drought (Frampton et al. 2000). However, it should be noted that ecosystems established in semifield tests are more representative of ephemeral ecosystems, such as temporary ponds, rather than established ecosystems, and are largely composed of species with good colonizing ability or other invasive characteristics. Our ability to extrapolate from such a biased subset of natural ecosystems is therefore somewhat constrained. Moreover, the conclusions drawn from such experiments should be interpreted with caution, particularly when applying the findings to systems possessing long-lived fauna or highly evolved mutual dependencies among species.

Semifield tests offer the ability to observe or measure a suite of biological and physicochemical variables, as with the analysis of natural ecosystems. Usually, the focus is on structural measures such as the number of species and number of individuals per population (often lumped into higher taxonomic units), physicochemical parameters (e.g., dissolved oxygen, and pH including toxicant residues), and some microclimate parameters (e.g., evaporation in TMEs).

Stagnant ponds and lakes are more separated entities. Although they may be linked with adjacent terrestrial landscapes, they can be viewed as relatively closed systems with distinct communities, material cycling, and energy flow. Their ecosystem properties are more easily studied, and much of the knowledge on interactions of species populations and ecosystem functional processes stems from the study of lakes. Freshwater outdoor mesocosms (van den Brink et al. 1996), laboratory generic freshwater microcosms (Nabholz et al. 1997; USEPA 2002b, 2002c; Daam and van den Brink 2003), and site-specific aquatic microcosms (Sibley et al. 2001a) are artificial ecosystems that develop communities and ecosystem functions closely corresponding to those of natural ponds (but see also above). They have become an important tool for studying the fate and effects of toxic substances to communities.

If we compare the physical surfaces in aquatic model ecosystems, which could affect the fate of the chemical under investigation, there is a high agreement between the model and natural system. In both types of systems, all surfaces, except the water surface, are covered by a biological layer (e.g., periphyton or biofilm) or are part of the sediment surface. Therefore, adsorption and desorption processes will take place in the same manner as in natural systems. Likewise, in outdoor systems, degradation processes such as photolysis can take place. Although fate processes in aquatic systems appear quite easy to extrapolate, the situation in a terrestrial system is much more complex due to the high spatial heterogeneity, the size of the systems, and the lack of distinct boundaries. The distribution and dissipation of compounds here are influenced by a number of factors not easy to control in a model system. Among these factors are many microclimate factors such as humidity, temperature, protection from wind and exposure to sunlight, and so on. Terrestrial model ecosystems, such as the soil-core microcosm (Nabholz et al. 1997; USEPA 2002a; Knacker et al. 2004), are appropriate to model parts of a natural terrestrial ecosystem such as a column of soil including the soil fauna and flora, or some weed plants together with their herbivores and predators. Also, with the use of plots in a natural field site, only a small part of a terrestrial ecosystem or a definite function of the system can be studied. In terrestrial systems, therefore, the method of choice may be to conduct a (bio)monitoring program after a toxicant application.

Belanger (1997) quantitatively reviewed the relationship between the structure, size, and various experimental design attributes of flowing model ecosystem studies as they affected biological complexity. There was no apparent influence of test system size on biological complexity, and it was apparent that most investigators take caution in designing systems of sufficient complexity to test the influence of stressors on complex species assemblages.

A number of studies have been conducted to characterize the community structure and function of model ecosystems. Some examples are presented below, providing weight of evidence that some systems correspond closely to (parts of) natural field communities.

Williams et al. (2002) compared the plant and macroinvertebrate communities of small microcosms to those of natural ponds in Britain. Using multivariate analyses, they showed that the fauna of the microcosms were most similar to those of less common deep circumneutral ponds and did not closely mimic more commonly occurring shallow ponds. The microcosms lacked a wide variety of invertebrate groups and plants typical of shallow natural ponds that possess a littoral zone.

Eckert (2000) compared the communities of stagnant 5000–L freshwater mesocosms with 10 natural ponds located around the mesocosm facility (Aachen, Germany). A littoral zone was lacking, and the model system represented pelagic and benthic assemblages rather than littoral ones. This type of mesocosm is characterized by a rich assemblage of phytoplankton and zooplankton as well as benthic invertebrates settling on the wall of the artificial pond and sediment surface, including chironomids, which fed on the periphyton, and the predatory midge larvae of *Chaoborus* spp. It turned out that only 4 of the 27 chironomid taxa found in the natural ponds were found in the mesocosms. The same occurred in 2 of the natural ponds located in the pastureland. The communities here appeared to be degraded, and macrophytes were lacking. The emerging insects and the zooplankton assemblages of these 2 natural ponds showed the highest degree of correspondence with the mesocosms. The emergent insect species belonged to ubiquitous species groups and were found in almost all of the remaining natural ponds. The majority of the natural ponds contained a more diverse insect community (including mayflies, caddisflies, and odonates) and vertebrates.

Dyer and Belanger (Dyer and Belanger 1999; Belanger 2003) presented information on the relationship between a specific flowing model ecosystem located in Ohio and its parent or source river. The system was rich in macroinvertebrates with several hundred taxa represented in the model ecosystem. Direct comparisons of assemblages sampled at the same time in both the model and the natural river demonstrated that there was a community similarity of 75% to 80% and that there were no statistically discernible differences detected from comparisons spanning over a decade.

The time schedule of aquatic model ecosystem experiments and the size of the system itself favor population growth of small organisms with short generation times (*r*-selected species) such as algae, protozoans, rotifers, small crustaceans, and benthic invertebrates with more than 1 generation per year. Macrophytes and longer generation species (e.g., odonate larvae, water beetle larvae, amphibians, and fish) are sometimes lacking, although it is recommended to add macrophytes to microcosms and mesocosms (Giddings et al. 2002). Because *r*-selected species are more common in these aquatic model ecosystems, the study of recovery from toxicant effects is limited to species groups exhibiting this type of life history strategy. Emergent insects from control ponds or neighboring natural ponds can easily deposit eggs on previously treated ponds, initiating a complete recovery of the populations (allogenic recovery) when the toxicant residues fall below the critical threshold. Zooplankton and phytoplankton species can build up new populations from resting stages in the sediment (autogenic recovery). *K*-selected macroinvertebrate species (1 or fewer generations per season) can easily be introduced into the model systems, but if they have emerged or been damaged by the toxicant, they are unlikely to contribute to the recovery process.

Another type of comparison can be based on functional processes (e.g., comparing process rates of primary production, consumption, or decomposition). Functional processes related to community metabolism are often measured in model and natural systems. If functional redundancy exists within the model ecosystem community, an extrapolation of toxicant effects is straightforward. This has been corroborated by studies of Peterson et al. (2001) on nitrogen export and processing within small streams using N-tracer additions. Eleven headwater streams representing a wide array of biomass throughout North America and one model ecosystem from Ohio were studied to determine nitrate and ammonium uptake and export. The driving force dictating nitrogen uptake length in all streams, including the model ecosystem, was stream discharge and the river from which the model ecosystem drew water. Further analysis of this data set by Dodds et al. (2002) indicated ammonium uptake rates were universally higher than expected and were not saturated at even high concentrations. The model ecosystem fit well within the boundaries established by the natural streams in the data set, strongly supporting that universal ecosystem functions can be extrapolated.

4.5.2 INDIRECT EFFECTS

Ecosystems are inherently complex, and a fundamental source of this complexity is the interactions that occur between organisms within and between different trophic levels. In this chapter we define complexity in the sense of "architectural complexity," referring to the product of species diversity and species connectance. Indirect effects are "changes in abundance of a population resulting not directly from the action of a causal agent (such as a toxicant) but indirectly through the effects of the causal agent on other species" (DeAngelis 1996, 25 p). The basic concept is illustrated in Figure 4.6, based on Van den Brink et al. (2002b).

Extrapolation between the population and community levels of biological organization and between community and ecosystem levels may be greatly confounded by the occurrence of contaminant-induced indirect interactions between organisms, leading to a high level of uncertainty in model predictions. In practice, the most

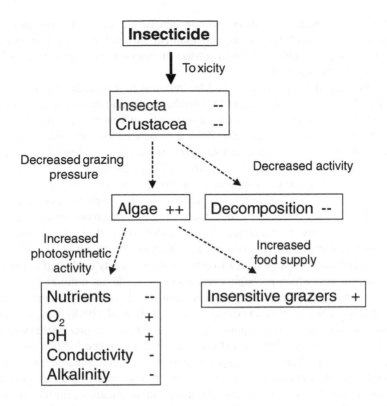

FIGURE 4.6 Schematic overview of direct (solid arrow) and indirect (dashed arrow) effects of toxicants on an aquatic ecosystem. *Source:* Redrawn from van den Brink et al. (2002b).

common approach to address this uncertainty has been to conduct semifield experiments (see Section 4.5.1). Indeed, the ecotoxicology literature is replete with examples of indirect effects caused by chemicals, with the most common examples occurring in micro- or mesocosm studies (Brock and Budde 1994). For example, herbicides directly affect the biomass or abundance of phytoplankton or plants, which may cause indirect declines in zooplankton abundance because the phytoplankton serves as a food source for the zooplankton (Jenkins and Buikema 1990; Kasai and Hanazato 1995a; Cuppen et al. 1997; Wellman et al. 1998). This in turn may cause an increase of insensitive grazers (Figure 4.6). It can be easily calculated by a mathematical model that top carnivores, even if not studied in the model system, will become extinct because not enough energy reaches this trophic level (Hommen 1998). Functional redundancy (less sensitive algal or plant species in higher numbers) could prevent or reduce this effect in the natural system. Conversely, a chemical that directly affects zooplankton populations often results in increases in phytoplankton populations because grazing pressure declines (Brock et al. 1992; Webber et al. 1992; Boyle et al. 1996; Sibley et al. 2001a, 2001b; van den Brink et al. 2002a; Wendt-Rasch et al. 2003; Friberg-Jensen et al. 2003). Also here, due to the higher functional redundancy in the natural system, the effect in the natural system might be smaller. Kersting and van den Brink (1997), for instance, found much smaller indirect effects on physicochemical parameters of similar concentrations of the insecticide chlorpyrifos in outdoor mesocosms compared to Brock et al. (1993) in indoor microcosms.

Theoretically, indirect effects could be manifest in a cascade effect across several trophic levels, reflecting changes in both the abiotic and biotic components of the ecosystem. Such cascades may occur via top-down or bottom-up manipulations or changes to the ecosystem. Hambright (1994) conducted a top-down biomanipulation experiment in a lake in which piscivorious fish were introduced into a lake that reduced the smaller planktivorous fish, which, in turn, led to an increase of zooplankton. The increase in grazing pressure of zooplankton led to decreased chlorophyll-a. A bottom-up example is often observed in microcosm studies when herbicides or nutrients are administered. In the case of herbicides, the decrease of macrophytes may lead to an increase of less sensitive or adapted phytoplankton and a decrease of macroinvertebrates due to habitat loss (Cuppen et al. 1997; van den Brink et al. 1997). However, interactions across more than 2 trophic levels are rarely evaluated in micro and meso-cosm studies, which probably contributes a significant amount of the uncertainty associated with extrapolating effects across levels of biological organization.

In a metapopulation-modeling exercise, Spromberg et al. (1998) found that perturbations that directly affect a given population may spread to populations that were never exposed to the chemical. They termed this "action at a distance," which they defined as the process by which impacts on a population in a given area can be transmitted by emigration from the contaminated patch to other patches within the metapopulation. The results of this study have 2 potentially important implications for the risk assessment process. First, from a practical standpoint, based on the potential for connectivity between the impacted and adjacent or nearby reference sites, these authors concluded that, ideally, reference sites should not be connected in any manner to test sites. This problem could be overcome by using a before–after control impact (BACI) design, although this approach has been criticized because it typically includes few reference sites (Underwood 1992). Spromberg et al. (1998) also suggested that reference sites established too far away from the test site would be unlikely to bear much physical or biological similarity to the test site and would thus also be inappropriate. In reality, the latter may or may not be true and could potentially be overcome by using approaches such as the multivariate reference condition in which multiple reference sites are characterized and against which the impaired site is compared (Reynoldson and Rodriguez 1998; Bailey et al. 2004). At a minimum, it clearly underscores the importance of being judicious in the selection of references sites. Second, in terms of extrapolation practice, the results suggest that, if ecologically relevant predictions of risk to a species or assemblage of species are to be made, sole consideration of individual tolerances and/or responses may not be sufficient (Preston 2002).

4.5.3 Extrapolating from Effects on Structure to Effects on Function

The relationship between structure (the composition and abundance of ecosystem components) and function (the processing of materials and energy flow) in ecosystems has constituted a central focus of ecological research for several decades

(Tilman et al. 1997; Naeem 1998; Hector et al. 1999). Despite these investigations, our understanding of the structure–function relationship today remains rudimentary, especially with respect to the response of ecosystems to natural or anthropogenic disturbances. Empirical investigations attempting to assess the relationship between structure and function are difficult, so it is not surprising that there is considerable uncertainty associated with extrapolations between these 2 entities. A key reason why structure–function relationships have not been adequately addressed is because experimentation rarely incorporates a sufficiently long temporal component. Functional responses often occur over much longer time scales than do structural endpoints (e.g., Kersting and van den Brink 1997), and the relation between responses in structural and functional endpoints is not linear. For the purposes of risk assessment, structural attributes have long been used as the primary measurement endpoints to assess ecosystem quality even though systems ecologists have studied functional aspects of ecosystems for more than half a century (Pratt and Bowers 1992). This is clearly reflected by the dominance of structure-based endpoints in most standard test methods (e.g., USEPA 2000b; ASTM 2003). The predominant use of structural endpoints reflects the prevailing philosophy among ecotoxicologists that initial signs of environmental stress are usually observed at the population level, affecting especially sensitive species (Odum 1992), and, conversely, that functional (process-oriented) endpoints are likely to respond only after damage to an ecosystem has already been sustained (Pratt and Cairns 1996). Pragmatically, structural endpoints are also generally easier to measure. This has resulted in widespread implicit acceptance among risk assessors that protecting the structural components of an ecosystem will protect its functional capacity. There is, of course, much debate about the scientific validity of this assumption.

The most accepted relationship between structure and function is, however, that structure is more sensitive than function. This hypothesis has considerable empirical support and is explained by the occurrence of functional redundancy, the idea that many species perform similar functions such that the loss of 1 or a few does not impair the overall functional integrity of the ecosystem (Walker 1991; Tilman et al. 1997; Naeem 1998; Hector et al. 1999). This hypothesis has its origin in ecosystem theory, which suggests that structural components of an ecosystem will exhibit departures from dynamic equilibrium conditions before functional measures, reflecting the inherent functional redundancy that exists within an ecosystem (Tilman et al. 1997). It must be noted that the recovery can take place to a stable state that is very different from the original one, depending on the original state of the system and the type and strength of the stressor (Scheffer et al. 2003). In the context of risk assessment, therefore, functional endpoints could provide a stronger and more practical basis upon which to monitor changes associated with ecosystem recovery following perturbation.

Opposing the view is the idea that function is more sensitive, or at least responds earlier, to perturbation than structure. Although there is less evidence to support this hypothesis, one can imagine situations in which the collective physiological capacity (function) of species within the ecosystem is impaired without a corresponding elimination of species (e.g., structure) per se (Pratt and Cairns 1996). For example, as described earlier, van den Brink et al. (1997) and Cuppen et al. (1997) found that community metabolism responded more clearly in response to a

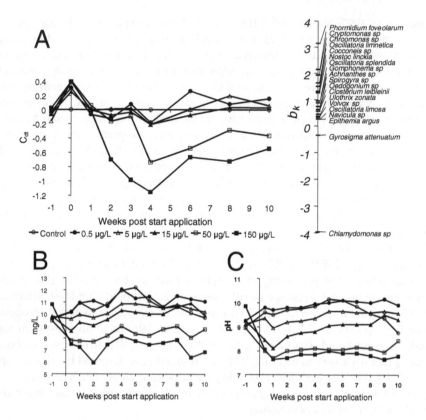

FIGURE 4.7 Figures showing the effect of a chronic application of the herbicide linuron on structural (A) and functional (B and C) parameters. *Note*: Figure A shows the principal response curves resulting from the analysis of the phytoplankton data set. Of all the variances, 47% could be attributed to the sampling date; this is displayed on the horizontal axis. Thirty percent of all variances could be attributed to the treatment level. Of this variance, 23% are displayed on the vertical axis. The lines represent the course of the treatment levels in time. The species weight (b_k) can be interpreted as the affinity of the taxon with the principal response curves (c_{dt}). Taxa with a species weight between 0.25 and −0.25 are not shown. A Monte Carlo permutation test indicated that a significant amount of the variance explained by treatment is displayed in the diagram ($P = 0.034$). Figures B and C show the changes in DO and pH in the different treatments in time. *Source*: Redrawn from data from van den Brink et al. (1997).

photosynthetic-inhibiting herbicide in aquatic microcosms compared to structural parameters. The principal response curves diagram (see van den Brink and Ter Braak [1999] for details), as shown in Figure 4.7A, shows that the structure of the phytoplankton community was affected at the two highest treatments levels. After 2 weeks of exposure, the highest concentration starts to diverge from the control; after 3 weeks also the second highest treatment also begins to diverge. Several species are indicated to have suffered from the treatment (they have a positive weight with the diagram), and an increase was indicated for *Chlamydomonas* sp. (it has a negative weight in Figure 4.7A). Macrophyte biomass was also significantly affected at the two highest concentrations (van den Brink et al. 1997). The functional parameters, however, were affected at the 4 highest concentrations (Figures 4.7B and 4.7C). So the reduced photosynthesis at the intermediate concentrations did not result in significant changes in macrophyte biomass or the structure of the phytoplankton community.

A third relationship may be that structure and function are so intimately related that a change in 1 will, in an approximately proportional fashion, lead to a change in the other. This is the premise of the "rivet-popping" hypothesis introduced by Ehrlich and Ehrlich (1981), which proposes that the structural components of an eco-system can be compared to the rivets of a plane; much in the same way that a gradual loss of rivets from a plane may weaken its structure, the gradual loss of species from an ecosystem gradually weakens its function, leading to reduced performance and, with sufficient perturbation, eventual failure. This view is supported, in part, by Ghilarov (2000, 408 p), who defined ecosystem function as "the energy transforma-tion and matter cycling resulting from the combined activity of living organisms." He argued that ecosystem function varies directly as a function of species diversity and that the latter cannot be diminished without a corresponding loss in ecosystem function. This statement is, however, not supported by experimental observations (i.e., it depends on what species are affected first). So species identity is key in this context, with mesocosms offering the possibility to study them jointly.

A final relationship has been suggested by Finlay et al. (1997), who argued that, in the case of microbial diversity, there is no discrete relationship between struc-ture and ecosystem function. They found that microbial activity and diversity in ponds are both a part of, and inseparable from, ecosystem function and suggested that concepts such as redundancy of microbial species and the value of conserving biodiversity at the microbial level have little meaning. Finlay et al. (1997) argued that ecosystem functions such as carbon fixation and nutrient cycling are governed by complex reciprocal interactions between chemical, physical, and microbiologi-cal factors. For microbial communities, these interactions continuously create new microbial niches that are quickly occupied by microbes originating from the resident pool of rare or cryptic microbial species. In this sense, the microbial ecologist is not necessarily concerned about species diversity per se because most groups of bacte-ria can be found in most environments as long as the environmental conditions and resources are appropriate (Ghilarov 2000). Finlay et al. (1997) suggested that the microbial diversity in an ecosystem is never so impoverished (e.g., infinite redun-dancy), even in the face of extreme changes in environmental conditions, that the microbial community cannot play a full role in biogeochemical cycles.

Evidence suggests that structure and function are intimately related, but the precise nature of the relationship remains to be fully elucidated. Although some research clearly shows that structure responds before function to perturbation, other evidence indicates that the functional group component of an ecosystem (grassland and plant systems) may be a greater determinant of ecosystem processes than the spe-cies component of diversity (Tilman et al. 1996, 1997). They conducted experiments in which species diversity (number of plant species), functional diversity (number of functional groups), and functional composition (different combinations of func-tional groups) of various plant species were manipulated in experimental field plots (Tilman et al. 1996, 1997). They found that the number of functionally different roles represented in an ecosystem was a stronger determinant of ecosystem processes than the total number of species per se. Importantly, these authors concluded that factors that produce changes in ecosystem composition are likely to strongly affect ecosystem processes and that the loss of species with certain functional traits may have a greater impact on ecosystem processes than the loss of other, less functionally important species.

A significant source of uncertainty in relating structure and function, and hence in extrapolating between each, is the notion of hidden treatments (Huston 1997). In this context — experimental manipulations that have multiple components, but only 1 or 2 are identified as an experimental treatment — the probability that incorrect conclusions regarding cause and effect will be drawn increases because the actual cause of the response is unknown or ignored. Sources of error may result from the interaction of undefined biotic and abiotic variables, thus complicating the interpretation of ecological experiments (Huston 1997). In addition, indirect effects (actions at a distance) of stressors on either structure or function can obfuscate specific relationships that may be under investigation, increasing prediction uncertainty.

Reliance on structural endpoints has been criticized as lacking ecological relevance, in part because of the incongruity between the scale at which the variables are typically measured (e.g., subcellular to organism) and that at which the entities to be protected occur (e.g., populations and communities). Although functional endpoints can be less responsive and practical to measure, they provide key information on ecosystem-level processes that may be manifested across all levels of structural complexity. Thus, to enhance ecological relevance in the ecological risk assessment process, and reduce uncertainty in the practice of extrapolation, some have advocated that greater effort must be directed toward developing functional endpoints and evaluating their performance in the context of changes in structure-based endpoints (Cairns et al. 1992; Pratt and Cairns 1996; Clements 1997).

4.5.4 COMMUNITY, ECOSYSTEM, AND LANDSCAPE MODELS

Community and ecosystem models are mathematical expressions that are intended to describe ecological systems composed of interacting species. Such models are typically spatially aggregated (include a minimum number of large habitat components within which the model assumes a homogeneous distribution of state variables) and represent biotic and abiotic structures in combination with physical, chemical, and ecological processes in aquatic or terrestrial systems (Bartell 2002). In ecotoxicology, they are often used to model bioaccumulation in food chains and to estimate the consequences of chemical stress on food webs, including indirect effects (Koelmans et al. 2001; Carroll 2002b). The advantage of these models over simpler ones is that they enable the integration of fate, bioaccumulation, and effects of chemicals at the ecosystem level in conjunction with information on food-web characteristics such as biomass, abundance and richness of component species, trophic structure, species interactions, and nutrient cycling. In this way, it is possible to extrapolate effects observed on populations to the community or ecosystem level.

Food-web modeling holds great promise in analyzing and predicting effects of chemical stressors (Baird et al. 2001) and has been developed for both terrestrial

(De Ruiter et al. 1994) and aquatic systems (Park 1999 [AQUATOX]; Traas et al. 1998, 2004 [C-COSM]). Food-web modeling also enables a direct way of linking structure and function. There is a large number of food-web models describing the flow of energy, carbon, or nutrients and their interactions in ecosystems (De Ruiter et al. 1995; Ågren and Bosatta 1998). Application of such models to ERA has taken very different routes. Whereas De Ruiter et al. (1995) applied classical stability analysis of food-web matrices (in the sense of May 1974b), a different approach is taken with dynamic food-web models to predict the time dependence of toxicant effects. A coupled differential equation model for pesticide effects in microcosms adequately described the dynamics of some primary and secondary effects (Traas et al. 1998), although this required considerable effort for parameter estimation. Besides their integrative nature, they force conceptual thinking and, as such, greatly transcend the information and predictions derived from models based on individual experiments and information derived from lower levels of biological organization. The primary disadvantage of ecosystem-based models is their complexity and associated uncertainty. In addition, they are poorly amenable to validation (Baird et al. 2001), often lack sufficient data, and may not account for a wide range of nonmodeled interactions (e.g., those resulting from pheromonal, gustatory, olfactory, physical, mechanical, or behavioral interactions) in the real world (Health Council of The Netherlands 1997). These kinds of approaches are part of the broader "ecological network analysis." Ecological network analysis is based on input models of energy or material flows (e.g., carbon compound flows) through a trophic network (e.g., a food web describing which species eat which other species; Johnson et al. 2003). Two main software packages have been developed to perform ecological network analysis: NETWRK4 (Ulanowicz 1987) and ECOPATH (Christensen and Pauly 1992). These tools offer approaches to assess the effects of perturbations across levels of organization on a functional level (e.g., nutrient and energy flows; Johnson et al. 2003).

A number of community and ecosystem models can be adapted for application at the landscape level; however, there are also a number of landscape models whose specific purpose is to model ecological phenomena, including effects of stressors, across landscapes (Mackay and Pastorok 2002). Landscape models are spatially explicit (i.e., they provide detailed information on the locations of organisms or the pattern of a landscape), may include information from several ecosystems, and typically use endpoints such as the spatial distribution of species, individual abundance of species within trophic levels or guilds, biomass and productivity, and trophic structure (Mackay and Pastorok 2002). Although predominantly applied in relation to ecological disturbance (e.g., forest succession and hydrologic changes), landscape models have been developed for the assessment of chemicals and watershed and land use disturbance (Kelley and Spofford 1977; Voinov et al. 1999; Nestler and Goodwin 2000).

Most recently, a new perspective on complex systems, one that recognizes the important role of thermodynamics in understanding the structure, function, and flow of energy through ecosystems, has begun to emerge (Nielsen 1992). A key feature of a thermodynamic perspective of ecosystem function that has potentially important implications for risk assessment is the principle of exergy. Exergy is a measure of the maximum capacity of the energy content of a system to perform useful work as it proceeds to equilibrium with its surroundings. In other words, it represents the quality of energy, measuring the distance between a system and its environment with respect to the entropy state of a system (Nielsen 1992). The ability to calculate and model exergy today can be traced back to the founding work of Boltzmann (1981), who investigated entropy as a function of microstates using statistical thermodynamics. Exergy can be used to explain differences between and within several levels of biological organization, structure, and order. Exergy can also be analyzed and seen as the mechanism behind the self-organizing properties of ecosystems (Nielsen 1992). The computational models behind exergy are derived from cybernetics and economic models.

The application of thermodynamic principles as a basis for understanding ecosystem dynamics has been restricted primarily to ecological studies (Ray et al. 2001; Jorgensen 2002; Jorgensen et al. 2002; Ludovisi and Poletti 2003; Zhang et al. 2003). This approach has been applied in a very limited capacity in the area of environmental toxicology (Xu et al. 2002) and will require considerable scientific scrutiny, and the full force of falsification (Popper 1959), before it can be applied as an interpretive and regulatory tool in ecological risk assessment.

4.6 DISCUSSION

In recent years, considerable effort has been expended on developing predictive (correlation and/or cause–effect) relationships between levels of biological organization as a basis for reducing extrapolation uncertainty in ecological risk assessment. Most of this work has focused on relationships between proximal levels of organization because these are easier to measure. If uncertainty associated with extrapolation between levels of biological organization is to be reduced, along with the existing heavy reliance on the use of uncertainty factors by regulatory agencies to ensure that the protection goals of legislation and regulatory authorities are achieved, greater effort must be directed toward developing quantitative relationships between less proximal levels of biological organization (Attril and Depledge 1997).

Biomarkers are used to detect exposure to chemicals, and observed responses can often readily be explained from environmental concentrations of chemicals. However, their predictive value for higher levels of organization is poor. Although actual biomarker effects are measured on a biochemical or physiological level, these effects can only be extrapolated as potential effects at the individual level. To date, biomarkers have failed to realize their early promise because many are not truly diagnostic of single stressors, and few studies have been able to demonstrate unequivocal ecological consequences arising when biomarkers are induced.

On a higher level, laboratory single-species tests can provide useful information to predict effects on natural populations because they describe how toxicological effects are translated into ecological currency (maturation, reproduction, and mortality). Many major taxa, however, remain poorly studied (e.g., soil fauna and snails), and the role of tested species for the ecosystems to be protected is generally unknown.

Population models can be a valuable tool to extrapolate data obtained from lab experiments to the field, but they are difficult to apply without context (i.e., the consequences of a stressor are contingent on local conditions and depend on population state).

Micro- and mesocosms offer the possibility to study effects on populations and communities (including indirect effects), as well as on some ecosystem functions (van den Brink and Ter Braak 1999), although complex food webs are difficult to create in experimental systems, due to spatial and temporal constraints. This suggests a requirement for some spatial-scaling experiments to improve the robustness of such approaches. Similar arguments can be made regarding temporal constraints. On the modeling side, food-web models are often found to be too complex and cumbersome, and their results are too difficult to communicate for application in generic risk assessment. Developing models that are complex enough to incorporate realistic food-web relations and simple enough to be manageable and communicable would be needed for their incorporation. The construction of a generic "risk assessment ecosystem" is, however, a difficult task. The development of "risk assessment scenarios" covering a variety of ecosystems possibly at risk may be a step forward. Also, site-specific food-web models should not be too complex; they should at least represent the actual ecosystem present at the site rather than a generic one (Traas et al. 2004).

Evidence from recent studies in ecology suggest that functional redundancy exists within some natural communities, but the relative contribution of different species to specific functions is far from equal, and the disappearance of certain species (i.e., dominants and keystone species) can have disproportionately large effects on overall system function and the occurrence of other species. For this reason, the precautionary principle advises that species loss be considered at least as a cause of incremental damage. Landscape-level models offer the possibility to improve our interpretation of local phenomena (e.g., species disappearance) in a wider spatial context.

In essence, successful extrapolation is linked to a mechanistic understanding of system structure and function, where one can have confidence in the existence and importance of causal structures. However, there is always a temptation to base approaches on untested assumptions (e.g., species equivalence in SSDs). For example, simply because A influences B does not mean that B will always respond to a change in A. We still lack a coherent framework in ecotoxicology that suitably integrates ecological theory. Although a clear understanding of toxicological effects (e.g., mode of action) is important for developing a mechanistic understanding of ecological responses, the primary domain of ecotoxicologists should be to understand their ecological consequences.

5 Mixture Extrapolation Approaches

Leo Posthuma, Sean M. Richards,
Dick De Zwart, Scott D. Dyer, Paul K. Sibley,
Christopher W. Hickey, and Rolf Altenburger

CONTENTS

5.1 INTRODUCTION

Organisms in the environment are rarely exposed to single stressors. However, most toxicological and ecotoxicological studies focus on exposure and effects of single compounds (Yang 1994a; Krishnan et al. 1997). Consequently, most environmental regulations are based on single-substance data, with the exception of effect-based diagnostic bioassay tools such as the whole-effluent toxicity (WET) approach. Quantitative data on mixture toxicity for relevant assessment endpoints (exposed target species or species assemblages) are needed to assess risks and to support regulation and risk management of chemical mixtures. With little data and numerous mixture possibilities, however, regulators have to make decisions based on single-substance data in combination with mixture extrapolation techniques.

Methods to extrapolate both exposure and effects data of chemical mixtures are applied for many reasons in risk assessment. These range from basic research questions to regulatory actions for mixtures of known and unknown compositions (Table 5.1). In this context, one may distinguish 2 basic objectives when examining mixture effects:

- A mechanistic, analytic, and experimental focus with the intention of understanding
- A prognostic focus, with the intention of developing methods for effect and risk prediction

The objective of experiment-based mixture studies is to understand organism or community responses when there is exposure to complex mixtures. This chapter reviews the results of such studies, and discusses practical implications for risk prediction. Using the set of experiment findings, there are 2 forms of utilization for risk assessment and management. For prognosis, the emphasis is on deriving environmental quality criteria (EQC) for mixtures that occur frequently in the same composition (e.g., mixtures of PCBs or PAHs), and the prediction of mixture toxicity before chemicals are released (Vouk et al. 1987). Both require making a purely predictive assessment of mixture toxicity. For diagnostic or retrospective risk assessment, emphasis is on site-specific assessment of spills and of otherwise multiple contaminated ecosystems for evaluation of remediation priorities or risk management decisions.

Mixture extrapolation is used to protect specific target species. Mixture extrapolation includes laboratory-to-field extrapolation, extrapolation from test species to target species, extrapolation between different matrices, and extrapolation from mixture A to mixture B. Mixture extrapolation is also used to protect specific ecological communities (species assemblages) and may be applied in the diagnostic sense. The latter involves extrapolations for the species as well as extrapolations across levels of biological organization (i.e., from species test data to the species assemblage level).

TABLE 5.1
Objectives of environmental mixture studies

Objective	Focus	Intention
Understanding		
Analysis of joint action	Exposure interactions	Characterization of joint exposure scenario
	Toxicant–target interactions	Characterization of biological sites of action
	Toxicant–co-solute interactions	Characterization of toxicological modes of action
	Species level experimentation	Understanding mixture responses in species
	Community-level experimentation	Understanding mixture responses in communities
Risk assessment — prognostic		
Predicting exposure	Site-specific exposure levels	Exposure aspects of mixture risk
Description of joint effect	Joint effects assessment	Effects aspects of mixture risk
Quantifying joint risk	Safe ambient concentrations	Derivation of environmental quality criteria for mixtures
		Decisions on release of new compounds on market
Risk management — diagnostic or retrospective		
Quantifying site risk	Calculation of exposure and effect	Knowing risk levels for targets 1, 2, and 3
1) Ranking components	Effective contributions	Prioritization of chemicals
2) Ranking sites	Expected site damage	Prioritization of site cleanup or prevention
3) Ranking species	Expected species group damage	Planning species protection measures
Quantifying site effects	Effects measurement of, for example, wastewater	Effects reduction

Most regulatory methods for the management of chemical compounds are based on single-substance risk evaluations, but this information, when used in conjunction with some basic toxicological noninteraction models (Loewe and Muischnek 1926b; Bliss 1939; Plackett and Hewlett 1952), may also be used to predict the joint effect of chemical mixtures on a single species. The models in question have been rigorously tested at the species level (see Deneer 2000; Warne 2003) and are generally found to yield reliable descriptions of test results for mixture studies testing high doses of a few constituents. However, most real-world environmental exposures involve low concentrations and a more complex range of chemicals than is usually investigated in the laboratory (Groten et al. 2001). Extrapolation from single-species mixture toxicity to in situ risk for an assemblage of species adds complexity. The nature of the chemicals in the mixture, the variability of exposure routes, and the ranges of sensitivities of the receptor organisms are all crucial factors that may determine the type and intensity of responses. The theoretical developments for multiple-species risk assessment are weak, and the collection of experimental data is even weaker. The development of validated methods for mixture risk assessment for species assemblages might be beyond the reach of current research capabilities. However, a relative ranking of contaminated sites and of the most affected group of species, as well as identification of the most hazardous compounds, can be sufficient for informed risk management decisions.

In view of the assessment problems and the limitations mentioned above, the aims of this chapter are to introduce, explain, and evaluate mixture toxicity extrapolation techniques at the species level and for species assemblages. These extrapolation techniques are discussed, their mechanistic and methodological foundations considered, and the quality of the data that support them assessed. Existing reviews on species-level mixture studies were used as a basis on which to address the quantitative aspects of mixture extrapolation, and to derive species-level extrapolation protocols. Species-level data and a novel combination of conceptual and pragmatic views are combined to explore, discuss, and define technical procedures for estimating the risks of complex toxicant mixtures for multispecies biological systems. Protocols for mixture toxicity extrapolations are provided. Limitations, uncertainty issues, technical feasibility (input data), and options for validation are explored. It is assumed that the reader has a broad knowledge of concepts, approaches, outcomes, and interpretations of mixture toxicity studies in ecotoxicology.

5.2 CONCEPTS FOR MIXTURE EXTRAPOLATIONS

Any attempt to extrapolate the toxicity of a mixture has to address exposure, matrix, and effect issues. For exposure, the main question is whether the contributing components are known or not, which in the following text we will call a defined or an undefined mixture, respectively. Matrix and media issues are treated in detail in Chapter 2 of this book. The focus in this chapter is thus restricted to combined effects.

Predicting the combined effects of a mixture from knowledge of the effects of its components requires a reference model of what to expect for a mixture. Reference models used in mixture extrapolation practice are typically based on pharmacodynamic assumptions on the type of interaction between a chemical and a biological system.

5.2.1 PHARMACODYNAMIC CONCEPTS FOR MIXTURE EXTRAPOLATION

The expectation of combined effects from mixture exposure is most often founded in the basic principles of toxicology and pharmacology (Loewe and Muischnek 1926a; Bliss 1939; Plackett and Hewlett 1952). The first strictly pharmacological ideas formulated (Loewe and Muischnek 1926a) were supplemented by biometrical considerations. Later, Bliss (1939), a biologist and a biometrician, provided the first consistent framework, as depicted in Table 5.2 (Plackett and Hewlett 1952). In this framework, the main ideas focused on the presence or absence of interactions (commonly referred to as interactive and noninteractive joint action) with respect to responses observed in test organisms, and the presence of the same or a different mode of action.

TABLE 5.2
Four possible mechanisms of joint action for mixtures as defined by Plackett and Hewlett (1952), with associated models

Effect	Mode of action	
	Similar	Dissimilar
Noninteractive	Simple similar action Model: concentration addition	Simple independent action Model: response addition
Interactive	Complex similar action Model: empirical	Dependent action Model: empirical

For noninteractive types of joint action, it is assumed that the chemicals in the mixture do not affect the toxicity of one another. Two different reference models are available for the analysis of noninteractive joint action, depending on the mode of action of the chemicals in the mixture. The modeling approach commonly known as concentration addition (CA) is used for mixtures with 2 or more compounds with a similar mode of action. The modeling approach called response addition (RA) is used for mixtures with 2 or more compounds with different modes of action.

For the analysis of interactive joint action (for either similar or dissimilar modes of action), no general models are available, but empirical descriptions may be used. The concentration addition and response addition models are often used in the analysis of experimental data from mixtures with compounds having known but different modes of action.

Various techniques, such as graphic illustrations (e.g., isobolograms), mixture toxicity indices (e.g., an additivity index), formulas, or fully parameterized models, exist for predicting an expected combined effect based on concentration addition or response addition (for review, see Bödeker et al. 1990). The quantitative relationship between the expected combined effect calculated according to concentration addition or response addition depends (in addition to other factors) primarily on the steepness of the concentration response relationship of the individual components (Drescher and Bödeker 1995). Concentration addition predicts a higher combined effect as compared to response addition when the mixture components have steep concentration response relationships, whereas the opposite is true for flat concentration response relationships of the mixture components.

In single-species risk prediction for individual toxicants and toxicant mixtures, the effect is expressed as the proportion of an exposed population that is likely to be somehow affected by toxic action (quantal responses), or as a reduction in performance parameters such as growth, clutch size, and juvenile period (continuous responses). Both concentration addition– and response addition–based methods are commonly applied for both response types. Assemblage-level risk prediction has only been introduced more recently (e.g., De Zwart and Posthuma 2005) and is founded on similar principles while focusing on the fraction of species that are likely affected by mixture exposure.

5.2.2 CONCEPTUAL LIMITATIONS AND SOLUTIONS

When a model (e.g., concentration addition or response addition) is considered appropriate for describing the mixture effects observed in experiments, it can serve predictive purposes (such as formulating a scientific null hypothesis for an experiment), or for practical extrapolation and for risk assessment. There are, however, limitations associated with the concepts and the associated models based on pharmacodynamic reasoning. These limitations were first recognized by Plackett and Hewlett (1952), yet have mainly gone unnoticed by followers of the mechanistic school of mixture toxicity. Three main limitations are identified, and extrapolation solutions are provided.

First, the 4 options of Table 5.2 can only be distinct for a limited set of mixture types, that is, mainly binary mixtures or mixtures for which the assumptions of either the concentration addition or the response addition model are valid. However, mixtures occurring in the environment are usually more complex in the sense that they contain groups of compounds that share the same mode of action *and* groups of compounds with different modes of action. Only recently have experimenters and theory developers started to investigate how this may be combined (known as conceptual "mixed models") to facilitate the analysis of complex mixtures (Ankley and Mount 1996; Posthuma et al. 2002a; Traas et al. 2002; Altenburger et al. 2005). Also, the question of what pharmacological or toxicological information exactly is necessary to define the mode of action of mixture components as similar or dissimilar and therefore to decide which concepts should be used is under debate (Grimme et al. 1996; USEPA 2000d; Borgert et al. 2004).

Second, interactive mixture toxicity is currently not predictable by the available mixture models. Various levels of possible interactions may be distinguished:

- Chemical–chemical interactions (e.g., reaction or precipitation) during mixture exposure
- Chemical interactions at the uptake, distribution, metabolism, or excretion levels in the organism
- Chemical–biological receptor interactions
- Biological interactions during propagation processes

Most of these interactions are not clearly distinguishable in ecotoxicological studies, because the investigations do not focus on molecular targets, as in toxicology and pharmacology, but rather on whole individuals or species in ecosystems. The ecotoxicological response in organism or assemblage studies is often the outcome of an array of interactions of compounds in various organs and tissues within individual organisms, and of interactions amongst organisms in assemblages. The difficulty arises in trying to determine which interactions are present, and how to address these extra interactions in experimental analyses and extrapolation.

Ashford (1981) developed ideas for mixtures in the framework of human pharmacology, which can be applied to ecotoxicological problems. Ashford focused on both the effects that modulate exposure as well as the factors that eventually determine the overall observed response. In his original work, Ashford proposed to study subsystems (e.g., separate organs) within the study entities (individuals). This implies that a set of mathematical formulae can be envisaged, based on the number of components in the mixture, the number of subsystems, and the number of sites of action. Physiologically based pharmacokinetic (PBPK) models have tried to address some of these aspects by mathematically calculating organ and target available concentrations of compounds (Yang 1994b; Haddad and Krishnan 1998; Andersen and Dennison 2004). General applicability of all the approaches has been rather low and has been restricted to human toxicology, due to the high level of data requirements.

A third limitation of the mechanistic approach is associated with the fact that observed effects in ecotoxicological studies are often composed of (unknown) sets of target-receptor responses, (1981). Consequently, experimental observations are often misinterpreted. For example, many authors have declared that compound mixtures (mechanistically) act in a concentration- or response-additive manner (implying simple similar action or simple independent action as molecular mechanisms), whereas the results were merely numerically similar to predictions from either of the 2 models. Rarely do ecotoxicological studies show the primary molecular interactions that occur at the target sites of intoxication. This implies that results of such studies can only be given in terms of statements like "The response is more than expected from concentration addition" or "The response is less than expected from concentration addition" rather than in absolute terms of "synergistic" or "antagonistic," respectively. Thus, for most ecotoxicological studies, the experimental data do not provide scientific proof to support mechanistic claims in explaining the experimental data.

5.2.3 Quality Issues

Many studies offer extrapolation approaches when addressing the issue of mixtures. However, when evaluating mixture papers for their support for mixture extrapolation procedures, study quality is a critical consideration. A set of criteria applies to evaluate the quality of the data (Altenburger et al. 1990), and these criteria can be applied to estimate whether and to what extent the data support the chosen extrapolation option.

First, there are experiment requirements. Organization of treatments is a major critical element of any mixture experiment, as mixture design will necessarily rely on previous experiments and may therefore be prone to error. An example is running the control group of a single-compound toxicity test asynchronous to the test exposure, which is a violation of a basic experiment design rule. Previous experiments should not be used, unless sufficient care is taken to randomize or control error sources. The type and quality of data that are used in a mixture assessment model are critical. For example, index-based approaches such as the toxic unit approach, or the use of toxic equivalence quotients, reduce full dose response curves to just 1 point each (e.g., the EC50), obviating the utility of other effect levels. The findings of a study may or may not be applicable to other exposure and/or effect levels, especially when the interest is in the "tails" of the curves (protection target), whereas experimental data pertain to EC50 levels. It is crucial to clarify the objectives of the study (mechanistic understanding, testing quantitative prediction accuracy of models, etc.) and to design the study in such a way that these targets are reached. For example, if one wants better mechanistic understanding, one should measure target concentrations rather than ambient exposure concentrations.

Second, there are biometrical requirements. Various exposure response models may be used and compared. The models need to be clearly defined, and goodness of fit should be reported, both for the separate exposures as well as for the mixtures. Concentration addition, response addition, and mixed-model results may be compared as possible alternatives, especially when underpinning of mechanistic assumptions is weak. Results at one exposure level (e.g., EC50) do not necessarily predict results at other exposure levels due to different slopes and positions of the curves for separate compounds and the mixtures. Statistical tests should be executed properly to compare predicted and observed responses. If any statements about the significance of results are made, the methods of dose–response analysis need to be reported.

Third, sensitivity should be considered. In principle, all models allow for the detection of significant differences through the calculation of confidence intervals. In practice, however, there are few examples where specific mixture models have actually been demonstrated to apply (amongst other options) by this means (Altenburger et al. 1990; Chen et al. 2005). Most statements of significant difference are based on implicit assumptions about variances. Precise data collection and manipulation are needed to avoid false interpretation that mixture effects are interpreted as "significant." Models chosen should be pharmacologically adequate for the quantification of nonadditive types of effects. The experimental design should have observations near the concentration level of interest to maximize sensitivity for the region of interest.

Fourth, there is the issue of specificity. The capacity to differentiate between different types of combined effects can be achieved by all models after statistical and mechanistic validation. For general application in ecotoxicology, we propose that combined effects be differentiated only with clear reference to the additivity concept of the model used.

Fifth, approaches are limited regarding inferences that can be made. Evidence supporting combination effects for a given mixture is often limited for many models. It must be emphasized that an interpretation of the results using point-estimate-based models is restricted to mixture ratios that have been investigated experimentally. Extrapolations used to address a more general issue such as "global additivity" are plausible but not statistically defensible. Overcoming such limitations would require the employment of parametric models. Most often, rather than using the terms "antagonism" and "synergism," it should be considered to use the terms "more than" or "less than" when referring to concentration addition or response addition, respectively, because the latter terms better define the test criteria by specification of a null model.

Sixth, approaches should be introduced with pharmacological transparency. A clear understanding of the researcher's biometrical approach (mechanistic assumption) is essential if confusion in mixture toxicology terminology is to be eliminated. The concept of additivity as a basis for an evaluation should be explicitly described as concentration addition, response addition, or another (new) concept, but should not be undefined. We recommend models that facilitate clear interpretation of effects and explicit critical judgment during the processing of data.

Many studies (an inventory is given in Section 5.3.1) do not unanimously pass all of these quality criteria. For example, mixture studies do not usually show significance testing of the observations. That is, the observed mixture effect concentration (a mixture concentration that causes a certain response) is usually compared to the mixture concentration that is predicted to yield that response. Thereby, the fact that single-compound test results as well as mixture test results encompass natural variation often is neglected. When the observed mixture effect is found at concentrations of 1.2 times the predicted concentration, this absolute comparison of values (1.2 observed versus 1.0 expected) suggests less than concentration-additive effects. However, when both are estimated with error, this could show that this difference is not significant (e.g., observed and predicted effects of 1.2 ± 0.15 and 1.0 ± 0.15, respectively, are not significantly different). Note that these methods still do not provide mechanistic proof. Some examples of statistical testing strategies are given in terrestrial mixture studies (see Section 5.3.1).

To address the quality limitations for extrapolation, the available experimental data on observed mixture effects were evaluated with care, and a pragmatic approach for mixture extrapolation was followed. Although mechanistic understanding was often not the purpose of the experiments, the extrapolation approach is based upon mechanistic principles, that is, regarding the choice between mixture toxicity models. Conceptual considerations on biases and mathematical characteristics of the models were included (see Section 5.3.3).

5.3 DATA FOR MIXTURE EXTRAPOLATION

5.3.1 REVIEWS ON EXPERIMENTAL EVIDENCE

A short review of mixture data, mostly taken from reviews on mixtures in aquatic systems (due to their prevalence versus those regarding terrestrial systems), was executed to test the appropriateness of models in describing experiment data. That is, we tested whether the models (despite the fact that there is no proof that the underlying mechanisms are applicable) accurately described the observations in the mixture experiments.

5.3.1.1 Aquatic Data

The European Inland Fisheries Advisory Commission (EIFAC; 1980) reviewed the toxicity of 76 binary mixtures of common effluent pollutants to fish. Mixture effects occurred at 0.4 to 26 times the exposure concentration expected under concentration-additive toxicity, with 87% of the data ranging between 0.5 and 1.5 times this concentration. Substances with concentrations lower than 0.2 toxic units (TU) appeared not to contribute to the toxicity of the mixtures. In contrast to the apparent lack of effects at low mixture concentrations, subsequent papers (Könemann 1981; Hermens et al. 1985; Deneer et al. 1988) showed that apparently equitoxic mixtures containing 8, 9, 11, 24, 33, and 50 organic chemicals at concentrations that were only small fractions of the individual EC50 values were indeed able to induce responses that agreed with the concentration-addition models. These formed the basis of an updated report (EIFAC 1987).

Wang (1987) reviewed the toxicity of heavy metal mixtures and concluded that combined toxicity cannot be readily predicted. However, 21 out of 37 mixtures showed an effect as expected under concentration addition, or an effect slightly less than concentration additive. Similar conclusions were reached in a review by ECE-TOC (2001) for the chronic toxicity of metal mixtures. The relative degree of unpredictability of the mixture response was attributed to the different modes of action of the metals. It was therefore recommended that concentration additivity should be assumed as the most balanced choice for a null model in mixture assessment, unless proven differently.

Deneer (2000) reviewed acute toxicity in aquatic organisms for about 200 pesticide mixtures with diverse modes of action. He found that combined toxicity in 90% of the cases differed from concentration additivity within a factor of 2. Two large studies on the chronic toxicity of binary pesticide mixtures (Faust et al. 1994; Altenburger et al. 1996) demonstrated that, in most cases (> 60%), the best fit of observed mixture effects was to the concentration addition model.

The largest reviews of mixture toxicity data conducted to date (Ross 1996; Ross and Warne 1997) examined the toxicity of approximately 1000 predominantly binary, tertiary, and quaternary mixtures. This analysis revealed that between 75% and 80% of the mixtures according to the concentration-addition model, 10% to 15% showed less response than expected under this model, and 10% to 15% showed a higher response than expected. Five percent of the mixtures had toxicity values that differed from concentration additivity by a factor greater than 2.5, and 1% of the mixtures had toxicity values that differed by more than a factor of 5.

In the vast majority of aquatic ecotoxicology mixture research, as indicated by the data in the compiled studies, the toxicity of the mixtures is compared only to the toxicity predicted by the concentration-addition model, irrespective of the (assumed) mode of action of the compounds in the mixture. More recently, however, authors have investigated whether the effects could be described by alternative models. For example, Altenburger et al. (2000) and Faust et al. (2001) clearly demonstrated that the toxicity of complex mixtures with assumed similarly acting chemicals, tested with luminescent bacteria and algae, was highly predictable with the concentration-addition model, which is in agreement with the reviewed studies. However, using the same toxicity tests with complex mixtures of assumed purely dissimilarly acting chemicals, it was shown that the response addition model produced the best prediction (Backhaus et al. 2000). These authors found that concentration addition underestimated the median effective concentration (EC50; i.e., it overestimated toxicity) of these mixtures by a factor of less than 3. From the recent studies that have compared concentration addition and response addition, it is concluded that the most predictive mixture extrapolation model (concentration addition or response addition) can likely be chosen on the basis of mode-of-action considerations, even though exact mechanisms of toxic action are not entirely known.

Finally, there is a debate as to whether combined effects are to be expected from low doses of mixture components. Evidence in ecotoxicological effect systems challenges the widespread notion that combined effects do not occur for components occurring below certain effect concentrations. Studies with narcotics (non-specifically acting compounds, typically chemicals used as solvents) indicated that concentrations of multiple components individually equivalent to 5% (Broderius and Kahl 1985; Xu and Nirmalakhandan 1998), 2% (Könemann 1980; Hermens et al. 1984), and even as low as 0.25% of the individual EC50 (Deneer et al. 1988) still produced a significant mixture effect. For specifically and similarly acting compounds such as photosystem II–inhibiting herbicides (Faust et al. 2001) and chemicals that are able to disturb endocrine function (Silva et al. 2002), it has been demonstrated that individual concentrations clearly below detectable individual-effect concentrations can produce predictable and significant combined effects. Even sediment pollutants with diverse modes of action in mixtures containing no-observed-effect concentrations (NOECs) of the components induced significant combined effects (Walter et al. 2002).

5.3.1.2 Terrestrial Data

The amount of ecotoxicity data for terrestrial systems is substantially more limited compared to that for aquatic systems. Posthuma et al. (1996) provided an overview of the limited set of other literature sources on terrestrial test results, which included the studies of van Gestel and Hensbergen (1997), Sharma Shanti et al. (1999), and Weltje (1998). Jonker (2003) published a PhD thesis on mixture toxicity in the terrestrial environment. Korthals et al. (2000) reported that the joint effects of metals on a nematode community in soil could be predicted by mixture modeling. Mesman and Posthuma (2003) published a report on applying generalized mixture extrapolation rules, as derived from the available literature, in terrestrial ecotoxicology. In a numerical sense, the findings of these terrestrial studies corroborated the findings of the aquatic reviews. That is, many studies with species such as earthworms, enchytraeids, nematodes, and plants showed that observed mixture responses were near the prediction of concentration additivity. Mechanistically, however, terrestrial ecotoxicologists commented explicitly (such as Ashford 1981) on the existence of multiple interaction levels. Pleas were made by Mesman and Posthuma (2003) for specific attention to mechanistic processes other than the compound–target interaction (e.g., sorption to the matrix) and to the above-mentioned criteria to describe study quality. Finally, an extended array of mathematical models to statistically test whether mixture responses significantly differ from the responses expected under the null model of concentration additivity has been developed in terrestrial studies (e.g., Posthuma et al. 1997; van Gestel and Hensbergen 1997; Jonker 2003).

5.3.2 MIXTURE EXTRAPOLATION IS JUSTIFIED BY THE DATA

Given the reviewed data, the first issue relevant for mixture extrapolation is whether extrapolation can be justified at all over no mixture extrapolation, in which separate responses only are assumed. In our opinion, the single-species test data are sufficiently clear to suggest that mixture extrapolation is preferred over no extrapolation. Although some exceptions exist (e.g., low effect range and specific compound mixtures), the majority of studies (aquatic and terrestrial) generally concluded that concentration addition is a reasonable conservative approximation of mixture responses. Indeed, the species-level experimental data we have reviewed clearly suggest that the observed responses occur at a mixture concentration approximately equivalent to one expected from concentration-addition modeling. Neither half the expected concentration (which would indicate some mechanism that induces more effects than expected) nor doubled or higher concentration levels (which would indicate less effects than expected) occurred to a significant degree. When mixture issues do not play a role, the observed response should always be equal to the response predicted from the toxicity curve of the most potent compound. This would imply that mixture effects would occur at twice the expected concentration for binary mixtures with both compounds present at concentrations of equal toxicity (equitoxic mixtures), triple the expected level for tertiary mixtures, and n times the expected level for n compound mixtures. The data show that this is clearly not the case.

In light of the large number of interaction levels that may influence the mixture toxicity response at the sites of toxic action, especially in terrestrial systems, it may come as a surprise that we are able to draw the conclusion that mixture extrapolations appear to be justified by the data. Although the interactions may all be relevant and may all require extrapolations according to the methods explained in the other chapters of this book, the net effect is apparently, and often relatively well, predicted by concentration addition.

In our opinion, the data are sufficiently clear to suggest that when it is not feasible to test the mixture in question, mixture extrapolation is the preferred option compared to no extrapolation. Indeed, all literature observations suggest that applying mixture extrapolation is to be preferred over not applying mixture extrapolation. Technical options for extrapolation are concentration addition, response addition, and the mixed-model approach, of which concentration addition is most often applied. Exceptions may apply in cases that are more specific. For example, when it is clear that 2 compounds precipitate (a situation of "no exposure" due to chemical interactions in the environment), one should acknowledge this prior to assessing mixture risks by mixture extrapolation approaches. When the data of a study allow, refined conclusions are possible. For example, when the study design is appropriate and the mathematical models are appropriate, researchers are able to discriminate between concentration addition and response addition, and (with sufficient experiment efforts) between these models and the mixed-model approach.

5.3.3 Considering Bias in Extrapolation Is Important

The different models show bias. Concentration addition implies that effects always increase with each additional compound (even when present at extremely low concentrations), and there is clear evidence that this occurs with similarly acting compounds, but it is not yet clarified with dissimilarly acting compounds (see reviews, above). Thus, applying concentration addition to mixtures of dissimilarly acting compounds might overestimate effects and risks, and this may be an undesired feature for various risk assessments (e.g., for retrospective, diagnostic risk assessments). Response addition implies that mixture effects only occur when at least 1 compound induces toxic effects. However, as cited in the review of data (Section 5.3.1), mixtures with the same mode of action can induce toxic effects even when the separate compounds do not. These biases imply that the mixed-model approach is conceptually sound in the sense that application of this model reduces both sources of bias. Thus, mixture extrapolation can be conducted using any one of the models (concentration addition, response addition, or the mixed-model approach). The final model choice for an assessment could differ depending on acceptance of bias in relation to the assessment target.

5.4 GENERAL PROTOCOL FOR MIXTURE EXTRAPOLATION

Mixture extrapolation can be complex due to the variety of model approaches, the optional tiers, and the different extrapolation targets. Therefore, apart from the technical mixture extrapolation protocols sensu stricto, a general stepwise protocol should be followed. Here, we propose a protocol of 5 steps.

5.4.1 First Step: Defining the Assessment Problem

The first step in mixture extrapolation for mixtures of known composition is the definition of the assessment problem. What exactly is the context of the mixture assessment? To address this problem, answers should be given to the following questions:

- What is the objective of the assessment: a specific assessment for a contaminated site, or a generic assessment that will derive a conservative degree of protection?
- What is the biological target of the assessment, a species, or a community?
- Which data are available to underpin the approach chosen in a mixture assessment? Are there specific data directly related to the problem, or only data that are peripherally related to the problem? What do the data suggest (e.g., a bias toward more or less than concentration-additive effects)?
- Which types and levels of interactions are present that might affect mixture toxicity, and how can these be appropriately addressed? For instance, is there an environmental interaction such as precipitation that would influence mixture effects through modification of exposure in a significant way? (*Note*: In the rest of this chapter, limited attention is paid to the issues of modifying effects brought about by nonchemical stressors, such as co-exposure to UV rays or pathogens. Those interactions are treated in the other chapters of this book.)

After these questions have been answered, the next step is to undertake a systematic analysis of the mixture itself, with the goal of identifying the toxicologically relevant components. For example, the concentration of compounds in the mixture could be compared to ecotoxicity data for those compounds to identify compounds most likely to contribute to mixture effects (exposure > sensitivity threshold), or grouping of compounds in the mixture according to modes of action. In the following steps, these issues are elaborated, starting with exposure assessment.

5.4.2 SECOND STEP: CONSIDERING THE ISSUE OF EXPOSURE EXTRAPOLATION

Matrix and media effects may modulate exposure concentrations. Methods for matrix and media extrapolation should be applied prior to modeling mixture risks according to the methods available in De Zwart and Posthuma (2005). The presence of mixtures of chemicals in the context of matrix and media extrapolation can result in particular effects. For example, 2 toxic compounds may, only when present together, form an insoluble complex, resulting in lower or more subtle mixture responses than expected. For example, various studies on terrestrial mixtures have shown that the bioavailability of a metal may be affected by the presence of another metal (van Gestel and Hensbergen 1997; Posthuma et al. 1997, 1998). In general, exposure may increase or decrease depending upon the nature of the compounds in mixed form. In accordance with the ideas of Ashford (1981), exposure assessments should be executed in mixture extrapolation before considering mixture effects at the target site of toxic action.

5.4.3 THIRD STEP: MIXTURE EXTRAPOLATION IN TIERED RISK ASSESSMENTS

Because there are various options to account for possible combined effects, including concentration addition alone, response addition alone, or concentration addition and response addition in concert, and motives from study quality, assessment endpoint, mechanistic features, and statistical characteristics and biases, there should be a way to logically choose amongst methods for mixture extrapolation. The logical and pragmatic way to choose amongst alternative approaches is to design and follow a tiered approach (see Chapter 1).

Tiering is often applied in risk assessment in order to reduce expenditures in time, money, and labor when the assessment requires only simple and possibly conservative output. Table 5.3 provides a suggested tiered approach in mixture extrapolation and is further described in the bulleted list below. The tiering is based on the way that mixture mechanisms are addressed in the approach. It is assumed that issues such as matrix and media extrapolation have been addressed according to the methods described in the pertinent chapters.

The following tiered approach is suggested for mixture extrapolation:

- **Tier-0 methods,** referred to as "no extrapolation" in Chapter 10, imply that mixture effects are assumed irrelevant for the assessment.
- **Tier-1 methods,** referred to as "simple generic approaches" in Chapter 10, consist of a nontoxicological way of addressing mixtures. For example, one may just define criteria for groups of structurally similar compounds or apply safety factors to account for possible combined effects, if mixture exposures are considered relevant. This approach has the advantage of needing no additional information but may be toxicologically meaningless if compounds are too different in their individual behaviors.
- **Tier-2 methods,** referred to as "moderately simple generic methods," assume a uniform mode of action for all compounds or a complete, nonuniform set of modes of action. On one hand, this implies application of concentration addition to address the mixture and likely results in

TABLE 5.3
Major tiers that can be distinguished in combined effect extrapolation

Tier	Model	Assumptions and/or model	Information required
0	No mixture extrapolation		
1	Additional safety factor, or Group criteria	Combined effect might be relevant	Occurrence of mixture exposure
2	Toxic unit summation	Point extimates on concentration-effect curves	Toxicological reference values for the mixture components, e.g., EC50, NOEC
3	Mixed-model approach	Similar and dissimilar action	Concentration response relationships for the components
4	Mixed model approach with emphasis on characteristics of receptor species	Similar and dissimilar action	Mode of action information, receptor species information

overestimation of risks (see Drescher and Bödeker 1995). On the other hand, it may imply using only response addition. The approach based on concentration addition is most often used. The advantage of the overall concentration-addition method is that it is simple: only summary toxicity descriptors such as EC50 or NOEC values are required, and all compounds may be expressed in terms of toxic units that are summed as dimensionless fractions of their effective concentrations (see protocols in Section 5.3). This method is based on point estimates from the concentration effect curves of all compounds present. Tier 2 methods are relatively simple, and should be applied in cases where the assessment problem is either simple or vaguely formulated (no clear, ecologically identified protection target), and where some conservatism in the assessment output is acceptable or desirable. An example might be the setting of quality standards for a clean environment.

- **Tier-3 methods,** referred to as "complex specific methods" in Chapter 10, allow for assessment of nonuniform modes of action, which may include the use of both concentration-addition and response addition models. This approach requires information on the concentration–effect relationship of the compounds, as opposed to only point estimates as in Tier 2.
- **Tier-4 methods,** referred to as "highly specific methods" in Chapter 10, address nonuniform mixtures from the perspective of using only not the assumed primary mode of action but also the assumed characteristics of the receptor species. Tier 4 methods can be applied when the assessment problem is defined specifically (regarding site, species, and compounds), and where an accurate result is preferred over a conservative one.

The tiers may be further refined and are further discussed in the sections on extrapolation protocols.

5.4.4 FOURTH STEP: EFFECTS EXTRAPOLATION ISSUES

In this step, the mixture extrapolation is planned. Decisions are made on the protocols to be used, depending on tier, exposed organism or assemblage, and so on. In this step, choices are made regarding the protocols for mixture extrapolation. In simple, generic analyses, one may want to work with point estimates from the concentration–effect curves. In more specific analyses, one may want to work with mixed-model approaches, acknowledging both the modes of action and the presence or absence of target sites of toxic action within different organism groups. If one wants to be more certain about probable mixture effects for a decision (i.e., whether predicted risks are below a threshold while being uncertain about the model), the concept of a window of prediction (i.e., use both concentration addition and response addition concepts to define the range and limits of expected combined effects; Walter et al. 2002) can be applied.

5.4.5 FIFTH STEP: EXECUTING AND INTERPRETING MIXTURE EXTRAPOLATION

The last step is to collect the appropriate models and their input data, to execute the mixture extrapolation, and to interpret the outcomes in view of the assessment target. The strength of the inferences that can be made can be explicitly reported (e.g., based on the amount and quality of the input data, based on the outcomes of the "window of prediction" assessment, or by elaborating on the uncertainties encountered in the stepwise procedures).

5.5 SPECIES-LEVEL COMBINED EFFECT PREDICTION

In the fourth step of extrapolation, specific mixture extrapolation protocols are needed. Below, some details on the theories and the associated protocols are given for concentration addition, response addition, and mixed-model approaches, and for the species and assemblage levels separately (this section and, next section, respectively).

5.5.1 PROTOCOL FOR CALCULATION OF CONCENTRATION ADDITION

Due to its conceptual and numerical ease of application, concentration addition is most commonly used as the null hypothesis in the assessment of mixture responses to known components. Thus, the concentration addition model is usually applied to predict the responses observed in single-species toxicity experiments in which the response to several compounds with assumed same modes of action is considered. Moreover, the joint effect of compound mixtures that have the same mode of action and that show no toxicological interactions will be quantitatively correctly described using the concentration addition model (Plackett and Hewlett 1952). Several concentration addition–related methods from graphic techniques such as isobolograms, through indices such as the "mixture toxicity index (MTI)," up to fully parameterized models have been developed to assess combined effects (Bödeker et al. 1990; Gentile et al. 1999). Concentration addition implies that the contribution of the individual toxicants to the overall effect can be added in the form of toxic units (TUs; Sprague and Ramsay 1965).

The concentration addition of a mixture can be described by the equation (Könemann and Pieters 1996) $TU = \sum d(A)/D(A) + d(B)/D(B) + \cdots + \cdots$, where d(A), d(B) are the actual exposure concentrations of the compounds A and B, and D(A), D(B) are the exposure concentrations of A and B that represent a common standard response such as a median lethal or effective concentration (LC50 or EC50) or a no-observed-effect concentration, and TU is expressed as dimensionless toxic units. The individual terms "d(A)/D(A)" and "d(B)/D(B)" may be called "toxic unit sum" (TUS). The use of NOEC values introduces an inaccuracy as an NOEC value is not a compound-independent measure of noneffect. It in fact quantifies a certain effect level in relation to experimental design issues.

Concentration addition implies that the common standard response is expected to occur at TU = 1. Because no concentration threshold exists for concentration addition, the concept predicts that a response may occur when organisms are exposed to a mixture of compounds where each of the constituents separately would not induce any effect. This is an important implication, as ecosystems often experience mixtures of chemicals at low concentrations.

The TU summation (TUS) approach for the concentration addition model suffers from a serious drawback. The method is based on point-estimate assessments taken from the whole of the concentration–effect relationship, because the standard response is often set at the LC50, the EC50, or the NOEC. If the slopes of the concentration–response relationship for all compounds in the mixture are not considered, there is no way to determine how far away a TUS value is from the effect of concern (Solomon and Takacs 2002). An improvement would be to use whole concentration–response functions for concentration addition modeling, which is possible when adequate effect information for the individual components of a mixture is available, as has been demonstrated by Faust et al. (2000). This can also be done using a probabilistic approach by applying the following protocol (De Zwart and Posthuma 2005).

5.5.1.1 Step 1

For the individual components of the mixture, estimate both slope and EC50 by fitting a log-linear regression model (or another appropriate continuous model, usually for sigmoid concentration response data; for reference, see Scholze et al. 2001) to the range of concentrations and effects observed in standardized single-species and single-substance toxicity experiments. Among the types of log-linear models, logistic, Weibull, or probit regression might be used for modeling a quantitative response (proportion of functional decline) as well as for a qualitative response (proportion of organisms affected). In practical terms, probit models come to the same conclusions as logistic regression, but have the drawback that probit coefficients are more difficult to interpret (there is no equivalent to logistic regression's odds ratios as effect sizes in probit). Both the cumulative standard normal curve used by probit as a transform and the logistic or Weibull (log odds) curve used in logistic regression display an S-shape. Though the probit curve is less flexible as it is symmetric about the midpoint, differences are small at median-effect levels.

A formula for the logistic regression is $\text{Logit}(P) = \ln(P/(1-P)) = (\log c - \log \text{EC50})/\beta$, and for the probit regression it is $\text{Probit}(P) = (\log c - \log \text{EC50})/\beta + 5$, where P is the probability (or proportion) of effect, c is the concentration that is supposed to cause this effect, and β and σ (standard deviation) are the respective slope coefficients of the regression. The ratio of σ over β is a constant: $\sigma/\beta = \pi/\sqrt{3} = 1.81$.

Probit values can be obtained from a probit table, or by applying the Microsoft Excel function NORMINV(P,5,1). As alternatives to this protocol, most graphic software programs contain a variety of nonlinear curve-fitting and regression procedures (including standard logit, Weibull, or probit models). As determination of concentration response functions is an iterative fitting procedure, fit algorithms typically need reasonable starting parameters that may be taken from inspection of the data in order to allow the convergence of functions within the set number of iterations.

5.5.1.2 Step 2

Calculate the sum of TUs for the mixture components, based on the concentrations and the estimated EC50 values for the individual components (as described with the TUS approach).

Calculate the combined effect according to concentration addition. To calculate the combined proportion of effect (P) in a log-logistic manner, substitute TU and the average of β in $P = 1/(1 + e^{-(\log TU/\beta)})$. For the log-probit evaluation, there is no analytical solution other than, for instance, to apply the Microsoft Excel function $P = \mathrm{NORMDIST}(\log(TU), 0, \mathrm{Average}(\sigma), 1)$, which provides the cumulative density function (CDF) of the normal distribution by Taylor series approximation. Using this protocol, the effect level to be expected from most of the realistic environmental mixtures can be predicted from combining single-species and single-compound effect data with the concentration addition approach.

Alternatively, you may utilize, if available, the slope information for the individual components by employing the inverse of the concentration response function to calculate expected concentrations for a defined mixture effect as follows. The concentration of each component in the mixture can be expressed as a fraction of the total mixture concentration. Consequently, a total concentration of the mixture, at which a certain effect is generated, can be calculated using concentration addition according to the following equation:

$$\mathrm{EC}x_{\mathrm{Mix}}\left(\sum_{i=1}^{n} \frac{p_i}{\mathrm{EC}x_i} \right)^{-1}$$

where $\mathrm{EC}x_{\mathrm{Mix}}$ is the total concentration of the mixture provoking $x\%$ effect; $\mathrm{EC}x_i$ is the concentration of component i provoking the $x\%$ effect, when applied singly; and p_i denotes the fraction of component i in the mixture. The calculation of total mixture concentrations for various effect levels leads to a complete iteration of an expected concentration–effect curve.

Although the TUs and the probabilistic approaches are useful in estimating the potency of different mixtures to a single species using concentration addition, they must be used with caution when extrapolating from 1 species to another (Compton and Sigal 1999). Species may not respond to the same set of toxicants in the same way (Parrott et al. 1995; van den Berg et al. 1998), which would invalidate the use of the same TUs or concentration response curves across species or species groups.

An example of this type of calculation is given in Box 5.1.

Substance	1	2	3	4	5	6	7	8	9	10	Total concentrat
Analysed concentration (µg/L extracted from 1 g sediment)	1,1000	0,3500	2,4700	3,8000	15,8000	1,6000	5,0000	5,0000	6,3400	7,6000	236,9575
Extract concentration (µmol/L)	6,1718	1,5331	10,9157	18,7877	72,0507	6,0790	33,1014	24,4774	26,2669	37,5717	
Molar fraction	0,0260	0,0065	0,0461	0,0793	0,3041	0,0257	0,1397	0,1033	0,1109	0,1586	

TIER 1   **Toxic Unit summation**

	1	2	3	4	5	6	7	8	9	10	Toxic Unit sum
required informatic EC50 (inhibition of algal reproduction; µmol/L)	2,84111	0,05942	0,04488	0,17253	0,15989	31,56605	3,34885	0,35565	0,06935	0,24369	1,00
TU-fraction for expected 1:1440 dilution + 50% effect (Exposure concentration / ECx)	0,00151	0,01792	0,16892	0,07562	0,31293	0,00013	0,00686	0,04779	0,26304	0,10707	

Expected EC50   0,16 µmol/L

TIER 2   **Response addition (or concentration addition)**

	1	2	3	4	5	6	7	8	9	10	Response Addi combined effec
required information dose-response model	Weibull	G-Logit	G-Logit	G-Logit	Logit	G-Logit	G-Logit	Weibull	Weibull	Weibull	
Parameter 1	-0,937	74,244	998,724	269,074	15,014	-38,418	-3,488	4,544	5,205	2,767	
Parameter 2	1,258	65,840	800,695	465,808	18,857	22,728	5,129	10,937	4,807	5,143	
Parameter 3		0,107	0,009	0,009		0,159	0,594				
Mixture conc. (µmol/L) => fractional effects of the individual components	0,750	0,000236	0,228727	0,078183	0,948235	0,000004	0,006393	0,000498	0,636585	0,130293	0,9890
	0,627	0,000137	0,133924	0,057240	0,500778	0,000003	0,005046	0,000213	0,501733	0,093051	0,9324
	0,524	0,000079	0,078415	0,041908	0,494071	0,000003	0,003863	0,000091	0,390860	0,060928	0,7926
Expected EC50	**0,44**	0,000046	0,045914	0,030682	0,183993	0,000003	0,003144	0,000039	0,281038	0,041202	**0,4988**
	0,367	0,000028	0,026883	0,022464	0,049485	0,000001	0,002481	0,000017	0,203138	0,027816	0,3225
	0,306	0,000015	0,015741	0,016446	0,011878	0,000001	0,001959	0,000007	0,144677	0,018737	0,2209
	0,256	0,000009	0,009217	0,012041	0,002768	0,000001	0,001546	0,000003	0,101970	0,012601	0,1574
	0,214	0,000005	0,005396	0,008816	0,000640	0,000000	0,001220	0,000001	0,071345	0,008467	0,1146
	0,179	0,000003	0,003160	0,006454	0,000148	0,000000	0,000963	0,000001	0,049965	0,005685	0,0645

Expected EC50   0,20 for CA-model

TIER 3   **Mixed models**

	1	2	3	4	5	6	7	8	9	10	Combined effec mixed model
required information mode-of-action with respect to the considered biosystem	unspecific	unspecific	unspecific	unspecific	reactive	unspecific	unspecific	unspecific	PS II inhibitor	unspecific	
Unspecifically acting components modelled by Concentration addition											
dose-response model	Weibull				Logit				Weibull		
Parameter 1	2,265				15,014				5,205		
Parameter 2	2,506				18,857				4,807		
Molar Fraction	0,5851				0,3041				0,1109		
Mixture conc. (µmol/L) => fractional effects of the individual components	0,68				0,8914				0,5618		0,9817
	0,58				0,6792				0,4423		0,9049
	0,49				0,3531				0,3386		0,7181
Expected EC50	**0,41**				0,1234				0,2537		**0,5033**
	0,35				0,0350				0,1871		0,3461
	0,30				0,0093				0,1364		0,2413
	0,25				0,0024				0,0986		0,1694
	0,21				0,0006				0,0708		0,1188

Observation   EC 50 for 10-component mixture   0,363 µmol/L

Substances

1 - Anthracene
2 - Benzo(a)anthracene
3 - Benzo(ghi)fluoranthene
4 - Fluoranthene
5 - N-Phenylnaphthylamine
6 - Parathiomethyl
7 - Phenanthren
8 - 2-Phenyl-naphthalene
9 - Prometryn
10 - Pyrene

$$Weibull\ E = 1 - Exp(-Exp(Parameter1 + Parameter2 * Log(Conc.\ [µmol/L])))$$
$$G\text{-}Logit\ E = 1 / (1 + Exp((-Parameter1) - Parameter2 * Log(Conc.\ [µmol/L])) \wedge Parameter3)$$

BOX 5.1 Example of a spreadsheet calculation of the expected combined defined effect for a multiple mixture using different amounts of information. *Note*: Tier-1 prediction relies on exposure and EC50 information (toxic unit summation), Tier-2 needs additional concentration response information for calculation of expected combined effects according to the reference models of response addition or concentration addition, and Tier-3 calculation (mixed models) requires information on the relevant mode of action. The sample is based on real analytical and effect data. *Source*: Redrawn from data from Altenburger et al. (2004).

5.5.2 PROTOCOL FOR CALCULATION OF RESPONSE ADDITION

The alternative concept to concentration addition is response addition, which was developed to address the issue of mixtures containing components with dissimilar actions. The reasoning is that toxicants may primarily interact with different molecular target sites but may lead to a common toxicological endpoint via distinct chains of reactions within an organism. Under these assumptions, the fractional effects of individual mixture constituents (e.g., 50% response) are expected to be independent from each other in a probabilistic sense. This means that the relative effect of a toxicant (e.g., 50% mortality) remains unchanged in the presence of another chemical. There is discussion as to whether responses against chemicals can be truly independent or whether they are typically partially correlated. Partial correlation of sensitivities, however, is not an extrapolative issue that has as yet been solved. Therefore, for predicting combined effects using response addition, the following approach applies, which assumes complete independence of responses:

$$E(c_{\text{Mix}}) = 1 - \prod_{i=1}^{n} (1 - E(c_i))$$

The effect at the total concentration of the mixture, $E(c_{\text{Mix}})$, is based on the effects of the components that they generate at the concentrations at which they are present in the mixture ($E(c_i)$). If the latter is expressed as a fraction of the total mixture concentration, it holds

$$E(c_{\text{Mix}}) = 1 - \prod_{i=1}^{n} (1 - E(p_i \times c_{\text{Mix}}))$$

This allows the calculation of an effect expected according to the concept of response addition for any concentration of the mixture. Again, the estimated individual effect may be taken from a concentration-response relationship derived on the basis of dose–response observations. It has to be noted that, in mixtures of many substances, the effects to be estimated for the individual contributors become rather small; therefore, a high-quality estimation of the concentration response, particularly in the low effect region, is needed. In such cases, it might be useful to consider models other than the standard probit or logit functions for description of the data.

An example of this type of calculation is given in Box 5.1.

5.5.3 PROTOCOL FOR MIXED-MODEL PREDICTION

As can be deduced from the review in Section 5.3.1, most studies on mixture toxicity have been conducted on single species exposed to binary mixtures. In the case of studies with more complex mixtures, generally, either the toxicants have all been selected to operate by the same mode of action, or they have all been selected to represent different modes of action. In the complex situation of a multiple contaminated environment, it is likely that the biota are exposed to a mixture where several modes of action (some similar, some dissimilar) are represented by a variety of toxicants (Teuschler et al. 2004; Chapin 2004). Predicting the toxicity of a complex mixture of toxicants could build on observations that both concentration addition

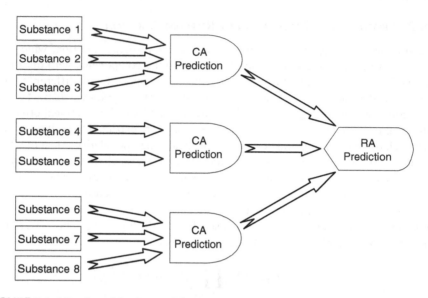

FIGURE 5.1 Mixed-model mixture risk assessment approach. *Note*: This illustrates the calculation of steps for combined effects of mixtures with similarly (e.g., Substances 1 to 3) and dissimilarly (e.g., Substances 1 to 3 versus Substances 6 to 8) acting components.

and response addition could be useful to predict the overall response, depending on the assumed mode of action of the mixture constituents (Altenburger et al. 2000). The literature now provides methods to predict the combined toxicity of such mixtures (Posthuma et al. 2002a; Altenburger et al. 2004, 2005; De Zwart and Posthuma 2005).

A proposed stepwise protocol for calculating an expectation for the combined effect of a mixture with components that act similar and groups or components that act dissimilar is presented in Figure 5.1. In the first step, evaluation of the concentration addition responses to individual modes of action is required. This calculation needs to be performed for a dilution series, which can subsequently be fitted to an expected concentration response function for the groups of similarly acting compounds. In the third step, the protocol requires evaluation of the response-additive effect of different modes of action. In the mixed-model case, the protocols for concentration addition are applied within groups of compounds that share the same mode of action, and response addition is applied across these groups.

The proposed protocol is a logical extension of the evaluation scheme provided by Plackett and Hewlett (1952) for the problem of predicting effects of complex mixtures of known composition. However, there are as yet only few data available to verify its predictive capacity (Altenburger et al. 2005). Also, ecotoxicity databases usually do not contain the full concentration effect data, which means that reevaluation of existing studies will be laborious.

Box 5.1 gives examples of single-species mixture effect calculations for different tiers of risk assessment.

5.5.4 INTERACTIVE JOINT TOXICITY

Recently, some models have been derived to analyze the occurrence of interactive joint action in binary single-species toxicity experiments (Jonker 2003). Such detailed analysis models are well equipped to serve as null models for a precision analysis of experimental data, next to the generalized use of concentration addition and response addition as alternative null models. However, in our opinion these models are not applicable to quantitatively predict the combined toxicity of mixtures with a complexity that is prevalent in a contaminated environment, because the parameters of such models are typically not known. Recently a hazard index (Hertzberg and Teuschler 2002) was developed for human risk assessment for exposure to multiple chemicals. Based on a weight-of-evidence approach, this index can be equipped with an option to adjust the index value for possible interactions between toxicants. It seems plausible that a comparable kind of technique could be applied in ecotoxicological risk assessments of mixtures for single species. However, at present, the widespread application of this approach is prevented by lack of available information.

5.6 ASSEMBLAGE-LEVEL MIXTURE EXTRAPOLATION

Experimental data or field observations on mixture toxicity and the responses in species assemblages are rarely available, with some exceptions (Korthals et al. 2000; Backhaus et al. 2004; Arrhenius et al. 2004). Nonetheless, risk assessment and legislation often focus on the protection of community-level endpoints in both prospective and retrospective risk assessments.

For addressing multispecies risk of mixture toxicity, we propose the following same procedure that is followed in general mixture studies in single-species ecotoxicology. That is, one must consider exposure, look at exposed species groups when necessary in view of the assessment endpoints, consider the mode of action of the components, and apply either of the sets of models based on this information. The practical protocols for mixture risk assessment that stem from this choice are worked out and discussed in the section below.

5.6.1 SPECIES SENSITIVITY DISTRIBUTIONS (SSDS) AND MIXTURE EXTRAPOLATION

5.6.1.1 SSDs and the Meaning of msPAF

Species sensitivity distributions (SSDs) are used for both prospective and retrospective risk assessments (Posthuma et al. 2002b). In prospective risk assessments, the concept is used to derive hazardous concentrations (e.g., HC5), which are used to derive environmental quality criteria. In retrospective risk assessments, the SSD approach is used to determine the local "toxic pressure" in terms of the potentially affected fraction (PAF) of species for each compound separately. Subsequently the multisubstance (ms)PAF, or optionally the combi-PAF, for the local mixture can be calculated. Originally, the combi-PAF concept was developed by Hamers et al. (1996) and assumes that only compounds exerting narcotic effects are addressed by concentration addition, whereas all other compounds are handled by response addition (mixed-model approach).

Subsequently, the msPAF was introduced by Posthuma and Traas and coworkers (Traas et al. 2002; Posthuma et al. 2002a), and also applies the mixed-model approach, but in this case all compounds that share the same mode of action (not necessarily narcotic effects only) are grouped and addressed (within such groups) by concentration addition. Thereafter, response addition is used to aggregate over the numbers obtained. In this case, the PAF can be estimated not only from an SSD for all tested species but also for subgroups of species that are sensitive to a particular class of compounds (e.g., insects are sensitive for insecticides, and the SSD is constructed from insect data only; see Posthuma et al. 2002a).

The meaning of the "toxic pressure" of an environmental compartment to the exposed species assemblage, as quantified by PAF, combi-PAF, and msPAF, is best explained via a hypothetical experiment. Consider a contaminated site where the PAF is estimated as 25% (with a 95% confidence interval [CI] of 20% to 30%), based on an SSD constructed from EC50s. This PAF value means that exposing a randomly selected subset of the tested species would induce 50% or more effect to 25% (CI 20% to 30%) of those species. Expressed alternatively, the probability of effects for a randomly chosen species (PES) from the population would be 25% (20% to 30%). Evidently, the toxic pressure that is imposed by the environment to any randomly chosen subset of the set of tested species would be interpreted as clearly different when the PAF for a set of sites ranges between (for example) 5% and 85%. This range is interpreted as a low versus a high probability of impacts on any species at these 2 sites. Evidently, the ecological "reality" of PAF estimates improves when the tested species are representative for the local community.

Thus, although the PAF is a statistical parameter summarizing a characteristic of the environment rather than a parameter with an ecological meaning per se, the relative rating of sites, affected species groups, and compounds is nonetheless feasible and can provide meaningful information for decision making.

5.6.1.2 Protocol

The risk of exposure to individual chemicals as calculated using the SSD method is based on the same mathematical principles used in the derivation of concentration–response curves in single-species toxicity evaluations. As for individual species, both the concentration addition and response addition models can conceptually be applied in ecological risk assessment for species assemblages exposed to mixtures of toxicants, which are now being formulated probabilistically (Traas et al. 2002; Posthuma et al. 2002a; De Zwart and Posthuma 2005).

The protocols to aggregate compound-specific PAF values to a single risk estimate for a mixture of compounds are derived from common toxicological theories on joint effects of compounds. As already proposed by van Straalen and Bergema (1995), these protocols may be applied after corrections for differences in bioavailability among test media and the actual field conditions have been made, if the necessary information is available. This adaptation to reflect actual exposure is conceptually motivated by the common theory of molecule–receptor interactions, which is founded on the principle that the concentrations at the target sites of action should be addressed rather than external concentrations (Ashford 1981). This adaptation is further numerically motivated by observations that matrix interactions cause major differences in availability between substrates (Chapter 2). Based on the evidence obtained from the single-species studies, the following protocols are proposed for the assessment of risks for species assemblages exposed to mixtures.

5.6.1.3   Data Sources for SSD Construction

An analysis of the assessment question should lead to the selection of the assessment approach in which the user chooses between single-compound PAF, combi-PAF, or msPAF approaches and in which (optionally) sensitive species groups are identified. In either case, data are needed to construct the SSDs. Data can be obtained from the USEPA's ECOTOX database (USEPA 2001) and the RIVM e-toxBase (Wintersen et al. 2004). The RIVM e-toxBase contains specific information on mode of action of compounds for specific organisms and also allows the user to select species groups on the basis of taxonomic relationships. For example, the selection of EC50 data for a compound can be made by focusing on a group of organisms such as "all insects" or "all crustaceans." The selected data are then used to construct SSDs for the separate compounds and/or for the separate species groups. Software for SSD calculations is provided by, for example, van Vlaardingen et al. (2004).

5.6.1.4   Basics of Concentration Addition Extrapolation

At the species level, the TU approach — a point-estimate approach — has been used to express the toxicity of one compound as a fraction of another with the same mode of action. Transfer of the TU principle to species sensitivity distributions (SSDs), by scaling compounds in a similar way, results in hazard units (HUs). The scaling is done on point estimates taken from an SSD, such as the HC5 or the HC50, or on any other point estimate that is considered relevant for the assessment (such as legal quality criteria).

Alternatively, when using a whole-curve approach for compounds with allegedly the same mode of action, a single SSDs can be derived using (relative) concentration addition quantified by hazard units. This SSD represents the separate compounds and any mixture of these compounds. It is assumed that the mode of action in this case applies to all species from which the SSD is derived. The msPAF calculations for concentration addition (msPAF$_{CA}$) are performed according to the protocols given in Section 5.6.2 (below). These protocols require toxicity data and SSDs for all components of the concentration-additive mixture for a variety of species.

5.6.1.5   Basics of Response Addition Extrapolation

Multiple-species risk for independent combined effects in terms of the potentially affected fraction of species can be assessed using models that are essentially the same as for the prediction of response-additive effects in single species. The underlying assumption in the application of a response addition model for compounds or groups of compounds with different modes of action is that correlation of species sensitivities to the different constituents of the mixture is again considered absent. The calculation of the msPAF for independent joint action or response addition (msPAF$_{RA}$) is performed according to the protocol described in Section 5.6.3 (below).

5.6.2 Protocols for Calculation of Concentration Addition

Two protocols are presented, the first for baseline toxicity, and the second for groups of compounds with the same assumed primary mode of action. The protocols are more broadly explained in De Zwart and Posthuma (2005). Example spreadsheet calculations can be found in Box 5.2.

	Column B Taxon	Column C Imaginary aquatic SSD information	Column D Anthracene	Column E Atrazine	Column F Benzene	Column G Cadmium	Column H Malathion	Column I Toluene
Row 4	Algae	Number of species	4.00	29.00	5.00	4.00	4.00	5.00
Row 5		Average of log acute EC50 (mg/L)	-1.13	-0.62	2.47	-0.51	1.12	2.05
Row 6		Standard deviation of log acute EC50	1.06	0.64	0.53	0.71	0.63	0.60
Row 7		Assumed Toxic Mode of Action	NPN	PSI	NPN	CA	NPN	NPN
Row 8	Crustacea	Number of species	3.00	9.00	8.00	12.00	29.00	6.00
Row 9		Average of log acute EC50 (mg/L)	-1.11	0.80	2.02	-0.91	-0.90	1.81
Row 10		Standard deviation of log acute EC50	1.07	0.53	0.53	0.80	1.75	0.61
Row 11		Assumed Toxic Mode of Action	NPN	NPN	NPN	CA	AChE	NPN
Row 12	Insect larvae	Number of species	4.00	3.00	7.00	12.00	65.00	3.00
Row 13		Average of log acute EC50 (mg/L)	-1.27	0.44	2.07	0.99	-1.31	2.04
Row 14		Standard deviation of log acute EC50	0.73	0.51	0.52	0.83	0.75	0.82
Row 15		Assumed Toxic Mode of Action	NPN	NPN	NPN	CA	AChE	NPN
Row 16	Fish	Number of species	3.00	19.00	18.00	18.00	71.00	13.00
Row 17		Average of log acute EC50 (mg/L)	-1.46	1.19	1.52	-0.02	0.15	1.80
Row 18		Standard deviation of log acute EC50	0.90	0.68	0.52	0.98	0.82	0.58
Row 19		Assumed Toxic Mode of Action	NPN	NPN	NPN	CA	AChE	NPN

	Assumed Toxic Mode of Action	Abbreviation
Row 23	Non Polar Narcosis	NPN
Row 24	PhotoSynthesis Inhibition	PSI
Row 25	Cadmium Action	CA
Row 26	Acethyl Cholinesterase Inhibition	AChE

	Imaginary concentration information	Anthracene	Atrazine	Benzene	Cadmium	Malathion	Toluene
Row 30	Total concentration in pond water (mg/L)	0.012	0.3	0.05	0.012	0.005	0.005
Row 31	pH of pond water	5.40	5.40	5.40	5.40	5.40	5.40
Row 32	Dissolved Organic Concentration in pond water (mg/L)	5.00	5.00	5.00	5.00	5.00	5.00
Row 33	Bioeffective concentration in mg/l (See Chapter 2)	0.009	0.011	0.045	0.009	0.004	0.004

	Taxon	Calculated Hazard Units	Anthracene	Atrazine	Benzene	Cadmium	Malathion	Toluene
Row 37	Algae	HU	0.12141	0.04608	0.00015	0.02912	0.00030	0.00004
Row 38		Example formulae	DS33/10^D5	ES33/10^E5	FS33/10^F5	GS33/10^G5	HS33/10^H5	IS33/10^I5
Row 39		Assumed Toxic Mode of Action	NPN	PSI	NPN	CA	NPN	NPN
Row 40	Crustacea	HU	0.11711	0.00176	0.00043	0.07293	0.03187	0.00006
Row 41		Assumed Toxic Mode of Action	NPN	NPN	NPN	CA	AChE	NPN
Row 42	Insect larvae	HU	0.16592	0.00399	0.00038	0.00091	0.08078	0.00004
Row 43		Assumed Toxic Mode of Action	NPN	NPN	NPN	CA	AChE	NPN
Row 44	Fish	HU	0.25733	0.00072	0.00136	0.00944	0.00280	0.00006
Row 45		Assumed Toxic Mode of Action	NPN	NPN	NPN	CA	AChE	NPN

	Taxon	Assumed Toxic Mode of Action	Sum TU	Avg StDev	(ms)PAF single TMoA
Row 49	Algae	NPN	0.12190	0.71	0.10
Row 50		Example formulae	D37+F37+H37+I37	AVERAGE(D6,F6,H6,I6)	NORMDIST(LOG(D49),0,E49,1)
Row 51		PSI	0.04608	0.64	0.02
Row 52		CA	0.02912	0.71	0.02
Row 53	Crustacea	NPN	0.11935	0.68	0.09
Row 54		AChE	0.03187	1.75	0.20
Row 55		CA	0.07293	0.80	0.08
Row 56	Insect larvae	NPN	0.17032	0.69	0.13
Row 57		AChE	0.08078	0.75	0.07
Row 58		CA	0.00091	0.83	0.00
Row 59	Fish	NPN	0.25947	0.67	0.19
Row 60		AChE	0.00280	0.82	0.00
Row 61		CA	0.00944	0.98	0.02

	Taxon	msPAF multiple TMoA	Formula
Row 65	Algae	0.13	1-(1-F49)*(1-F51)*(1-F52)
Row 66	Crustacea	0.32	1-(1-F53)*(1-F54)*(1-F55)
Row 67	Insect larvae	0.19	1-(1-F56)*(1-F57)*(1-F58)
Row 68	Fish	0.21	1-(1-F59)*(1-F60)*(1-F61)
Row 69	All Taxa avg	0.21	AVERAGE(C65:C68)

BOX 5.2 Example of a spreadsheet calculation of toxic risk (msPAF) for a species assemblage in an imaginary aquatic pond as the result of exposure to a mixture of toxicants with diverse and species-dependent toxic modes of action. *Note*: Overall risk values (msPAF) per species group were calculated assuming concentration addition within common modes of action and response addition between modes of action. This example only serves to demonstrate the method of calculation. The SSD information on the mixture constituents as well as the total and bioeffective concentrations in pond water were randomly selected by realistic expert judgement. The gray cells contain examples of the formulas applied.

5.6.2.1 Method A: Unspecific Modes of Action (Baseline Toxicity)

Baseline toxicity or narcosis is considered to be associated with processes equally affecting the functioning of cell membranes in all types of organisms. Full baseline toxicity occurs in mixtures of organic compounds that exhibit only baseline toxicity, which pertain to about 60% of all organic industrial compounds (Warne and Hawker 1995). An earlier report from Bol et al. (1993) allocated about 30% of classified OECD high-production-volume chemicals (HPVCs) into the group of inert or less inert chemicals, for which toxicity is assumed to be only baseline. According to the funnel hypothesis (Warne and Hawker 1995), responses near baseline toxicity will also occur in mixtures of all organic compounds with extremely low concentrations.

5.6.2.1.1 Step 1

Toxicity data with the same type of evaluation criterion, either chronic NOEC or acute L(E)C50 values, or any EC_x, are obtained from an appropriate data set for each compound. Depending on data availability, the sets of species included in the SSDs for different toxicants may or may not be matched. For various assessment problems, Forbes and Calow (2002) argued that care should be taken to include species of different trophic levels and to include endpoints with comparable relevance to population success. For some toxicants, the toxicity data may be limited to the set of species prescribed in standard testing protocols (e.g., for the OECD: algae, crustaceans [*Daphnia*], and fish). The set of toxicity data may be extended by applying relative methods to extrapolate unknown species sensitivity from known species sensitivity (von der Ohe and Liess 2004). Usually, the existing databases contain mainly classical criteria, such as chronic NOEC and acute EC50 or LC50 values, not EC_x or raw concentration effect data. In general, SSDs for deriving environmental quality criteria are based on chronic NOEC data that are relatively scarce compared to acute data. Forbes and Calow (2002) argued therefore that if chronic data are used, the resulting SSDs are likely less representative of the sensitivity of natural species assemblages. Further, NOEC exceedance does not give a clear indication of the type and magnitude of effects to be expected. Whether the chronic NOEC is the appropriate criterion to use in ecological risk calculations depends on the assessment problem itself. An influential factor for the choice of acute or chronic criteria is the level of contamination that is to be assessed. If the contamination level is very low, NOECs might be good for ranking — though the meaning of msPAF as output is relatively unclear. If it is high, as in the case of spills or highly contaminated sites, EC50 values might be more discriminative for ranking.

5.6.2.1.2 Step 2

Toxicity data are scaled into dimensionless hazard units, preferably based on bioavailable concentrations. A hazard unit is defined as the concentration where the effect criterion (e.g., NOEC) is exceeded for 50% of all species tested, that is, the median of the toxicity data of the whole data set, $HU_i^j = NOEC_i^j / \overline{NOEC_i}$, for $i = 1$ to n compounds and for $j = 1$ to m species, with HU_i^j = the scaled NOECs in dimensionless hazard units (mg·L^{-1} / mg·L^{-1}), and $\overline{NOEC_i}$ = the median NOEC for substance i. The SSDs for each compound are obtained by fitting a log-logistic or log-normal model to the log toxicity data in hazard units. For the log-normal procedure, the SSDs are fully characterized by the slope of the cumulative distribution function that equals the standard deviation (σ) of $\log(HU_i^j)$; the median of the distributions is zero by definition. If a log-logistic procedure is applied, the slope of the curve (β) is given by $\beta = \sqrt{3} \times \sigma/\pi$.

5.6.2.1.3 Step 3

For each compound present in an environmental sample, the bioavailable exposure concentration is again recalculated to hazard units, HU_i = Bioavailable concentration$_i$ /$\overline{NOEC}_i$. The HU values are added (non-log transformed) for substances with corresponding modes of action and corresponding slopes, $HU_{TmoA} = \Sigma_i HU_i$. The log-normal concentration addition model gives the toxic risk for mixture constituents with the same mode of action by applying the Microsoft Excel function NORMDIST: $msPAF_{CA,TMoA} = NORMDIST(\log(\Sigma HU_{TMoA}), 0, \sigma, 1)$. Using the log-logistic concentration addition model, toxic risk for mixture constituents with the same mode of action can be calculated by

$$msPAF_{CA,TMoA} = \left(1 + e^{\frac{-\log(\Sigma HU_{TmoA})}{\beta_{Tmoa}}}\right)$$

5.6.2.2 Method B: Specific Modes of Action

For compounds with a specific mode of action, the situation is more complicated. Some species experience the same type of effect due to specific interactions at targets sites that only occur in a fraction of the species, and other species only experience narcosis, due to a lack of specific target sites for the compound. This holds, for example, for organophosphate (OP) biocides. Each OP biocide acts by acetylcholinesterase inhibition, and thus all compounds in this group can be seen as dilutions of each other for those organisms that have a receptor for this compound (e.g., insect larvae and crustaceans). In such cases, concentration addition modeling (Method A) should be applied for the specifically acting OP biocides, with an SSD for sensitive species. However, species or groups of species that lack the target receptor are not sensitive for OP exposure (e.g., bacteria and algae), and will only experience narcotic baseline toxicity, or a secondary level of toxicity (Solomon and Takacs 2002). This means that concentration addition for a specific mode of action is only appropriate within a single SSD, if the SSD consists of species having the specific receptor. For the organisms without the receptor, it may be more useful to consider narcotic or nonspecific concentration addition according to Method A. This requires splitting SSD curves according to groups of species with different types of receptors, as proposed by Posthuma et al. (2002a).

5.6.3 PROTOCOLS FOR CALCULATION OF RESPONSE ADDITION

The $msPAF_{CA,TMoA}$ values for the different modes of action in the mixture are calculated according to the procedures in Section 5.6.2 describing the multiple-species risk model for concentration addition, even if a mode of action is only represented by a single substance. The combination effect for compounds with different modes of action is again calculated analogous to the probability of 2 nonexcluding processes. For the present use in SSDs, it is assumed that sensitivities are uncorrelated in response addition. For more than 2 chemicals or groups of chemicals with different modes of action, this leads to $\text{msPAF}_{RA} = 1 - \Pi_{TMoA}(1 - \text{msPAF}_{TMoA})$, for mode of action = 1 to n substances or modes of action, with msPAF_{RA} representing the multisubstance potentially affected fraction of various (groups of) compounds calculated by response addition. Example spreadsheet calculations can be found in Box 5.2.

5.6.4 MIXED MODELS ALLOWING SPECIES-DEPENDENT MODES OF ACTION

Various msPAF protocols for complex, known mixtures can be conceptualized on the basis of the protocols of Section 5.5.

- Mixed model with concentration addition only for narcotics: The msPAF protocol originally proposed by Hamers et al. (1996) applied concentration addition only to compounds with a narcotic mode of action, whereas all other compounds were treated following the response addition approach.
- Mixed model with concentration addition for all compounds that share 1 mode of action: In current practice (e.g., Traas et al. 2002; Mulder et al. 2004; De Zwart and Posthuma 2005), concentration addition is also applied to mixtures of nonnarcotic compounds with the same mode of action, such as photosynthesis inhibition or acetyl-cholinesterase inhibition. The protocol is shown in Figure 5.2.
- Mixed model with assignment of exposed subgroups: When the mixture contains compounds with highly specific modes of action that also differ for the different groups of species, it is possible to generate mixed-model (concentration addition and response addition) msPAF values (again) for the individual groups of species — provided that sufficient toxicity data are available. In this approach, a single assessment for a contaminated ecosystem yields various msPAF values for different groups of species (Box 5.2) gives a hypothetical mixed-model example calculation for the msPAF of a mixture containing different modes of action for different species groups, where the species groups have equal weight in the analysis.

As a last step in the aggregation process, an overall msPAF can be calculated by averaging over the msPAFs generated for different groups of species. This average msPAF may be weighted, amongst others, according to the relative representation of groups of species in the field (Posthuma et al. 2002a; Forbes and Calow 2002). The weighting may also be motivated in different ways, for example, numerically based, biomass based, or trophic position based. Nonweighting relates to the fact that the SSD approach handles species as equal entities (numbers), with each given the same weight. Nonweighting implies that each species group is considered to be equal in (ecological) importance. Weighting can be preferred when some species groups are considered more important than others, such as in the case of a subgroup containing so-called "keystone species" or "ecosystem engineers." The weighting itself may be helpful to tailor the basic (numerical) outcome to the problem under investigation. As yet, no universally applicable and validated method has been derived.

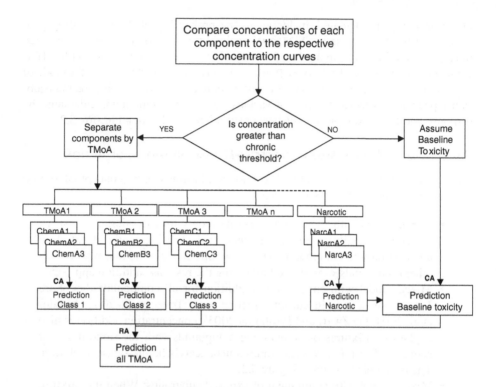

FIGURE 5.2 Schematized presentation of the mixed-model approach for assemblage-level extrapolation. *Note*: A similar approach is followed for species-level mixed-model extrapolation. The system can be simplified by assuming response addition for all extrapolations except the baseline toxicity assessment (approach of Hamers et al. [1996], yielding combi-PAF). The system can also be more complex when predictions for compound class effects are made for different species groups.

5.7 MIXTURES OF UNDEFINED COMPONENTS

5.7.1 OVERVIEW AND TYPOLOGY

In the case of complex mixtures of unknown composition, the application of extrapolation techniques using the approaches described in Sections 5.5 and 5.6 is possible only after the components and other stressors have been identified. Thus, the main issue when starting with an ill-defined mixture, or a mixture of completely unknown toxicants, is to isolate the agents or factors that cause biological stress.

There are various types of issues for unknown mixtures. An assessor might first wish to know that effects do occur. An example of an approach to address such a problem is the internationally common whole-effluent testing approaches (Chapman 2000) or the Dutch pT-monitoring approach (De Zwart and Sterkenburg 2002), both of which attempt to quantify impact.

Next, an assessor may wish to know the identity of the most influential compounds within the effluent or mixture, or the possible role of mixtures when a set of stress factors is present. If the concentrations of a set of toxic compounds are established, the mixture extrapolation techniques introduced in the previous sections can be applied, but the extrapolation in such cases represents a retrospective diagnostic approach. That is, extrapolation is undertaken to improve understanding of the effect causation as much as possible, to be able to make informed risk management decisions such as defining remediation priorities. The assessor follows a causal–analytical approach to decide on the most effective risk management option. By chemical and physical means, the assessor can try to partition the mixture into separate compounds, or groups of compounds, and test these mixtures.

Finally, extrapolation of unknown mixtures could consist of investigations on a large number of cases of environmental contamination with mixtures, where an investigator wants to predict the effects of unknown mixtures for nonstudied sites.

An important first question for risk managers trying to assess complex contaminated sites or samples and the cumulated risk for target populations or assemblages is whether the mixture is of a stable composition in time (Vouk et al. 1987). For unstable exposure situations such as effluents with greatly varying composition of toxicants, there are currently no mixture extrapolation techniques available, and thus only toxicity-monitoring techniques are suitable. In contrast, for mixtures of constant composition, as is often found at contaminated soil, sediment, or groundwater sites, mixture extrapolation techniques can be applied in solving the environmental problems after the identification of major causative toxicants.

5.7.2 Diagnosis and Identification of Relevant Compounds

5.7.2.1 Biologically Directed Diagnosis

Biologically directed approaches, which include bioindication, biotesting, and biomonitoring, have become important tools used to measure spatial and temporal responses of ecological systems to environmental disturbance (Markert et al. 2003). Implicit in the use of living organisms to assess environmental quality is that they integrate the effects of all stressors to which they are exposed. Primary emphasis in biologically directed approaches is placed on the use of living organisms as indicators of ecological condition. Such approaches have been particularly well developed for assessment of aquatic systems (Cairns and van der Schalie 1980; Suter et al. 1999a; Barbour and Yoder 2000). This approach typically involves measuring individuals, populations, or communities and using this information in 1 or more (multimetric) indices of relative environmental quality. Such studies include comparisons of community structure upstream and downstream of contaminant discharges, between reference and polluted sites (including the reference condition approach), and along concentration gradients of an environmental stressor.

5.7.2.2 Causal Emphasis

The composition of complex mixtures that occur in effluents or contaminated sediments is often unknown. In such cases, it is difficult to establish cause–effect (concentration–response) relationships aimed at identifying specific contaminants to which ameliorative efforts can be applied. To identify specific contaminants that are responsible for the toxicity observed in contaminated field samples containing mixtures of chemicals, it may be necessary to use chemical methods (fractionation) capable of separating the mixture constituents. Two approaches that can be applied are toxicity identification and evaluation (TIE) and bioassay-directed fractionation (BDF). BDF and TIE are diagnostic tools that use both toxicity testing and chemical methods in an iterative, directed approach to arrive at the identification of chemicals in sediments or waters and to link these to effects determined in toxicity tests (Mount and Anderson-Carnahan 1989; Norberg-King et al. 1991; Mount et al. 1993; Burgess 2000; Ho et al. 1997). Guidance has been published on stressor identification techniques and approaches to eliminate noncritical stressors from the matrix (USEPA 2000c).

TIE was originally developed to identify chemicals in effluents but has since been adapted for use in sediments using pore water extracts and bulk sediments (Schubauer-Berigan and Ankley 1991; Ankley and Schubauer-Berigan 1995; Doe et al. 2003). Bioassay-directed fractionation is a more directed form of TIE. BDF combines toxicity tests with fractionation using organic solvent extracts. Thus, BDF is generally limited to organic toxicants (Brack 2003). For both TIE and BDF, isolated chemical fractions are used in bioassays to determine the presence and magnitude of toxicity. In TIE for aquatic systems, common test organisms include *Daphnia magna* and *Ceriodaphnia dubia*. In BDF approaches, sample volumes are frequently small, so toxicity tests have been developed using small organisms with rapid life cycles, including bacteria (e.g., *Vibrio fischeri*), rotifers, and algae.

Both approaches are most frequently applied at higher tiers of the retrospective risk assessment process due to the cost and effort that are required and the more mechanistic nature of the information they provide. For example, TIE and BDF have been applied after initial screening using whole-effluent testing (De Vlaming et al. 2000). The primary limitations of BDFs and TIEs include insensitivity to the detection of both toxicity and chemical toxicants due to the chosen nontarget analysis. Therefore, the methods are typically applied at established or suspected hotspot contaminations. Recently, methods for the application of in situ BDFs and TIEs have been developed and have been shown to be more sensitive than lab-based TIEs at detecting sediment contamination from pore waters (Burton et al. 2002). Overall, TIE and BDF methods offer a logical solution for relating effects and exposure assessments and therefore could provide a more accurate and comprehensive characterization of risk (Simon 2003; Brack 2003). Inherent to any TIE or BDF approach is the use of mixture extrapolation techniques, as the relative roles of identified toxicants need to be confirmed regarding their contribution to the overall toxicity of the original sample that has to be explained (Grote et al. 2005).

The concept of ecological disturbance evaluations (EDEs) is another approach to address causation. EDE is conceptually comparable to TIE and BDF. As with TIE and BDF, which are used to identify chemicals responsible for toxicity in whole effluents and contaminated sediments, an EDE is applied after an impact has been detected in a community. The EDE uses an integrated index (e.g., index of biotic integrity), or other metric, allowing the researcher to extract information that might represent a response signature for the suspected cause of the impact.

This approach was applied by Fore (2003), who developed impact-specific multimetric indices using benthic invertebrates, which were used to identify specific causes of impacts due to metal contamination in rivers flowing through hard-mining areas in Colorado.

A recent example of fish assemblage stress analyses by means of multivariate gradient analyses that involved the use of local toxic stress quantifications (msPAFs) is provided by De Zwart et al. (2006). The targets of this study were both impact quantification and the assignment of probable causation. A specific approach in this respect has been termed biological response signatures (BRS) and is based on the use of biological data to discriminate between different stressors causing impairment (Yoder and Rankin 1995; Simon 2003). At present, assigning causes and sources of impairment in biologically directed approaches is perhaps best achieved by applying a weight-of-evidence approach that incorporates information from multiple lines of evidence drawn from response indicators such as biological response signatures or biocriteria. An important component of such studies is to develop a conceptual model of the range of stressors that may be affecting various communities. Once a decision has been made using the weight of evidence, and ameliorative actions have been instituted, biologically directed approaches can then be used to monitor improvements in biological condition.

5.7.3 Predictions of Toxicity for Unknown Mixtures

In contrast to the list of prediction methods that can be constructed from the literature on extrapolation of known mixtures, there is no specific technique that can be applied in practice as a method for mixture extrapolation for mixtures of unknown composition. Even if one has data (e.g., from a large series of WET observations at a certain location), the use of any extrapolation approach to predict the toxicity of a new WET test for that location may be inappropriate, especially when the situation of concern is unpredictable.

When a problem is, however, relatively well known and predictable from monitoring data sets collected in the past, and when risk managers have to handle this recurring situation, mixture extrapolation may be useful. Such a situation exists in The Netherlands, where risk managers have to make decisions on the deposition of slightly contaminated sediments from rural areas on adjacent land. Ongoing research here focuses on the categorization of sediments into 3 classes (Posthuma, unpublished results): sediments that can always be put on land because mixture risks are always acceptable; sediments that can never be deposited on land due to persistent, nonacceptable risks; and sediments for which the site-specific risk is sometimes acceptable and sometimes not. The "sometimes" category exists because the risks that a sediment may pose on a local scale will depend on various factors such as the propensity of the compounds to adsorb to local soils. The categories are defined using centiles (e.g., the 95% and 50% protection levels) to discriminate between "always" and "sometimes" and between "sometimes" and "never," respectively. The possible levels of mixture risks from all possible combinations of compounds and soil types (scenarios) are calculated by means of the mixed-model approach. Pattern recognition statistics are subsequently used to identify the proxy parameters that influence risks most. Because of this latter step, the local mixtures for which the risks eventually need to be handled are "unknown" (i.e., not precisely known). They are represented by the few selected proxy parameters. For sediments with unknown mixtures (but known proxies), the msPAF scenario values can be used to decide on the categorization of a sediment into 1 of 3 categories for daily practice decision making. The use of proxies introduces a new kind of uncertainty as the trade-off of ease of use, but the choice of regulatory acceptance criteria can be made relatively conservative to avoid "false positives." In short, a practical way to predict risks of imprecisely defined mixtures from the risks of known mixtures (scenarios) has been developed to support repetitive risk management questions.

5.8 APPLICATIONS IN RISK MANAGEMENT PRACTICES

5.8.1 OVERVIEW

Risk management problems for known mixtures have existed since the beginning of ecotoxicology. Hence, various practical ways to handle mixture problems have been designed in the past, and are used in various regulations. This section describes selected applications of the various parameters (e.g., TU and SSD) described in the previous sections in the context of risk management. The approaches are presented and described (Sections 5.5 and 5.6) using a tiered approach, as shown in Table 5.4.

5.8.2 APPLICATIONS OF SPECIES-LEVEL MIXTURE EXTRAPOLATION

5.8.2.1 Tier-0: No Extrapolation

The absence of mixture extrapolation is common. In many studies, the issue of possible mixture effects is mentioned, but not addressed. In various cases, mixture assessments are only partial, for a selected subset of compounds with assumed similar modes of action. For other compounds with different modes of action, no mixture assessment is made.

5.8.2.2 Tier-1: Application of Substance Group Criteria or Additional Safety Factors

Various approaches are used for which the occurrence of mixture exposures has led to the formulation of regulatory standards for whole groups of chemicals of similar structure or use (e.g., ground- and drinking-water quality standards in the EU). In The Netherlands, substance groups for which group criteria are used are given by Traas (2003). Some environmental quality criteria for individual compounds, such as the EU water quality objectives or the Dutch Target Value (to identify "clean soil" with negligible risk) have been formulated using uncertainty factors that explicitly account for combined effects from mixture exposures. The most often used extrapolation methods in species-level mixture extrapolation are based on this model. The modeling makes use of acronyms such as TUs, TEFs (toxic equivalence factors), and TEQs (toxicity equivalent quotients) (e.g., van den Berg et al. 1998). The TEF and TEQ approaches have been especially developed for assessing risks of specific mixtures (e.g., of PCBs and PAHs) for target species groups (i.e., birds and mammals).

TABLE 5.4
Overview of mixture extrapolation approaches used in practice

Level of mixture consideration	Approaches	Details
Interactions between compound and exposure medium	Matrix and media chapter	
Mixture exposure modeling	Matrix and media chapter	See discussion for future options
Combined effect modeling for species	Tier-1: point estimates	TU, MTI, AI, TEF, TEQ, and isobols
	Tier-2: concentration response function based	Concentration addition and response addition
	Tier-3: mode of action based	
Combined effect modeling for species assemblages	Tier-1: point estimates	CCU, HU, and EQC
	Tier-2: concentration response function based	SSDs
	Tier-3: concentration response function based, mixed models	SSDs, Combi PAF, msPAF, spatiotemporal variation, and EPC pie diagrams
	Tier-4: concentration response function based, mixed models with attention for receptor species type	Specific SSDs and msPAF$_{NOEC}$

Recently, methods for the evaluation of mixtures in contaminated sediments have been developed, including the use of TU modeling, development of numerical sediment quality guidelines (SQGs), and application of weight-of-evidence approaches. Where the contaminants in a sediment are known to act in a concentration-additive fashion, one approach is to simply apply a sum of toxic units using either pore water or organic-carbon-normalized (OC-normalized) TUs. The sediment–TU approach has been applied successfully to metals. When applied in conjunction with measurements of acid volatile sulfide (AVS), a strong binding phase for metals in anaerobic sediments, and simultaneously extracted metals (SEMs), this approach provides a good geochemical basis for extrapolating the cumulative effects of metal mixtures in anaerobic sediments (Swartz and Di Toro 1997).

The TU approach, combined with equilibrium partitioning (EqP) and (Q)SAR modeling, was also used by Swartz and DiToro (1997) to develop the ΣPAH model to predict the toxicity of sediment-associated PAH compounds. A (Q)SAR and EqP method is also presented by Swartz and DiToro (1997) for modeling narcotic chemicals in sediments. In this approach, the sediment quality guideline for a mixture of narcotic chemicals that exhibit additive toxicity could be expressed as the sum of the fraction of the OC-normalized sediment concentrations divided by the SQG for each chemical. A calculated SQG for the mixture that exceeds a value of 1 indicates that the total concentration of the narcotic-acting compounds in the sediment is expected to exceed the level that will protect 95% of the species from acute effects (determined by interpolating the 5th percentile of the species sensitivity distribution for toxicity endpoints for narcotics). A chronic SQG could be calculated by assuming an acute-to-chronic ratio (ACR) of 10 (Swartz and Di Toro 1997).

In the USEPA guidance document on the confirmation step in toxicity identification evaluation (Mount and Anderson-Carnahan 1989), different steps are proposed to confirm that the causes of toxicity are correctly identified. In one of these steps, the correlation approach, which especially applies when mixtures of toxicants are present, the authors recommended the utilization of a TU calculation. However, being aware of the problem arising from the application of TU calculations of nonconcentration additive compounds, they stated that great care must be taken to understand the interactions of the toxicants. They stated that "if two or more toxicants are strictly non-additive, only the major one (the one present in the most TUs) should be included in the data set" (Mount and Anderson-Carnahan 1989: Section 2 to 3). Grote et al. (2005) pointed out that in TIE and EDA, by using TU summation approaches as a confirmation strategy, there is substantial bias potentially leading to the risk of overlooking major toxicants or unresolved toxicity.

It has also been noted that species-related approaches must be used with caution when extrapolating from one species to another (or amongst species groups) as these species may not respond in the same way (Parrott et al. 1995; Compton and Sigal 1999). As a practical question, risk assessors face the problem of judging mixtures in which the compounds have different modes of action. Conceptually, the concentration addition approach would not be applicable. However, based on the ease of application of the models that only require EC50 values for the mixture components, the assessor might consider which approach would introduce the least bias: choosing nonextrapolation (and viewing each compound separately), or choosing a conceptually wrong model that nonetheless provides a numerical approximation of risk. As stated in Section 5.3.2, it is usually more beneficial to extrapolate with a less than ideal model than to extrapolate and judge each compound separately. The latter might yield the undesired situation of a high level of "false positives" — that risks (per compound) are not unacceptable — but the mixture as a whole induces undesired effects.

In human toxicology, it has become a standard procedure to split the toxicologically acceptable dosage of a chemical for different routes of exposure, thus accounting for sequential multiple exposures. All these applications of mixture extrapolation do not necessarily require any additional information on the type or degree of combined effects. The approaches are mostly precautionary rather than evidence based.

5.8.2.3 Tier-2: Concentration Response Function-Based Approaches

As summarized in Section 5.3.1, the vast majority of aquatic mixture toxicity studies report that the actual toxicity of mixtures is very close to the toxicity predicted by concentration addition. The application of Tier-3 approaches is mostly restricted to the analysis of experimental data. These approaches are not used in predictive frameworks outside that field, due to the lack of available concentration response data. Unlike the issue of SSDs, single-species toxicity data comprising the raw data of mixture studies are not systematically collected, which hampers the use of existing data sets to generate predictions for response addition in general and for concentration addition at other than EC50 or NOEC levels. An important improvement over using point estimates would be to use concentration response functions for the assessment, which is possible when adequate information for the individual components of a mixture is available. This can be done in a probabilistic way, both for compounds that share the same mode of action and for more complex mixtures. Predicted mixture effects are usually generated within experiments, from single-compound treatment group responses, using concentration addition or response addition as null models based on the assumed mode of action. Recently, an advancement was proposed for the establishment of causal links between chemical contamination and observed toxic effects of environmental samples (Grote et al. 2005). In this effect-directed analysis, use is made of tools for the assessment of mixture toxicity that accounts for unknown modes of action and heterogeneity of concentration response (Grote et al. 2005). For this purpose, toxicants were identified in sediment extracts, and subsequently all identified compounds were tested individually as pure compounds as well as in mixtures at ratios equal to those found in the sediment extracts. The observed extract toxicity was then compared with the expected combined effects calculated according to the models of concentration addition and response addition as well as with the observed toxicity of the synthetic mixture. An Index of Confirmation Quality was introduced, providing a quantitative measure of confirmation over a range of different effect levels.

There are various mathematical models that can be used to describe and analyze experimental data (Scholze et al. 2001). In addition to these curve-fitting approaches, response surface models are also available (e.g., Greco et al. 1995), but these are suitable primarily for the analyses of experimental data, rather than for predictive purposes. As an example, Altenburger et al. (2004) applied both concentration addition and response addition and observed that the combined effect of a 3-compound mixture out of 10 identified sediment toxicants was sufficient to explain the observed combined effect of the more complex mixture. For identifying remediation priorities in site-specific assessment of complex contamination, this approach has great potential.

5.8.2.4 Tier-3: Mode of Action-Based Approaches and the Mixed Model

As in Tier-2 approaches, Tier-3 approaches use concentration response information but in addition try to acknowledge particular mode of action information. This leads to the use of concentration addition as a model to predict cumulative effects within groups of compounds with the same mode of action, and response addition to predict the overall cumulative effects (over those groups and the compounds with a mode of action that is unique in the investigated mixture). For identifying remediation priorities in site-specific assessment of complex contamination, this approach has great potential.

5.8.2.5 Tier-4 Approaches

Tier-4 approaches do not exist for the single-species level. The Tier-4 approach is defined by the presence of more than 1 exposed species, which is not the case for single-species mixture extrapolation.

5.8.3 APPLICATIONS OF ASSEMBLAGE-LEVEL MIXTURE EXTRAPOLATION

5.8.3.1 Tier-0: No Extrapolation

There are numerous examples of no extrapolation. The clearest example is the common practice of setting environmental quality criteria for each compound separately. Conceptually, it is difficult to imagine how a single quality objective could be formulated for every possible mixture in terms of maximum permissible mixture concentration, unless one could develop a criterion that is formulated as a maximization of impact or risk.

5.8.3.2 Tier-1: Point Estimates and Concentration Addition

Transfer of the TU principle to the community level results in hazard units or cumulative criterion units (CCUs). The most common types of hazard assessment use this concept.

Toxic unit–type approaches applied at the community level have been used in a limited number of field studies investigating metal-contaminated streams in North America (Clements et al. 2000; Fore 2003) and New Zealand (Hickey and Clements 1998). Clements et al. (2000) found significant correlations between CCUs and 10 out of the 16 biological metrics that they tested to assess benthic community structure. Hickey and Clements (1998) found that abundance and species richness of mayflies; number of taxa in the orders Ephemeroptera, Plecoptera, and Trichoptera (EPT); and total taxonomic richness were the best indicators of heavy metals in New Zealand streams. Further New Zealand studies with macroinvertebrates exposed in mesocosms for 34 days to a mixture of Cu and Zn found strong relationships with CCU. Variance in community structure was best explained by 3 quantitative variables: total mayfly abundance, a mollusk (*Potamopyrgus antipodarum*) abundance, and a summary parameter (QEPT, or the quantity of EPT individuals). These studies concluded that the CCU model could represent a useful approach for monitoring the toxicity of metal mixtures in aquatic environments. Indeed, the CCU approach is similar to the total toxicity of a mixture (TTM), a method that is currently used in a regulatory context in Australia and New Zealand (ANZECC ARMCANZ 2000; Norwood et al. 2003). As with CCUs, the TTM is determined by summing the ratio of the toxicity of the individual metals to their water quality criteria.

When using TU-based approaches, specific attention should be paid to the definition of the units that are used. Commonly, these units are the environmental quality criteria that are locally applicable. In The Netherlands, however, a total of 3 CCUs does not necessarily imply that there are ecosystem risks. This relates to the fact that the unit values in the regulations are the lowest of a set of subordinate risk limits, being the human risk limits for the set of compounds and ecological risk limits for those compounds (Swartjes 1999; Sijm et al. 2002). Exceedance of a quality criterion may thus imply the presence of human risks, ecological risks, or — at high exposures — both.

5.8.3.3 Tier-2: Whole-SSD Curve-Based Approaches, Concentration Addition, or Response Addition

SSD-based approaches are used for monitoring and trends analyses regarding toxic mixtures per se, and for the analysis of situations of multiple stressors. However, none of the SSD-based methods assumes a priori that only concentration addition or response addition would apply. Even the earliest applications recognize the relevance of both models. These uses of SSDs are described under Tier-3 methods.

5.8.3.4 Tier-3: Whole-SSD Curve-Based Approaches, Concentration Addition, and Response Addition (Mixed Model)

The earliest SSD-based approach for monitoring and trends analyses yielded the so-called combi-PAF (Hamers et al. 1996) and, by aggregating over a package of substances used in The Netherlands, the indicator known as I-tox. The latter parameter has been used to investigate the changes over space and time in overall toxic pressure in The Netherlands for soil and water data in various "State of the Environment" reports of RIVM. Such analyses have suggested, for example, that I-tox values have declined in the larger rivers over the last few decades.

A site-specific example of mixture risk analysis was provided by Traas et al. (2002). Spatial and temporal distributions of mixture risk (msPAF) were analyzed for a major estuary basin, for which potential adverse influences were suspected from major industrial activities upstream. The analyses showed spatiotemporal variation in toxic pressure in the estuary, and these data were used to support risk management through the identification of the periods and the sites of the highest expected impacts, and the major compounds most likely contributing to the risks. Various authors elaborated on this idea further to identify hotspots of highest risks along a river stretch, to identify the added risks of introducing new chemicals into the system (Verdonck et al. 2003), and to make GIS maps of pesticide impacts in The Netherlands (see Figure 5.3). This approach was called "georeferenced probabilistic risk assessment."

A very specific way of applying estimates of the ecotoxicity of mixtures at the community level is the application of the msPAF approach in life-cycle assessments (Huijbregts et al. 2002). In this application, the target is to assess ecological impacts of mixtures in a single value, and to compare this value to the predicted impacts of other stressors (e.g., ozone depletion and energy expenditure).

Various examples exist of the use of msPAF in multiple-stress analysis to acknowledge the relative role of toxicant mixtures in shaping ecological communities. Mulder et al. (2004) studied the decline of butterfly populations in a nature reserve in The Netherlands. It appeared difficult to establish associations between decline and the major environmental parameters, such as pH and water relationships.

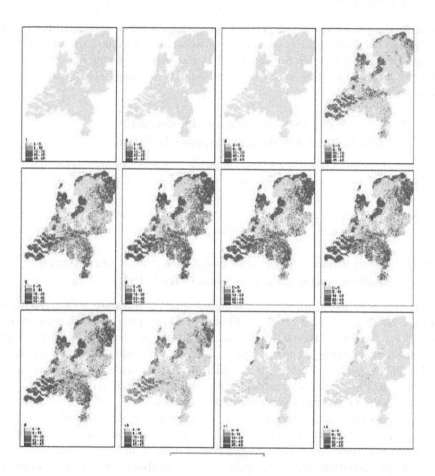

FIGURE 5.3 Predicted ecotoxicological risk ($mPAF_{NOEC}$) of pesticide use in field ditches. *Note*: The maps from left to right and from top to bottom represent the development of pesticide risk for 4-week periods throughout the year 1998. Darker colors indicate higher risk, up to the maximum level of 51% mPAF. Original data redrawn from De Zwort (2005).

By using mPAF as a surrogate parameter for the toxicant mixture (metals), it was found that the decline of the species was associated with metal stress, but mediated by the direct effects of the compounds on the host plants. In this example, ecological (trophic) relationships between species were an explicit part of the mPAF-mediated analyses. In a further example that merges ecological and ecotoxicological methods, De Zwart et al. (2006) used mPAF methods in concert with ecological modeling (RIVPACS) to identify the relative role of toxicant mixtures as compared to other human-induced stress and natural variability in shaping local fish communities in Ohio surface waters. Based on a large database of abiotic (physical and chemical) and biotic (fish counts for 100 species) variables in Ohio rivers, these authors applied the mixed-model mixture extrapolation technique to quantify the likely impact of the local mixtures of metals and household chemicals. The analyses resulted in so-called "effect and probable cause" (EPC) pie diagrams, the pie size for each sampling site representing the magnitude of impact (compared to the community that is locally expected under reference conditions through ecological modeling, or RIVPACS) and the slice sizes indicating the relative contributions of the (statistically) probable causes. The strength of the method in such diagnostic analyses is that 1 msPAF can be used to represent toxic stress instead of a large set of measured compound concentrations, which strongly improves statistical power.

5.8.3.5 Tier-4: Whole-SSD Curve-Based Approaches for Species Level with Mode of Action and Exposed Species Analysis

A possible criticism to the Tier-3 approaches is that one single compound can have multiple modes of action in the case of community-level exposure. For example, an insecticide that acts on the nervous system of insects may also elicit responses in plants and other nontarget organisms (such as narcotic effects). As a solution, for specific assessment problems, one may acknowledge the presence of different groups of ecological receptors (see Box 5.2). This approach reduces the potential for disparities between data and SSD that can occur, for example, when nonhomologous groups of ecological receptors are used to create an overall SSD.

Higher tier mixture extrapolation approaches such as this are used in practice, by calculating site-specific msPAF values, across a broad range of assessment questions. A practical example is provided by De Zwart (2005), who studied the impacts of pesticide use in The Netherlands. In this study, the specific mode of action of the pesticide was taken into account, as shown in Box 5.2. This analysis resulted in spatiotemporal indicators of relative toxic pressure across The Netherlands (Figure 5.3).

From a legal standpoint, Tier-4 methods might constitute the best option for defining a problem. When environmental laws are violated, and the "violator" is prosecuted, very specific questions are asked regarding likely local impacts. This relates to the fact that larger impacts result in higher penalties. Legal cases asking for quantitative mixture risk assessment have occurred in The Netherlands (RIVM, unpublished results). In such cases, depending on the exposed species types, Tier-3 methods could be the best option to address the question of seriousness of impact and penalty.

5.8.4 ABSOLUTE VERSUS RELATIVE INTERPRETATIONS

In single-species assessments, the interpretations of mixture assessments tend to be mostly absolute. Hence, risk assessors often focus on particular species and particular compound groups (e.g., risks of PCB mixtures for birds), allowing them to interpret and explain their experimental data to the best of their abilities. On the other hand, many risk assessors apply mixture extrapolation methods to address risks for communities. The applications of SSD-based methods for this evolved fast and now cover a wide set of approaches, ranging from ecological multiple-stress analyses to overall approaches such as life-cycle assessment. Especially in the latter set of approaches, the risk assessor can often allow the method to only yield relative rankings of risks (e.g., to identify sites or compounds of highest concern). The parameter msPAF can serve those purposes, although an absolute, ecological interpretation of PAF remains to be established.

5.9 UNCERTAINTIES AND FUTURE ISSUES IN MIXTURE EXTRAPOLATION

5.9.1 GENERAL

We have reviewed current conceptual and modeling approaches in mixture ecotoxicology as well as current experimental evidence to derive practical risk assessment protocols for species and species assemblages. From the review of conceptual approaches in mixture ecotoxicology, it appears that there is a difference between a mechanistic view of joint action from a compound mixture and a probabilistic perspective on combined toxicity and mixture risk. A mechanistic view leads to emphasis on the distinction of modes of action and physicochemical properties first, then on the choice of the appropriate joint toxicity model, followed by a comparison of the models' prediction with experimental observations. A probabilistic orientation leads to the observation that concentration addition often yields a relatively satisfactory quantitative prediction of observations for the integral level of effects as observed in individual organisms or populations. In these applications, concentration addition is frequently connected with a slight bias to conservatism, especially for compounds with different modes of action (Backhaus et al. 2000, 2004; Faust et al. 2003).

The different viewpoints frequently cause misunderstanding in debates between scientists adopting the causal–analytical viewpoint (for research) and those developing operational instruments (for practical use in risk assessment). In this chapter, we merge some of the issues by proposing mixed-model methods for complex mixture risk assessment based on probabilistic modeling, taking into account mode of action knowledge at higher tier extrapolation.

5.9.2 MIXTURE EXPOSURE

5.9.2.1 Improvements on Exposure Extrapolation

Ashford (1981) and Mesman and Posthuma (2003) emphasized the issue of exposure assessment within the context of mixture modeling (i.e., estimating the actual dose at the target site of action). In ecotoxicology, this practice is still under development. Regarding the issue of environment–chemical interactions, there are various compound-specific interactions between the compound and matrix and, superimposed on that, interactions between the compounds (such as nonsoluble complexes). As a rule, the latter are likely to be less important than the former. For metals, for example, the range of matrix influences (e.g., pH) on separate metals can be substantial, resulting in orders of magnitude of exposure variance (Janssen et al. 1997). Van Straalen and Bergema (1995) have highlighted the effects of changing soil acidity on the ecological risks of metals to soil organisms as predicted by SSDs. They showed that risks steeply increased at lower pH levels. Screening of typical compound–compound interactions should be part of the mixture extrapolation steps to avoid missing critical interaction data. Focus should be on expected special interactions, such as those rendering a compound either substantially less or more mobile in the presence of the other (e.g., by formation of insoluble precipitates).

Regarding the issue of internal redistribution of multiple compounds over tissues (toxicodynamics), there is a general lack of practical approaches and data. The concept of physiologically based pharmacokinetic modeling might provide specific solutions in the future. PBPK models were suggested for application in assessing human health risks of contaminants in drinking water (Ashford 1981) on the basis that most interactions between organic substances occur as a result of induction or inhibition of metabolism. There are some practical examples. In practice, however, the option to execute a risk assessment of mixture toxicity based on application of PBPK models in ecotoxicology is currently very limited. Although conceptually attractive, the use of PBPK modeling as a basis for estimating mixture toxicity is far from being operational for ecotoxicity assessments due to lack of data. Hence, the exposure extrapolation issues can currently be assessed in practice only by applying matrix and media extrapolation approaches as discussed in Chapter 2.

When the assessment concerns a community-level risk assessment, it should be considered that each species explores its habitat in its own, specific way, so that in the higher mixture assessment tiers (where species groups might be separated out for deriving specific SSDs for them) the assessor may need to consider that these different groups would need different exposure extrapolations. In Posthuma et al. (2002a) and Vijver (2004), the issue of differences between different soil organisms was conceptually worked out in an example for metals. In soil, where many species have soft integuments, the dominant exposure route is through the skin (as compared to passage through the alimentary system), such that contaminant uptake is highly dependent on the pH of the soil. However, for soil species with a hard integument, exposure is better correlated with the total metal level in the food. This relates to the regulated conditions (e.g., pH for metals) in the alimentary canal. Applying species-group–dependent corrections for exposure adds realism in the assessment when executed well and with enough input data, but data limitations may oppose this expected increase in confidence. That is, when the available data are split into 2 subgroups, the conceptual improvement trades off in the format of increasing size of the confidence intervals of subgroup results.

5.9.2.2 Improvements on Toxicant Detection

The major challenge for dealing with mixtures of unknown composition is to generate meaningful information on the contributing toxicants. Current approaches look at property estimations, using physicochemical parameters such as chromatographic distribution behavior. Refined protocols of toxicant identification are being developed, based on fractionation schemes for complex contaminated matrices. The use of effect assays as toxicity detectors in such approaches is being pursued, but confirmation of the identified effect contribution and quantification of unresolved effects are vital for useful application (Brack 2003). The practical need for high-throughput effect assays, however, trades off with the need to also focus on ecologically relevant effects at prolonged exposure times.

5.9.2.3 Simultaneous versus Sequential Exposure

All mixture extrapolations considered so far have their focus on cases of simultaneous exposure to a set of toxicants, whereas organisms or species assemblages in the environment frequently face sequential exposure. Although extrapolative methods are available for sequential exposure to single compounds, this is not the case for mixtures.

5.9.3 MIXTURE EFFECTS

5.9.3.1 Modeling

Based on a review of the data, 2 main models were encountered most frequently: concentration addition and response addition. These models are regarded as the 2 biologically most plausible and useful reference concepts suitable to calculate an expected mixture effect based on effect information from the components of the mixture under consideration. The appropriate selection of the reference concept to be applied is believed to depend on the similarity of the mode of action of the components. This is, however, an ambiguous form of information (Grimme et al. 1996), because the mode of action of a compound is codetermined by the receptor. One is therefore often left with the dilemma of 2 possible ways to assess the same data. Consideration of the quantitative difference in the expected combined effect or the modeling of a prediction window using both concepts might be more helpful than selecting 1 by chance.

Despite the fact that concentration addition and response addition are completely different in mechanistic backgrounds and modeling, the predicted effects may numerically be similar. Evidence for this is presented by Drescher and Bödeker (1995) and Deneer (2000). At a slope of log-logistic beta (slope) of about 0.4 and a log-normal standard deviation of about 0.8, concentration addition and response addition methods yield similar risk predictions. The degree of similarity of predicted effects between the concentration addition, response addition, and perhaps mixed-model approach is, in part, dependent on the mathematical model that is used to describe the data (log-normal, log-logistic, Gaussian, etc.) on the slope divergence, on the position on the curve for which the extrapolation is needed (e.g., lower tail), and on the number of components. The models themselves may equally fit experimental data in many cases, so that the different models yield equivalent prediction values when used in extrapolation in such cases! This means that mixture risk prediction might be more robust than often thought.

The existence of mathematical features of the models used in mixture extrapolation does not mean that it is appropriate to apply unchecked models to all assessment problems. Most data for which the models were tested pertain to the higher exposure (e.g., EC50) level. There may be mathematical features of the models that introduce bias when applied to lower exposure levels. The models should be applied with a clear understanding of the potential biases that may occur. A model might be discarded for a certain use because of its bias. For example, using concentration addition for all mixtures, even when there is clearly a case of different modes of action, would typically overestimate mixture risks when assessing risks at a contaminated site, which would be an undesired feature for this type of assessment.

In the assessment and ranking of contaminated sites, one tries to obtain a realistic view, not a conservative one. On the other hand, a model might be preferred if the bias is favorable for the assessment target. The concentration addition approach is biased toward overestimation of risks at lower exposures, but this is desired in cases when environmental protection is the assessment endpoint.

The issue that matters most is the presence of data at the concentration where the models are to be applied. For example, many experiments focus on the concentration level of maximum response (i.e., the inflection point of effect curves, e.g., the EC50), whereas for the objective of setting EQC one usually applies NOECs (low concentration range). Various authors have shown that there is a concentration-dependent pattern in similarity between observed and predicted values using concentration addition (Posthuma et al. 1997; van Gestel and Hensbergen 1997; Sharma Shanti et al. 1999; van der Geest et al. 2000; Jonker 2003). Mixture experiments do not usually have observation points at low concentrations that are precisely measured so as to clearly distinguish concentration addition from response addition. This is caused by a general lack of data in the tails of the curves, yielding low statistical power. The underpinning of mixture extrapolation is thus weakest in the low exposure range, as is true for all dose–response extrapolations.

If there are doubts about the mode of action in relation to the studied ecological receptors and the concept of the receptor sites, a good approach may be to use a reasonable worst-case assumption, which in many cases seems to be concentration addition. A better approach may, however, be the generation of a "prediction window" (Walter et al. 2002). A prediction window shows the outcomes of alternative modeling options, when model choice (concentration addition, response addition, or mixed model) is not supported by the available knowledge. Prediction windows are produced by applying alternative null models. The location of the prediction window (bordered by the outcomes of the alternative models) as compared to the chosen protection endpoint may yield insights that are sufficient for decision making. For example, Posthuma et al. (2006) applied this concept in the design of a decision support system for deciding on the deposition of slightly contaminated sediment on land. In cases where the prediction results of alternative models all suggested that the mixture risks were below the (policy's) acceptance threshold (in this case, the use of a 95% protection criterion as a critical level for unacceptable risks, whereby less than 5% of the species is supposed to be exposed at a level higher than their no-effect level), the decision can be made that sediment could be deposited on land without causing unacceptable risks.

5.9.3.2 Improvements at the Toxicological Level Sensu Stricto

Although most mixture studies in ecotoxicology focus on integral effects (i.e., vital characteristics such as growth and reproduction), various molecular assays exist for assessing or predicting mixture effects on gene or protein induction, inhibition, or alteration. It can be argued that the reactions that occur in vital characteristics as a result of these physiological changes may be more valid endpoints for extrapolation toward ecologically relevant mixture effects, and that molecular reactions can provide an "early-warning" signal. Recent technical advances in the

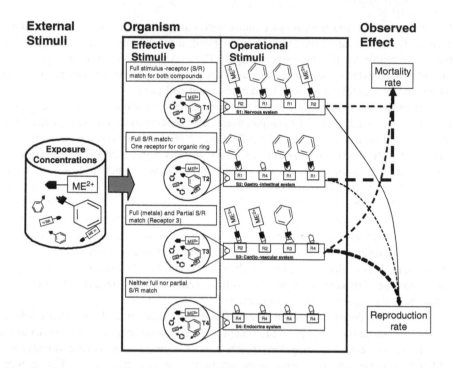

FIGURE 5.4 Theoretical concept of the joint action of chemical compounds according to Ashford (1981), adapted to describe ecotoxicological effects at the single-species whole-organism level. *Note*: Ashford's external stimulus is the mixture (e.g., of an organic compound and a metal-containing compound). Ti is the concentration at target site i. Rj is a receptor for toxicant j. Sk is a physiological subsystem k. The arrows of variable thickness leading to mortality and reproduction are indicating the relative contribution of underlying processes to the overall effects studied in ecotoxicology. *Source*: Ashford (1981).

field of eco(toxico)genomics might provide the options for empirical test strategies for addressing mode-of-action issues and mixture modeling. A challenge here is to develop "early-warning" effect tools that are also suitable to detect and quantify combined effects for ecologically relevant endpoints.

A neglected area in ecotoxicology has, as yet, been the internal redistribution of compounds within organs and cells within an organism. Ashford (1981) developed such ideas in the framework of human pharmacology, and these ideas could be applied to ecotoxicological problems (Figure 5.4). In ecotoxicology, external stimuli (ExS, the toxicant exposure concentrations outside the organism) are most commonly used as surrogates of exposure. However, the use of effective stimuli (ES, toxicant concentrations that enter the tissues) and operational stimuli (OS, concentrations at particular target sites of action [Ti]) would be much more appropriate for describing the true physiological (internal) activity of a contaminant. The ES and OS depend on internal distribution processes amongst tissues, and the OS is not necessarily the same at all sites of action in different tissues. A receptor (Rj) can be present in one or more physiological subsystems (Sk). The action can be a toxic interaction, where the compounds in the operational stimulus and the receptors are fully complementary (full stimulus–receptor [S–R] match; function is fully inhibited) or a partial interaction, whereby the stimulus inhibits the normal function, but only partially. Compounds may affect different subsystems, with different combinations of Rj. Figure 5.4 shows mechanistic effects at target sites of toxic action in exposed subsystems and overall effects at the level of the organism (e.g., on mortality and reproduction rates). The latter effects are studied in ecotoxicology, and the former in pharmacology. The final toxic effect (mortality, altered reproduction, or altered growth rate of a species) is the net effect of all subsystem-level effects, including their interactions.

Each subsystem may have various active sites for the toxic action of compounds. However, between subsystems, the same mixture is likely to cause quantitatively different responses. Because an organism is composed of an array of subsystems, the outcome of a mixture study with compounds with distinctly different modes of action on reproduction effects may demonstrate a response that is numerically similar to concentration addition, whereas the underlying responses in all subsystems (e.g., endocrine, energetic, and metabolic systems) may differ mechanistically. Although all this is theoretically plausible, the challenge still is to provide empirical evidence that the models derived accurately predict mixture effects in exposed species.

5.9.3.3 Tiering

Current efforts to extrapolate mixture effects are dominated by TU-based approaches, which result in prediction error when the models are used for situations where the concentrations deviate from the original effect level that is used to define TU. Provided that the data are available, mixture extrapolation at the species level may improve by using the proposed higher tier protocols. It should be acknowledged, however, that the data needed for such an enterprise at the species level are not systematically stored in databases, as is the case for the databases available to construct SSDs (see Section 5.6.1). For a significant advancement, researchers therefore should strive for full-curve modeling over point-estimate models (i.e., to model at Tier-2 and Tier-3). The major requirement would therefore be to not only produce but also report systematically on concentration response functions for individual compounds, as this would allow prediction of any yet untested mixture for the same biological response.

5.9.3.4 Suggestions for Improvements at the Assemblage Level

5.9.3.4.1 Models
This chapter proposes the use of SSD and mixture toxicity models in ecological risk assessment of species assemblages by calculating the multisubstance potentially affected fraction of species on the basis of measured or predicted (biologically active) concentrations of toxic compounds in the environment. The msPAF method has been scrutinized for its conceptual basis. To address this scrutiny, we cite the human toxicology work of Ashford (1981) as a cross-link.

Indeed, the ideas of Ashford (1981) could be introduced productively into mixture ecotoxicology. Ashford's ideas can further be seen as a prelude to the interactions between multiple toxicants and multiple species, as a next level of added complexity and improved relevance of mixture studies. In this added level, species can be considered as active sites of mixture effects. They might be called "ecological receptors" of the compounds, as equivalent to the toxicological target sites of action (i.e., the toxicological receptors). The subsystems might be envisaged as species groups that show interactions (e.g., within and between functional groups in a food web). The concept of an "ecological receptor" was originally introduced by van Straalen (1992). An ecological receptor is an ecological entity at a higher level of biological organization than the toxicological receptor. Regarding the mixture literature, the concept would in practice often relate to the species level. For example, the primary ecological receptors of insecticides are insects, as they possess unique physiologies or specific molecular targets for insecticide action. Any other species that responds similarly to another insecticidal compound could also be considered an ecological receptor for insecticides. The response may be exclusive for that species or species group, and is dependent on the way the species (or group) explores its niche and/or on its molecular and physiological characteristics. At the community level, it is conceptually problematic to maintain the idea that "each compound can be assigned a single mode of action." In this case, a compound can have more than one mode of action, and likely even many modes of action. For example, an organophosphorus insecticide has been designed to kill insects whereby the mode of action is a chemical interaction with particular target sites of toxic action in the nervous system of insects. The same insecticide likely has other types of (inter)actions in the tissues of other, noninsect organisms. In other species, it may only cause baseline toxicity. Posthuma et al. (2002a) called these phenomena "target," "partial," and "minimum" toxicity to designate the apparent effects of the slope and position of species sensitivity distributions. Data inspection, as done by De Zwart (2002), might indicate the presence of different ecological modes of action (EMoAs) of a compound. EMoAs are defined here as the equivalents of the concept of mode of action, but the concepts pertain to a higher level of biological organization than the molecular target.

The proposed mixed-model approach for assemblages, preceded by an exposure analysis, is in line with Ashford's ideas; the ecological interactions need further attention. Whether Ashford's ideas can be fully worked out conceptually, tested experimentally, and applied in a validated predictive framework remains to be solved by mixture ecotoxicologists.

5.9.3.4.2 Tiering

A proposal for tiering was suggested, based on an array of simple and more complex modeling approaches, consisting of an uncertainty factor–based approach (Tier-1), point-estimate–based extrapolation (Tier-2), and full-curve–based extrapolation (Tier-3) to curve-based extrapolation with special emphasis on the link between mode of action and ecological receptors (Tier-4, only for assemblages).

The TU-based approaches (Tier-1) have to date most widely been used in practice. A critique of this approach is that it operates under the assumption that a linear relationship exists between risk and criterion exceedance or concentration, and that the slope of the concentration–risk relationship is equal for the different substances in the mixture. This may not be the case, because the sensitivity distribution (SSD) is typically characterized by a nonlinear bell-shaped response relationship that may have a different width for different substances. Further, the individual criterion concentrations (c_i) for each compound will be those determined by a particular regulatory agency and may vary between jurisdictions and with periodic revisions.

Specific critiques of the TU-based method have been listed by Solomon and Takacs (2002). A specific problem is that investigators limit the application of this approach to compounds with an assumed similar mode of action, because they have mechanistic-based objections against broadened applications. However, especially in the case of site assessment, the goal is to know the risk of the complete mixture, not only the risk of compounds with an assumed similar mode of action. This could result in the underestimation of risks and "false positives" wherein the risk assessment made for the limited set of compounds could suggest "absence of risks" although the total mixture would in reality induce risks. To prevent "false positives," one could sum the TUs for all compounds, thereby removing the objections. The numerical similarity between predictions generated by concentration addition and response addition can be seen as numerical support for this idea.

5.9.3.4.3 Validity
For msPAF to be a useful parameter in risk management, it is crucial to obtain values of msPAF that are relevant to predict community risk under field conditions, either in an absolute sense, where risk is related to ecological effects, or in a relative sense, where the method is useful for ranking contaminated sites. However, the ecological meaning of msPAF is still rather unclear. The results of msPAF evaluations can be validated by analyzing combined physicochemical and biological monitoring data or with mesocosm experiments. Observing effects in the field may be difficult due to high variability and a lack of reference sites. The expected effects may be masked if less sensitive species outcompete the affected species, or effects may be larger than expected due to interactions between species not predicted by single-species toxicity tests (Klepper et al. 1999; Mulder et al. 2004). Ecological interactions are an important point to consider when interpreting SSD-based output. In practice, different natural stressors or any other nontoxicological environmental stressor may influence the effects of toxicants either directly by affecting the species' viability or through interactions with the compound. This is a factor that is not taken into account when calculating mixture risk in the proposed way.

When interpreting SSD-based risk predictions, one should note that the SSD approach solely focuses on direct effects of toxic compounds on species assemblages. Indirect ecological effects resulting from changes in species interactions are an aspect not reflected in the msPAF risk modeling proposed in this chapter. It is likely that indirect effects can only occur when direct effects are present and that their importance increases at increasing levels of direct effects. When msPAFs can be calculated, such as in Box 5.2, ecologists might be able to identify some likely indirect effects (e.g., when msPAF values for primary producers are high, there is an increased probability of indirect effects on grazers). Confirmation studies could shed light on the implications associated with increased levels of toxic risk.

There is quantifiable uncertainty in the steps associated with SSD and msPAF assessments. Although there is as yet no universal procedure to develop a specific SSD-based ecological risk assessment, it is essential to take uncertainties into account. One of the options to present uncertainties is the use of confidence intervals. Although the confidence intervals of a single toxicant PAF can be calculated quite easily (Aldenberg and Jaworska 2000), the mixture risk approach uses additional concepts and parameters, for example, related to exposure and bioavailability issues. These uncertainties should be superimposed on those of the separate SSDs, implying that further uncertainty analysis of SSD-based models is needed, along with a clear ecological interpretation of the final output and its confidence interval.

5.9.3.4.4 Use

Various options were listed for practical application of the proposed protocols for risk prediction for species assemblages. One of the most useful features of the SSD concept is that its output in terms of dimensionless PAF values can be aggregated to reflect the risk posed by multiple substances, according to a set of protocols that are motivated by established toxicological principles and that are further supported by quantitative data on model behavior and model characteristics. Mathematical analyses of an array of mathematical response models are supporting the viewpoint that, in the case of lack of data on modes of action, reasonable robust risk predictions can be obtained, especially when slopes are approximately equal. If a mixture consists of compounds with moderate and similar slope values, then this would remove part of the necessity to have a debate on mechanistic principles in the absence of data to support either viewpoint. When the slopes diverge outside the given bounds or are different among substances, the outcome of the assessment could be characterized as likely overestimating or likely underestimating the direct risks as a consequence of mathematical model characteristics. It remains to be established how robust predictions of msPAF are when compared to field or mesocosm data, and how uncertainties proliferate through the whole protocol. Despite the profitable mathematical characteristics, it should be noted, however, that the principles of combination toxicology, derived from single species, are directly transposed to the level of multiple species, with the hidden assumption that joint effects transfer unchanged from the species level to the community level. Whether this assumption holds true remains an open question.

5.10 CONCLUSIONS

Mixture extrapolation is needed for many assessment purposes, but techniques to execute mixture extrapolations are not strongly validated by test data that are appropriate to the problem, such as the almost complete lack of data on assemblage-level mixture effects.

Typical problems prohibit the derivation of firm "mixture extrapolation laws" that are sound with regard to both conceptual underpinning and support by test data. Due to the lack of high-quality, appropriate data, mixture extrapolation is mostly founded on a few simple pharmacodynamic concepts of noninteractive joint action, commonly called concentration addition and response addition, and on the statistical features and biases of these models. Those are, however, well known, and they can thus be used for decision making.

This chapter provides some new views on theoretical concepts that are in use in mixture studies, such as the ecological equivalent of the mode of action, which is the ecological mode of action, and the introduction of the mixed-mechanism problems and mixed-mechanism models for cases in which various types of compounds affect the exposed biota. It is argued that the conceptual approaches in mixture studies should further address environmental–chemical interactions, toxicokinetic and toxicodynamic interactions, and ecological interactions.

Because these necessary concepts have not been widely applied in the existing experimental studies, the data that have been collected in the past were reviewed on the basis of existing reviews and the mathematical characteristics of mixture models. From that, it was concluded that the mathematical models that are used in the best-case studies do predict mixture responses relatively well, although the use of some models may not be mechanistically justified, and although the models have peculiar biases that need be taken into account in relation to the objective of the extrapolation.

A tiered system for mixture extrapolation is proposed, in which the main tiers are extrapolation based on substance co-occurrence (Tier-1); point-estimate extrapolation using toxicological summary information such as, for example, EC50 values (TUs, TEFs, CCUs, etc.; Tier-2), and whole-curves–based extrapolation relying on concentration response information (Tier-3), respectively. In the latter case, the approach may be further specified according to recognition of mode-of-action information or on identification of different ecological receptors (Tier-4). These basic tiers can be applied in predicting responses of species and/or assemblages, or in understanding monitoring responses. Targeted research can be designed to investigate the issues where extrapolation techniques are currently weakest (i.e., in community-level mixture assessments or interaction at exposure or effect propagation level).

6 Temporal Extrapolation in Ecological Effect Assessment of Chemicals

Theo C. M. Brock, Keith R. Solomon,
René van Wijngaarden, and Lorraine Maltby

CONTENTS

6.1 INTRODUCTION

Populations and communities integrate the effects of environmental conditions over different temporal and spatial scales. Natural communities are characterized as being spatially heterogeneous and temporally dynamic systems (Connell and Sousa 1983; Sousa 1984), and populations may show a high degree of stochastic variability, particularly when they are small and spatially isolated (May 1974a; Lande 1993). Furthermore, most ecological perturbation experiments are performed on spatial and temporal scales that are much smaller and shorter than the natural systems and time frames of interest (e.g., Englund and Cooper 2003), and the scale at which scientists work may have a large impact on their perspective of ecological and ecotoxicological phenomena. Although, in the extrapolation of ecotoxicological data, it is not always easy or desirable to separate aspects of spatial and temporal variability, this chapter has its focus on temporal extrapolation. The following chapter (Chapter 7) deals with aspects of spatial extrapolation in ecological effect assessments of chemicals, and also builds further on the data presented in this chapter by paying attention to the ecological effect assessment of chemicals at the landscape level.

Measurable properties of populations and communities are usually in flux because of diurnal and seasonal changes in key abiotic factors (e.g., light conditions, temperature, and hydrological conditions) or internal ecological mechanisms (e.g., the increase in biomass and species richness in the course of succession). Consequently, when extrapolating risks of toxicants in time, one should take into account the natural variability of factors that affect chemical fate, exposure, and ecotoxicological responses. When the intensity of chemical stressors is small, their ecological impacts may be hard to detect, especially when the effects fall within the "normal operating range" of the measurement endpoints (Maise 2001). At greater intensities of chemical stressors, however, the structure and functioning of ecosystems may change significantly.

In the effect characterization of contaminants, both lower and higher tier toxicity tests may be performed for comparative, retrospective, or predictive purposes. When one is primarily interested in ranking the toxicity of chemicals, highly standardized tests and/or (Q)SARs are used, and there is less focus on spatiotemporal extrapolation. For retrospective and predictive applications, however, temporal issues in the risk assessment procedure cannot be ignored. In the retrospective risk assessment of persistent toxicants, important research issues may be temporal changes in bioavailable fractions due to aging and breakdown, as well as temporal changes in ecological responses due to recovery and adaptation. Predictive risk assessment procedures, performed for the registration of new chemicals, initially adopt a "realistic worst-case approach" that intends to take into account possible uncertainties related to spatiotemporal variation in exposure concentrations and in species, populations, and habitats potentially at risk. At a more advanced stage (higher tiers) of the risk assessment procedure, the focus is on more environmentally realistic exposure regimes and organisms at risk. An important regulatory question at stake is whether higher tier tests performed in certain periods of the year and with certain exposure regimes can be extrapolated in time. It is, however, neither financially nor practically feasible to test all ecologically relevant exposure scenarios (e.g., pulsed, intermittent, or chronic) on a large number of species and communities during different time periods. Therefore, both extrapolation across time-varying exposure regimes and the temporal extrapolation of ecotoxicological effect data are important issues in ecological risk assessment.

This chapter aims to provide insight into 1) our current knowledge of temporal variability of ecological responses to chemical stress and 2) the tools that can be used to extrapolate across time. The focus of this chapter will be on aquatic organisms and ecosystems, and it is divided into 2 major sections:

- Extrapolation across time-varying exposure regimes: This section will focus primarily on the toxicological and ecological implications of different exposure regimes and the relation between the time frames of exposure and response.
- Temporal variability and ecotoxicological data extrapolation: This section has its focus on temporal variation in population and community structures and related differences in sensitivities, and on the ability of populations and communities to recover from, or adapt to, chemical stress.

6.2 EXTRAPOLATION ACROSS TIME-VARYING EXPOSURE REGIMES

6.2.1 EXPOSURE DURATION AND INCIPIENT TOXICITY LEVELS

The response of organisms to toxic chemicals depends on both exposure concentration and exposure time and duration. Consequently, when using laboratory toxicity values to estimate effects in the field, it is important to consider the relationship between exposure duration and the time needed to express the toxic effect (Figure 6.1). For example, there was a highly significant negative relationship between exposure duration and effect concentrations for aquatic amphipods exposed to the pyrethroids lambda-cyhalothrin and cypermethrin: 1-hour exposures were 18 times less toxic than 96-hour exposures (Maund et al. 1998). Effect measures such as LC50s normally decrease with time until an incipient level is reached, at which point toxicity stabilizes. Standard toxicity testing guidelines (e.g., those provided by the OECD) prescribe (minimum) test durations, usually 48 to 96 hours for acute toxicity

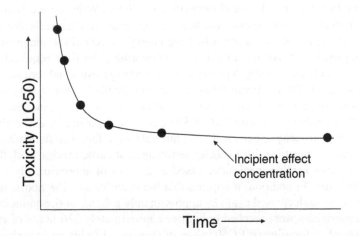

FIGURE 6.1 Illustration of change in toxicity over time until incipient effect (lethality in this case) is reached.

TABLE 6.1
Relation between sublethal (EC50) or lethal (LC50) toxicity values (µg/L) and exposure duration (in hours) for the microcrustacean *Daphnia magna,* the macrocrustacean *Gammarus pulex,* and the flatworm *Dugesia lugubris* exposed to the fungicide carbendazim

Time	*Daphnia magna* EC50	LC50	*Gammarus pulex* EC50	LC50	*Dugesia lugubris* EC50	LC50
48 hours	192	320	167	1041	178	876
96 hours	87	91	55	177	25	134
168 hours	61	61	47	53	12	22
336 hours	55	—	26	34	13	12
504 to 600 hours	44	—	16	16	12	14

Source: After van Wijngaarden et al. (1998).

tests with aquatic organisms. Reaching incipient levels for lethal effects, however, may require longer than 48 hours, depending on the mode of action of the chemical. For example, in toxicity tests exposing aquatic arthropods to the insecticide diazinon, it was observed that lethality increased, on average, by a factor of 4.4 between 48 and 96 hours (Stuijfzand et al. 2000). This increase in toxicity with exposure time may be smaller when focusing on more sensitive sublethal measurement endpoints (e.g., immobility). For most organophosphorus and pyrethroid insecticides, there is little difference in EC50 values for aquatic arthropods exposed for 48 or 96 hours; EC50s decreased by a factor of 1.4 for lambda-cyhalothrin ($n = 10$; Schroer et al. 2004a) and by a factor of 1.6 for chlorpyrifos ($n = 7$; van Wijngaarden et al. 1993).

Temporal effects may be more marked for other types of chemicals. For instance, a 96-hour exposure period may not be long enough to reveal the incipient toxicity of the fungicide carbendazim to aquatic invertebrates, the time required to reach incipient concentrations being dependent on the test organisms and the measurement endpoint selected (Table 6.1; van Wijngaarden et al. 1998). In Table 6.1 it can be seen that, for the crustaceans *Daphnia magna* and *Gammarus pulex*, incipient toxicity concentrations were not attained after 336 hours of exposure to carbendazim. In these long-term toxicity experiments, the animals were fed. The flatworm *Dugesia lugubris* was the most sensitive species tested against carbendazim; both the EC50 and LC50 values for this species decreased by a factor of approximately 7 between 48 and 96 hours. In addition, it appears that between 96 and 168 hours, the LC50 value for *Dugesia* decreased again by approximately a factor of 6, and incipient toxicity concentrations were reached only after approximately 336 hours of exposure. In contrast, when focusing on EC50 values of *Dugesia*, differences in toxicity values between 96 and 168 hours differed only by a factor of 2, and incipient concentrations were reached after 168 hours of exposure (van Wijngaarden et al. 1998). It can be concluded that, for certain receptor-specific modes of action, we may need longer

testing periods than 48 to 96 hours. The problem, however, often is lack of elucida-
tion of the toxic mode of action of the chemicals of concern.

The data presented above are for temperate species and show that the time needed
to reach incipient toxicity is dependent on the type of chemical and the species and
endpoint tested. However, the time needed to reach incipient toxicity values may be
different for species from different geographical regions subject to different climate
conditions (Chapman and Riddle 2003). In colder polar environments, metabolic
rates may be slower than in temperate regions, resulting in longer response times,
whereas in warmer climates, metabolic rates of organisms may be higher, resulting
in shorter response times. For Antarctic species exposed to zinc, it was observed that
an increase in exposure time from 4 to 14 days resulted in an increase in toxicity by
a factor of 3 to 22 (Ling et al. 1998). According to a review by Mayer and Ellersieck
(1986), temperature and toxicity are positively correlated for most chemicals. Elevated temperatures may increase respiration in the organisms and increase uptake
across membranes. At the same time, elevated temperatures may reduce exposure by
increasing the metabolism of toxicants, resulting in more rapid degradation within
the organism. In a study of the effects of temperature (test range 10°C to 26°C) on
cadmium toxicity to *Daphnia magna*, it was demonstrated that cadmium concentrations in tissue showed a greater rate of increase at higher temperatures and that the
time needed to express the effects of a certain exposure concentration decreased
(Heugens et al. 2003). The mathematical model DEBtox (Bedaux and Kooijman
1994) provided an adequate fit to the cadmium concentrations in the daphnids. In
addition, the survival of the daphnids was described well by the estimated tissue
concentrations. DEBtox is able to describe time-dependent toxicity data, which contain information about the dynamic aspects of the occurrence of effects.

Adequate extrapolation of results from standard laboratory toxicity tests to other
time scales of exposure and response requires observations on the time course of
toxic effects. These observations can then be used to construct time-to-event models,
such as the DEBtox model mentioned above. These models explicitly address both
intensity and duration of exposure to hazardous chemicals, and better use is made
of the data gathered from toxicity experiments. Diverse endpoints in time can be
addressed, and individual organism characteristics and/or environmental circumstances (e.g., temperature) can be incorporated as covariables. An overview of time-
to-event models and approaches and their use in the risk assessment of chemicals is
provided by Crane et al. (2002).

6.2.2 Reversibility of Responses and Time-Variable Exposure

Some responses, such as mortality, are irreversible. However, many sublethal
responses may be reversible, such as the impact of the photosynthesis-inhibiting herbicide linuron on macrophytes (Snel et al. 1998). Linuron inhibits photosynthesis
by disturbing electron transport in photosystem II. Table 6.2 presents the kinetics
of photosynthesis inhibition when shoots of macrophytes are placed in water with
50 μg/L linuron, and subsequent recovery when placed in uncontaminated water. The
EC50 values are remarkably similar between macrophytes, and half-life estimates
for inhibition and recovery are less than 2 hours (Table 6.2). Except for *Potamogeton*

TABLE 6.2
EC50 values (and 95% confidence limits) for the effects of linuron
on the efficiency of photosystem II in aquatic macrophytes

Plant species	EC50 in µg/L	Inhibition $t_{1/2}$ in hours	Recovery $t_{1/2}$ in hours
Elodea nuttallii	13.3 (10.7 to 16.8)	0.41	1.5
Myriophyllum spicatum	11.8 (6.9 to 20.5)	0.23	1.8
Potamogeton crispus	12.9 (9.6 to 17.4)	1.90	0.57
Ranunculus circinatus	13.2 (11.0 to 16.0)	0.16	1.00
Chara chlobularis	12.1 (7.9 to 18.5)	0.10	0.45

Note: Kinetics of the inhibition of photosystem II electron flow in these macrophytes by 50 µg/L linuron
and the subsequent recovery of inhibition by washing with uncontaminated well water are expressed as
half-life times ($t_{1/2}$).
Source: Adapted from Snel et al. (1998).

crispus, inhibition is more rapid than recovery; however, differences between species are relatively small. Not surprisingly, Chara chlobularis, with its small distance between chloroplasts and external medium, shows the most rapid kinetics. These data are in accordance with observations on other photosynthesis-inhibiting herbicides. For instance, recovery of algae from exposure to 50 µg/L of atrazine was nearly instantaneous, once the herbicide was removed from the overlying water (Klaine et al. 1997), and reversibility of the effect of metribuzin on the photosynthesis of periphyton communities could be demonstrated by placing periphyton in clean water (Gustavson et al. 2003).

Reversibility of effects in exposed individuals has also been reported for animals. For example, Little et al. (1993) exposed bluegill (Lepomis macrochirus) to the pyrethroid esfenvalerate and assessed aggressive interactions among these fish after 30, 60, and 90 days of continuous exposure to the insecticide, and after 21 days of recovery. There was a significant reduction in aggressive behavior in all treatment groups (0, 0.01, 0.025, 0.05, 0.1, and 0.2 µg/L), but after 21 days, aggressive behavior had only recovered in the 0.01 µg/L treatment group; hence, the reversibility of the effect on individuals was concentration-dependent. Pulsed exposure to esfenvalerate resulted in a significant decrease in the aggressive behavior of fish exposed to 0.1 µg/L or greater, but, in contrast to continuous exposure treatments, aggression was similar among all treatment groups 21 days after the final pulsed exposures. Therefore, reversibility of the effect was dependent on concentration, exposure type, and duration of exposure. These data demonstrate that the issue of reversibility of toxic effects cannot be ignored when evaluating the risks of pulsed and multiple exposures.

For some compounds, no difference in toxicity has been observed between time-varying or repeated and continuous exposures. This situation can occur when responses are related to peak exposure rather than time-weighted exposures, when exposures are cumulative (compounds with slow depuration rates), or when effects are only slowly reversible (organism "memory"; van der Hoeven and Gerritsen 1997).

For example, no difference in the toxicity of organophosphorus compounds to larvae of the midge *Chironomus riparius* was observed between pulsed and continuous exposures (Kallander et al. 1997). In the same study, however, two 1-hour pulses caused significantly fewer symptoms of intoxication than 2 hours of continuous exposure to carbamate compounds, when animals were placed in clean water for at least 2 to 6 hours between treatments (Kallander et al. 1997), suggesting that detoxification or elimination of the toxicant during the toxicant-free period can reduce the toxic effects of the earlier exposures.

Most laboratory single-species tests are conducted using standard times (48 or 96 hours) that may not be similar to realistic environmental exposures of shorter duration, for example, when environmental dissipation and/or degradation rates are relatively great, or when concentrations quickly decline due to dilution (e.g., pulsed exposure regimes in rivers). Several model ecosystem experiments, particularly in artificial streams, have demonstrated that the severity of ecological responses and/or threshold concentrations is lower for shorter pulse-exposure durations. Hose et al. (2002) studied the effects of 12- and 48-hour exposures to aqueous endosulfan in artificial streams. NOECs for this organochlorine insecticide with the most sensitive endpoint of the macroinvertebrate assemblages were 8.7 and 1.0 µg/L for the 12- and 48-hour exposure regimes, respectively. In other Australian studies, Pusey et al. (1994) observed that a 6-hour pulse to 0.1 µg/L chlorpyrifos did not result in effects on arthropod populations in experimental streams, whereas Ward et al. (1995) demonstrated pronounced effects on the macroinvertebrate assemblage in these test systems as a result of continuous exposure to the same concentration for 3 weeks. Similar relationships between exposure duration and ecological responses in experimental ecosystems are reported for herbicides. Adverse effects on biomass and cell densities of periphytic algae could not be observed in streams exposed twice for 24 hours (interval 14 days) to 100 µg/L atrazine (Jurgensen and Hoagland 1990). In contrast, in experimental streams, a 7-day exposure to 100 µg/L of this photosynthesis-inhibiting herbicide resulted in significant effects on periphyton (Moorhead and Kosinski 1986). In addition, a 21-day exposure to 10 µg/L also resulted in significant treatment-related effects on periphyton (Kosinski and Merkle 1984; Kosinski 1984). These observations were confirmed by model ecosystem experiments performed with another photosynthesis inhibitor. In artificial streams, a 12-hour pulse exposure to 2700 µg/L hexazinone elicited little effect on the structure and functioning of the aquatic community (Kreutzweiser et al. 1995), whereas in lake enclosures a single peak load of 100 µg/L (without renewal of water) caused pronounced effects on oxygen metabolism and densities of plankton (Thompson et al. 1993a, 1993b). The mechanism of action of these herbicides is inhibition of photosynthesis, which can be reversed (Jensen et al. 1977). This reversal explains the minimal impact that short-term pulse exposures to atrazine and hexazinone have on the structure of periphyton communities (see above).

Effects on organisms may be similar when they are exposed for a short time to a greater concentration or for a longer time to a smaller concentration, a phenomenon referred to as "reciprocity" (Giesy and Graney 1989). For example, a 4-day exposure at 4 µg/L may cause the same effects as a 1-day exposure at 16 µg/L or a 2-day exposure at 8 µg/L, an example of linear reciprocity. Similarly, Heming et al. (1988)

showed that short (2-hour) pulsed exposure of early life stages of rainbow trout to methoxychlor (580 μg/L) produced similar effects to static exposures to 30 μg/L. According to Giesy et al. (1999), reciprocity relationships may be used to estimate responses to shorter environmental exposures where they are less than those used in "acute" bioassays in the laboratory.

However, several experimental studies revealed that reciprocity did not adequately describe effects in other types of time-varying exposure regimes (e.g., Bailey et al. 1985; Schulz and Liess 2000; Naddy and Klaine 2001). Apparently, reciprocity only holds when dealing with an exposure that causes a certain degree of damage in the organism that is not, or is only to a small degree, repaired in the interval between repeated exposures. If the damage is repaired relatively quickly and the effect is reversible, the effect is predominantly caused by the highest concentration during the exposure period. If, on the other hand, elimination of the compound from the organism occurs relatively slowly, and/or irreversible binding of the compound to receptors occurs, then the effect is probably best described by some other, more integral measure of exposure, and not by the maximum concentration during the exposure period.

Intermittent and multiple exposures may allow not only for interexposure detoxification and/or excretion, but also for acclimation, behavioral adaptation, selection of more susceptible individuals, or population recovery (for organisms with short generation times). Population recovery and adaptation are discussed in detail in Sections 6.3.3 and 6.3.4. An example of acclimation is the induction of biotransformation enzymes such as cytochrome P-450-dependent monooxygenases and mixed-function oxygenases in fish (Sipes and Gandolfi 1986; Parrott et al. 1995; Burnison et al. 1996). These enzymes better allow exposed organisms to physiologically cope with the toxicants. An example of behavioral adaptation, avoidance of exposure to pulses of toxicants, is provided by Rajagopal et al. (2003). They studied the impact of continuous and intermittent chlorination (pulses of 4 hours of exposure followed by 4 hours of no exposure) on 3 mussel species (*Dreissena polymorpha*, *Mytilus edulis*, and *Mytilopsis leucophaeata*). Mussels shut their valves as soon as chlorine was applied and opened them only after chlorine exposure was stopped. Under conditions of continuous chlorination, mussels kept their shell valves shut and experienced 100% mortality after 3 to 7 weeks (dependent on species) of continuous chlorination at 1 mg/L. However, the 3 mussel species subjected to intermittent chlorination at 1 mg/L showed little or no mortality during the same periods.

6.2.3 MECHANISTIC MODELS TO PREDICT EFFECTS OF TIME-VARIABLE EXPOSURE

Mathematical models have been developed and used to extrapolate toxicity under pulsed exposure conditions (for an overview, see Boxall et al. 2002; Reinert et al. 2002; Ashauer et al. 2006; Jager et al. 2006). Some models consider concentration × time (Meyer et al. 1995); others, uptake and depuration (Mancini 1983) or damage and repair (Breck 1988). Several models are based on the concept of critical body residues, which integrates toxicokinetics and the effects of exposure time on toxicity (McCarty and Mackay 1993; Barron et al. 2002). This approach is promising because several studies showed that toxicity from pulse exposures is largely controlled by accumulation and elimination rates of toxicants by exposed organisms (Hickie et al. 1995). However, the main limitation of the use of the critical body residue approach in risk assessment is its dependence on age, sex, lipid content, and other variables of an individual within a species (e.g., Deneer et al. 1999).

In the mechanistic models used to predict toxic effects of time-variable exposure to organisms, a distinction can be made between 1-step models and 2-step models (Ashauer et al. 2006). One-step models only consider toxicokinetics, whereas 2-step models consider both toxicokinetics and toxicodynamics. One-step models try to describe the uptake and elimination of a given compound in an organism and relate the calculated internal concentration to the effect occurring. Usually, an average total body residue is calculated, assuming that the concentration at the actual site(s) of action will be linearly related to the total body concentration. In specific cases, it may be necessary to calculate the concentration at the site of action through the use of more refined multicompartment (PBPK) models.

Two-step models include, besides the toxicokinetic terms "uptake" and "elimination," terms addressing toxicodynamics such as "injury" and "repair." Ashauer et al. (2006) concluded that 2 approaches of such 2-step models are suitable to model the effects of time-variable exposure to aquatic organisms: one approach originates from the damage assessment model (Lee et al. 2002), and the other from the DEBtox concept (Bedaux and Kooijman 1994). Ashauer et al. (2007) combined these approaches to form the threshold damage model. The 2-step models mentioned above will be difficult to operationalize, calibrate, and validate without extensive research, which is partly due to the fact that species- and compound-specific parameterizations are necessary. The feasibility of extrapolation to other species and/or compounds has not yet been demonstrated. Nevertheless, as stated by Ashauer et al. (2007), parameterizing these models for different species could facilitate a better understanding of the causes for the distribution of species sensitivities toward toxicants, hence leading to new approaches for interspecies extrapolation of toxicity.

Where effects are known to be dependent on pulsed exposures, and the temporal nature of the exposures is modeled or measured, the exposures can be characterized using a tool such as the Risk Assessment Tool to Evaluate Duration and Recovery (RADAR), developed as part of the efforts of ECOFRAM (ECOFRAM 1999; Reinert et al. 2002). This tool provides information on pulse magnitude, duration, and inter-pulse interval, which is particularly useful for assessing likely effects on classes of organisms with known recovery times and time-exposure responses.

6.2.4 ACUTE-TO-CHRONIC RESPONSE

The extrapolation from acute responses to no-observed-effect concentrations or chronic responses is particularly important as chronic tests are more costly and time-consuming than acute tests. Traditionally, relationships between acute and chronic effects were estimated using a simple ratio, the acute-to-chronic ratio (ACR). Where acute and chronic effect measures are available for the same species, this ratio is used to estimate chronic responses in related organisms for which only acute data are available (Stephan and Rogers 1985). This approach is based on the assumption that there is a relationship between the responses in acute and chronic tests, an

TABLE 6.3
Acute–chronic ratios (acute L[E]C50 and/or chronic NOEC) from the ECETOC aquatic toxicity (EAT) database for all aquatic species

Group	Minimum	50th Centile	90th Centile	Maximum
Metals and organometals	0.3	28	192	1290
Other inorganics	2.9	8.4	20.1	69.3
Pesticides	1.3	12	83.7	371
Other organics	0.13	3.9	15.9	27.5

Source: From Länge et al. (1998).

assumption that is not necessarily true. For example, in an acute test the response observed may be lethality, whereas in the chronic test it is growth or reproduction.

ACRs (= acute L(E)C50 and/or chronic NOEC) were calculated for all aquatic species in the ECETOC database and presented according to specific substance classes (Länge et al. 1998). For 8 of the 28 species in the analysis, it was possible to calculate ACRs for 5 or more chemicals. The minimum, median, 90th centile, and maximum values for the acute–chronic ratios were determined for several classes of those chemicals (Table 6.3). Median and 90th centile ACRs were highest for metals and organometals, followed, in decreasing order, by pesticides, other inorganic substances, and other organic substances.

Although it has been concluded that a significant relationship exists between acute and chronic effects in some organisms and for some types of data, a high uncertainty factor may be necessary if predictions with a very high protection are required. For example, for pesticides, uncertainty factors between 100 and 800 may be necessary for aquatic invertebrates and 95th centile predictions (Elmegaard and Jagers op Akkerhuis 2000). The data for pesticides presented in Table 6.3, however, suggest that an uncertainty factor up to 100 may suffice for 90th centile predictions, and around 10 for 50th centile predictions.

Methods for acute-to-chronic extrapolations have been developed and are available as computer programs such as the acute-to-chronic estimation (ACE; Mayer et al. 1994; Ellersieck et al. 2003) software, which makes use of 3 methods — regression, multifactor probit analysis, and accelerated life testing — to consider the relationship between exposure concentration, degree of response, and time course of response (Mayer et al. 1994; Sun et al. 1995; Lee et al. 1995). All methods produce confidence intervals around the LC and/or EC percentage point estimate.

An analysis of regularities observed in species sensitivity distributions (SSD) fitted on acute and chronic aquatic toxicity data for a large number of organic and inorganic toxicants is provided by De Zwart (2002). The log-logistic sensitivity model he used is characterized by the parameter α, which is the mean of the observed log10-transformed L(E)C50 or NOEC values over a variety of test species, and β, a scale parameter proportional to the standard deviation of the log10-transformed

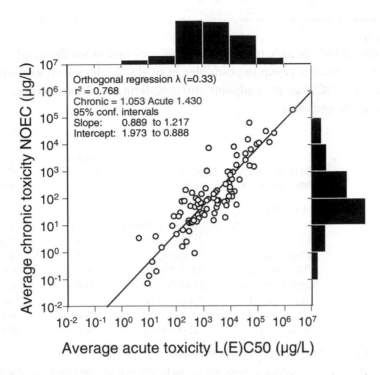

FIGURE 6.2 Comparison of average acute and chronic toxicities for aquatic species *Source*: Drawn from data from De Zwart (2002).

toxicity values. A regression of acute and chronic β values for a large number of chemicals reveals that the average acute toxicity is approximately a factor of 10 higher in concentration than the average chronic toxicity (Figure 6.2). Provided that enough data for individual species are available, regression analysis on the acute and chronic β values indicates that the factorial difference between the sensitivities of the most and least sensitive species tested is about equal for both acute and chronic tests. De Zwart (2002) suggested that the observed regularities may be used to assign surrogate SSD parameters in situations where appropriate sets of acute toxicity data are available but corresponding appropriate chronic data sets are lacking.

Maltby et al. (2002) and Van den Brink et al. (2006a) compared SSDs based on acute and chronic laboratory toxicity data for aquatic test species exposed to pesticides. The SSDs were constructed with toxicity data for the most sensitive taxonomic group, because of the specific toxic mode of action of the pesticides selected. The SSDs were used to calculate the hazardous concentration to 5% of the species (HC5) by means of a log-normal distribution model, and comparisons were performed for 2 insecticides and 7 herbicides (Table 6.4). The log-normal model did not fit the diuron (herbicide) short-term L(E)C50 data or the atrazine (herbicide) long-term NOEC data. Consequently, the L(E)C50 HC5 value for diuron and the NOEC HC5 value for atrazine should be interpreted with caution, as well as their acute HC5–chronic

TABLE 6.4
Estimates of HC5 in µg/L (with 95% confidence limits) for the most sensitive taxonomic group to pesticides in single-species short-term toxicity tests using L(E)C50 as the endpoint and long-term toxicity tests using NOEC as the endpoint

Pesticide	NOEC HC5 (95% confidence limits)	L(E)C50 HC5 (95% confidence limits)	L(E)L50: NOEC HC5 Median	L(E) C50: NOEC HC5 lower 95% confidence limit
Azinphos-methyl	0.038 (0.005 to 0.121)	0.044 (0.009 to 0.142)	1.2:1	1.8:1
Diazinon	0.094 (0.019 to 0.217)	0.286 (0.090 to 0.674)	3.0:1	4.7:1
2,4-D	5.1 (0.57 to 16)	71 (7.1 to 199)	13.9:1	12.5:1
Atrazine	3.0[a] (1.3 to 5.3)	13 (5.8 to 24)	4.3:1	4.5:1
Diuron	0.34 (0.04 to 1.0)	12[a] (7.6 to 16)	35.3:1	190:1
Linuron	0.50 (0.09 to 1.4)	5.8 (0.74 to 17)	11.6:1	8.2:1
Metribuzin	1.4 (0.20 to 3.3)	7.4 (4.0 to 11)	5.3:1	20:1
Pendimethalin	0.51 (0.03 to 1.8)	2.0 (0.20 to 5.1)	3.9:1	6.7:1
Simazine	6.4 (1.7 to 13)	52 (18 to 92)	8.1:1	10.6:1

[a] The log-normal model did not appropriately fit the data.

Note: The most sensitive taxonomic group was Arthropoda for insecticides (azinphos-methyl and diazinon), macrophytes for the herbicide 2,4-D, and algae and macrophytes for the other herbicides (atrazine and simazine).

Source: Data for insecticides according to Maltby et al. (2002) and those for herbicides after Brock et al. (2004) and van den Brink et al. (2006a).

HC5 ratios. Of the 9 pesticides investigated, 6 compounds showed a ratio of less than 10 between the median L(E)C50 HC5 and the median NOEC HC5 values. For the herbicides linuron and 2,4-D, this ratio was slightly greater than 10 (i.e., 11.6 and 13.9, respectively), and for the herbicide diuron this ratio was substantially greater (i.e., 36.3). The acute HC5–chronic HC5 ratios based on lower limits were somewhat lower to 5 times greater than those based on median values. The 2 selected insecticides showed the lowest ratios. The data presented in Table 6.4 suggest that, for all pesticides except diuron, a chronic HC5 may be estimated from an acute HC5 by applying an uncertainty factor of 10 to 15, whereas, for all compounds except 1, an uncertainty factor of 20 would suffice to extrapolate lower centile acute HC5 values to the lower centile chronic HC5 values. A pragmatic approach for a specific substance might be to adopt an uncertainty factor equivalent to the acute-to-chronic ratio observed in guideline tests with sensitive standard test species. However, the data presented in Table 6.4 are limited to 2 acetylcholinesterase-inhibiting insecticides (azinphos-methyl and diazinon), 1 auxin-simulating herbicide (2,4-D), 5 photosynthesis-inhibiting herbicides (atrazine, diuron, linuron, metribuzin, and simazine), and 1 herbicide that inhibits cell division and shoot elongation (pendimethalin).

More chronic toxicity data on compounds with other toxic modes of action are required before a more robust generalization can be offered.

A few aquatic micro- and/or mesocosm experiments allow for a comparison between ecological threshold concentrations for short-term and long-term exposures to pesticides. In experimental ditches, a short-term pulsed exposure to the photosynthesis-inhibiting herbicide linuron resulted in an NOEC of 5 μg/L based on the most sensitive structural endpoint (abundance of algae; Kersting and van Wijngaarden 1999; van Geest et al. 1999). The threshold concentration for long-term exposure was a factor of 10 lower; in indoor macrophyte-dominated microcosms, which were treated chronically for 4 weeks with linuron, the NOEC of the most sensitive structural endpoint was 0.5 μg/L (abundance of algae and biomass of macrophytes; van den Brink et al. 1997; Cuppen et al. 1997).

The same test systems were used to study the impact of the acetylcholinesterase-inhibiting insecticide chlorpyrifos. In the experimental ditches, a single application of chlorpyrifos resulted in a short-term exposure regime and an overall NOEC of 0.1 μg/L based on the most sensitive structural endpoint (abundance of arthropods; van den Brink et al. 1996; van Wijngaarden et al. 1996). In indoor microcosms, which were treated chronically for 4 weeks with an insecticide mixture of chlorpyrifos and lindane, slight and transient effects (class 2 sensu; Brock et al. 2000a) on 1 crustacean species were observed at treatment concentrations of 0.01 μg/L chlorpyrifos in combination with 0.3 μg/L lindane (van den Brink et al. 2002b; Cuppen et al. 2002). These data suggest that, at the ecosystem level, and when considering responses of the most sensitive endpoints to chlorpyrifos, the ACR of this compound will be around 10.

A single application of the acyl-urea insecticide diflubenzuron was studied in indoor microcosms and outdoor enclosures (Moffet et al. 1995). The acute exposure regime to diflubenzuron resulted in an NOEC of 0.3 μg/L and slight transient effects (Effect class 2, according to Brock et al. 2000a) on some arthropod populations at 0.7 μg/L (Moffet et al. 1995). In artificial streams, Hansen and Garton (1982) studied the ecological impact of chronic exposure (26 weeks) to diflubenzuron. They observed an NOEC of 0.1 μg/L based on the most sensitive endpoint of the invertebrate community and populations. Again, the short-term and long-term exposure studies with diflubenzuron reveal a ratio of less than 10 between the acute $NOEC_{ecosystem}$ and the chronic $NOEC_{ecosystem}$ in experimental aquatic ecosystems. These observations are in line with the ratios between median acute HC5 and median chronic HC5 values for pesticides presented above (Table 6.4).

In the scientific literature, relatively few examples of model ecosystem experiments can be found that compare the ecological responses between an acute and a chronic exposure regime for the same chemical. A larger database may be obtained when combining the results of experiments performed with compounds that have similar toxic modes of action and by expressing the effect concentrations in toxic units (1TU = L[E]C50 of the most sensitive standard test species). For example, this was done for insecticides by Brock et al. (2000b) and van Wijngaarden et al. (2005b) (see Figure 6.3). The probability of insecticide effects occurring in microcosm and mesocosm studies was calculated by analyzing the combined data set of all available insecticide studies using logistic regression; by making a distinction

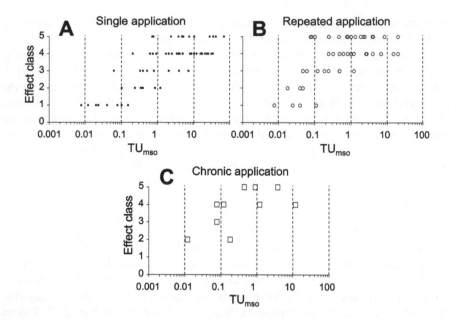

FIGURE 6.3 Responses of the most sensitive endpoints in aquatic microcosm and meso-cosm studies treated once (A), repeatedly (B), or chronically (C) with an acetylcholines-terase-inhibiting or pyrethroid insecticide. *Note*: The effects are expressed in toxic units (1 TU = L[E]C50) of the most sensitive standard test species (TU_{mso}) and classified (Effect class) according to magnitude and duration. 1 = no significant effect, 2 = slight effect, 3 = clear short-term effect (< 8 weeks), 4 = clear effect in short-term study (recovery moment unknown), and 5 = clear long-term effect (> 8 weeks). *Source*: Redrawn from data after Brock et al. (2000b) and van Wijngaarden et al. (2005b).

between single-application exposure, repeated-application exposure, and continuous chronic exposure; and by using the effects classes presented in Table 1.3 (Chapter 1) to describe responses. Table 6.5 presents the 5th, 50th, and 95th centile values for micro- and/or mesocosm effect concentrations (MEC; expressed in toxic units) with 95% confidence intervals when comparing no or transient effects (classes 1 and 2) versus clear effects (classes 3, 4, and 5) for the 3 exposure regimes. The confidence intervals of the 50th centile MEC values are relatively small when compared to the 5th and 95th centile values. Overall, the difference in MEC values between the single- and repeated-application exposure regimes was larger than between the repeated and chronic exposure regimes. It should be noted that the category "repeated applications" is rather heterogeneous because the number of applications between studies varied considerably. When comparing the median 5th, 50th, and 95th centile MEC values of no or transient effects versus those of clear effects (Table 6.5) between single-application exposure and chronic exposure, it appears that uncertainty factors of 12, 6, and 3, respectively, are needed to extrapolate threshold concentrations for acute insecticide exposure to those of a chronic exposure regime.

The insecticides used in the studies contributing to Figure 6.3 and Table 6.5 comprised mainly nonpersistent compounds. In microcosm–mesocosm experiments

TABLE 6.5
Microcosm and/or mesocosm effect concentrations (MEC) as calculated by means of logistic regression

		Estimate	95% confidence limits
No and slight effects versus clear effects			
Single	MEC5%	0.036	(0.007 to 0.198)
	MEC50%	0.261	(0.126 to 0.541)
	MEC95%	1.862	(0.502 to 6.914)
Multiple	MEC5%	0.023	(0.007 to 0.070)
	MEC50%	0.052	(0.032 to 0.085)
	MEC95%	0.119	(0.050 to 0.284)
Chronic	MEC5%	0.003	(0.000 to 4.868)
	MEC50%	0.043	(0.003 to 0.665)
	MEC95%	0.544	(0.010 to 29.01)

Note: MECs with 95% confidence limits are expressed in toxic units (1 TU = L[E]C50 of the most sensitive standard test species). MECs were expressed as 5, 50, and 95 percentages of probability of effects occurring on the most sensitive endpoints for acetylcholinesterase-inhibiting and pyrethroid insecticides (Figure 6.3). MECs were calculated for the scenario where no and slight effects are placed against clear effects (classes 1 and 2 versus classes 3, 4, and 5). Results were based on responses found in studies using single, multiple, and chronic insecticide applications.

Source: Adapted from van Wijngaarden et al. (2005b).

performed with more persistent water-soluble pesticides like atrazine (for an overview, see Brock et al. 2000b), the ecological threshold concentrations for short-term and long-term exposure regimes are more difficult to compare because most experiments were performed in lentic or recirculating lotic test systems, in which a single application already resulted in chronic exposure.

6.2.5 LATENCY OF RESPONSES

Delayed responses or latency must also be considered when extrapolating effects. In order to demonstrate latency (or lack thereof) in acute studies, observations must continue after the exposure is completed and the organism has been removed from the stressor (Figure 6.4). This is not commonly done in routine acute toxicity testing, and special tests may have to be designed to observe this response. In studies characterized by long-term exposure regimes, latency can be determined by observing the responses of organisms during their whole life cycle. It may even be required to make observations on the responses of the offspring.

A latent response in an acute test may occur because the action of the toxic substance causes its effects through a chain of irreversible reactions that take some time to occur. For example, in experiments in which grass shrimps (*Palaemonetes pugio*) were exposed for 6 hours to the insecticides azinphos-methyl and endosulfan,

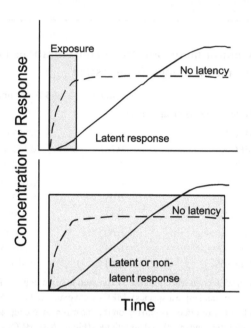

FIGURE 6.4 Illustration of latent (delayed) and nonlatent toxicity responses for short-term (top) and long-term (bottom) exposures.

the highest mortality was observed in the postexposure period when the test animals were placed in a pesticide-free medium (Moore et al. 1990).

The presence of a sensitive life stage or period may also be one of the causes of latency. For example, the effects of diflubenzuron (an insect growth regulator) on aquatic macroinvertebrates were not observed until molting began, some 2 to 4 weeks after a single exposure (Hurd et al. 1996). When larvae of the trichopteran *Limnephilus lunatus* were exposed for 1 hour to 0.1 µg/L fenvalerate, and subsequently placed in contaminant-free artificial streams, there were few mortalities, but a delay in development and emergence was observed (Liess and Schulz 1996). In aquatic ecosystems that periodically dry out in summer, such a delay in development time may have severe consequences for population survival (Liess 1998).

In chronic studies, latency may result from delays in the chain of events between exposure and expression of effects. In some cases this may be the result of an effect on a developmental process that only becomes apparent later in the life cycle of the organism, or even in its offspring. For example, selenium is an essential metal that is highly toxic at high concentrations, particularly to the early life stages of certain species of fish, ducks, and wading birds (Lemly 2002). Selenium is efficiently transferred from the female's diet to the eggs, where it can result in teratogenic deformities and death, effects that only become apparent upon hatching. Adults can survive and appear healthy despite the fact that massive reproductive failure is occurring (Lemly 2002). There is also evidence that the fitness of *Daphnia magna* neonates may be affected by certain chemicals. For example, a dispersant (naphthalene-sulfonic acid with formaldehyde) showed clear effects on the F1 and F2 offspring of *Daphnia*

(Hammers-Wirtz and Ratte 2000). The dispersant induced an increase in offspring number of approximately 50% that was coupled with a decrease in offspring quality (smaller size, higher mortality, and fewer F2 offspring; Hammers-Wirtz and Ratte 2000). In an experiment with the estrogen mimic diethylstilbestrol, reduced fitness in the second generation of daphnids was also observed (Baldwin et al. 1995). Studying the fitness of the F1 and F2 generations is not taken into account in the conventional *Daphnia* reproduction test. To evaluate the fitness of the F1 generation, either an additional test (starting with F1 neonates) is necessary, or a multigeneration population growth experiment must be performed (Hammers-Wirtz and Ratte 2000).

Latency of responses to toxicants may also become apparent in conjunction with other stressors (e.g., food shortage and temperature shifts) that cause increased metabolic demands resulting in increased use of stored body lipids and thus increased bioavailability of stored lipophilic toxicants. A well-known example is dieldrin toxicity in eider ducks triggered by the use of fat stores (Walker et al. 2001). Similar effects have been reported for toxic metals and fish. For example, selenium toxicity may be induced in certain species of warm-water fish (*Centrarchidae*) when lower water temperatures and shorter photoperiods in winter reduce feeding activities and consequently result in metabolic distress, a phenomenon referred to as the "Winter Stress Syndrome" (Lemly 2002). It is also reported that short-term exposure to chemical and natural stress conditions during and after the exposure may affect long-term population survival. For example, exposure to environmentally relevant concentrations of endosulfan for 96 hours (0.8 µg/L) resulted in significant effects within a cohort of *Litoria citropa* tadpoles. Short, pulsed exposure to this sublethal concentration in conjunction with fluctuating temperatures had long-term impacts on fitness. In addition, both variable temperatures and endosulfan increased a tadpole's subsequent vulnerability to predatory odonates when tested more than 3 weeks later. In other words, a short-term and pulsed exposure to a sublethal endosulfan concentration in natural conditions may have the potential to manifest fitness effects after cessation of the contaminant exposure due to multistress conditions (Broomhall 2002).

Knowledge about the mechanism of action of a substance, and of specific sensitive periods in the life cycle of organisms, may be all that is required to determine if the substance has latent effects or not. For substances where the mechanism of action is well known, the question of latency in the target organism can usually be deduced. However, in organisms that lack the receptor system, another mechanism may be responsible for toxicity, and this may or may not show latency. The only models that can be used to address extrapolation of latent toxicity at this time are physical models such as specially designed bioassays and tests. Latency should be addressed if, through analogy to similar substances or knowledge of mechanisms of action, it is suspected to occur. In cases where latency is known not to occur in compounds with a similar toxic mode of action, it might be disregarded. Uncertainty factors that can be used for these types of extrapolations have not yet been developed.

At an ecosystem level, complex interactions within and between populations and nonlinear biological dynamics may create a latency period between the exposure event and certain effects (Landis et al. 1996; Matthews et al. 1996), particularly when considering the indirect effects of chemical stress. Populations of organisms may be affected by toxicants in an indirect way when a reduction or elimination of toxicant-susceptible species results in a disturbance of biological interactions and processes. Ecosystem changes that follow and result from direct toxic effects are termed "indirect effects," or "secondary effects" (Hurlbert 1975). The taxa in which indirect effects are observed in different (model) ecosystem experiments subject to the same stressor may differ considerably, as well as the time frames in which these indirect responses can be recorded. For example, in 2 different types of macrophyte-dominated microcosms (with and without weekly nutrient additions) that were treated once with 35 µg/L chlorpyrifos, a fast decline (within a few days) in arthropod populations (e.g., daphnids and insects) was observed due to direct toxic effects. In both types of microcosms this decline was rapid (within 1 to 2 weeks), followed by an increase in periphytic algae due to the release from grazing by arthropods. In the test systems without weekly nutrient additions, an algal bloom, dominated by blue-greens, developed that persisted for approximately 8 to 10 weeks, followed by a bloom of filamentous green algae. Blue-green algae are not a preferred food source for nonarthropod grazers, so snails only increased in abundance 16 weeks after the insecticide application, which is when the periphyton was dominated by green algae (Brock et al. 1992). In contrast, in the microcosms enriched with nutrients, a periphyton bloom dominated by green algae and diatoms developed immediately and persisted for 4 weeks. This bloom was controlled by 2 species of nonarthropod grazers, the oligochaete worm *Stylaria lacustris* and the snail *Lymnaea stagnalis* (Brock et al. 1995). Notably, the abundance of *Stylaria lacustris* increased between 1 and 3 weeks post chlorpyrifos application, during which time the periphytic algae had also increased, and declined again 7 weeks post insecticide application, when densities of *Lymnaea stagnalis* started to increase. The increased numbers of snails may have contributed to the decline of *Stylaria* through competition for food. That *Lymnaea stagnalis* responded later to the increase in periphyton than *Stylaria* might be explained by its longer generation time. The 2 microcosm experiments described above clearly illustrate latency of indirect effects, as well as the context-dependent time frames of the indirect responses observed.

Extrapolation of latent indirect effects of contaminants in ecosystems can be addressed using physical models, such as specially designed microcosm and mesocosm experiments, or computer models that simulate food webs. These integrated ecosystem models, which are simplifications of ecosystems, can be used to predict effects of toxicants on the structure and functioning of ecosystems. Species in food webs may be aggregated in functional groups. In this case, the temporal dynamics in response to toxicant stress are predicted at the functional group level, and usually not at the level of individual populations. An overview of integrated modeling of eutrophication and organic contaminant fate and effects in aquatic ecosystems is given by Koelmans et al. (2001). Examples of aquatic food-web models are AQUATOX (Park 1999), CATS (Traas et al. 1998), and CASM (Naito et al. 2003). Integrated dynamic food-web models offer the advantage of studying the dynamics of toxicant effects in time, incorporating both direct and indirect effects. The downside of these models is the large number of parameters that they require. According to Traas (2004), further development of these models needs to focus on simplifying the estimation of population parameters, improving methods for calibration, and quantifying prediction uncertainty.

6.3 TEMPORAL VARIABILITY AND ECOTOXICOLOGICAL DATA EXTRAPOLATION

6.3.1 DEVELOPMENTAL VARIATION IN SENSITIVITY OF INDIVIDUALS AND POPULATIONS

In the environment, chemicals may stress populations at different periods of the year and during different phases of the populations of the development. Consequently, when extrapolating ecotoxicological data, it is important to be aware of the possible temporal variability in sensitivity associated with different life stages.

It is frequently reported that the developmental stage of an organism cannot be ignored when studying the relation between exposure concentration and effects. For example, in a study where the first and last larval instars of the trichopteran *Hydropsyche angustipennis* were exposed to the insecticide diazinon, it was shown that the younger life stages were up to 22 times more sensitive than the older ones. In addition, ranking the 48-hour LC50s available in the literature for this compound indicated that differences between species within a taxonomic group are often smaller than differences between instars of the same species (Stuijfzand et al. 2000). In several studies with other chemicals, including metals and organic contaminants, it was also found that younger stages of aquatic invertebrates are usually more sensitive than older ones (Connor 1972; Sanders 1972; Green et al. 1986; Mayer and Ellersieck 1986; McCahon and Pascoe 1988; McCahon et al. 1989; Ringwood 1992; Williams and Hall 1999; Meier et al. 2000; Schulz and Liess 2001). However, exceptions to this rule have been reported among invertebrates (Kiffney and Clements 1996; Hutchinson et al. 1998b; Handersen and Wratten 2000; Arizzi Novelli et al. 2002; Blanck et al. 2003). The higher susceptibility of early developmental stages may be due to their greater body surface area per unit body mass (for adsorption and uptake of contaminants), their relatively higher metabolic rates, or their less advanced development of detoxification or excretion mechanisms, or because they are in the process of growth and development, where chemical signaling is important.

Several compilations of vertebrate toxicity data have confirmed the consensus described above for invertebrates that adult full-grown individuals are less sensitive to toxicants than are individuals at early life stages, although exceptions also exist (Mayer and Ellersieck 1986; Hutchinson et al. 1998b). Based on fish NOEC data in the ECETOC database, fish larvae were more sensitive than embryos for 68% of the substances, whereas fish larvae were of greater than or equal sensitivity to juvenile fish for 85% of the substances. Based on fish EC50 data, juveniles were more sensitive than adults for 92% of the substances (Hutchinson et al. 1998b). Amphibians may be more or less sensitive to chemical stressors later in development. Premetamorphs of larval *Xenopus laevis* were less sensitive to chlorpyrifos than were metamorphs (Richards and Kendall 2002). The sensitivity of tadpoles of *Bufo woodhousii fowleri* to DDT increased as animals matured (Sanders 1970), whereas that of tadpoles of *Rana sphenocephala* to toxaphene decreased with age (Hall and Swineford 1980). In a study on the impact of α-cypermethrin on various life stages of *Rana arvalis* tadpoles, it was observed that individuals exposed to this insecticide in earlier life stages (as eggs or newly hatched tadpoles) metamorphosed earlier, whereas exposure throughout their whole development prolonged metamorphosis (Greulich and Pflug-macher 2003). Cyclical activity such as reproduction and molting may make organisms more sensitive at certain times of the year, or life cycle, than at others, and the spectrum of sensitivity may be confined to certain classes of substances such as the endocrine modulators (Tillitt et al. 1998; IPCS 2002).

Standard toxicity tests typically use early life stages, as it is assumed that they are more sensitive, and their use in effect assessment procedures will increase the number of species and populations that will be protected in ecosystems. The organisms used in standard toxicity tests are usually cultured in the laboratory, thereby minimizing temporal variability in the quality of the test individuals. Where the test individuals are obtained from the field, however, their history may affect sensitivity. Factors that may lead to variability in toxicity tests include the source and health status of organisms and the time of the year, both of which influence the physiological condition of the test organisms. Reduction in adult vigor due to pressures from the external environment has been shown to affect subsequent larval development and sensitivity. For example, adults of the marine bivalve *Mytilus edulis* subject to high temperatures, nutritive stress, and low salinity produced less viable gametes than unstressed adults did (Bayne et al. 1995), which may explain the large variation in sensitivity to copper between some broods of *M. edulis* larvae (Beaumont et al. 1987). In addition, extended periods of contaminant exposure in adult *M. edulis* have been shown to alter larval sensitivity via maternal and genetic processes in this cosmopolitan species (Beaumont et al. 1987; Hoare et al. 1995). These phenomena, however, may be context-and species-dependent. In a related bivalve species (*Mytilis galloprovincialis*), no seasonal differences in sensitivity in larval development tests to several inorganic (e.g., zinc and cadmium) and organic (e.g., sodium dodecyl sulfate and phenol) toxicants were observed (Williams and Hall 1999).

The arrangement and condition of individuals through time in a certain habitat are indicative of the local state of the population. This suggests that the sensitivity of populations may vary throughout the year, or that the habitat is only suitable for the organism at different times of the year. Chemical stresses that affect early developmental stages can potentially have serious consequences on population recruitment. In addition, failure to consider endpoints above the level of the individual often leads to an overestimation of risk, but in some cases may lead to an underestimation of risk (Forbes and Calow 1999). Extrapolating time- and/or development-related effects of chemicals on individuals can be done by building and using quantitative models. An overview of how population-level effects of stressors may be measured or projected from individual effects is provided by Maltby et al. (2001). Choosing an adequate population model for use in a risk assessment will depend on several factors. Primarily the model should address the seasonal variation in demographic structure of the population of interest. An overview of population models that might be used is provided by Bartell et al. (2003) and Pastorok et al. (2003). The models comprise scalar, life history, and individual-based population models. When the effects of toxic chemicals are age or stage dependent, life history models can be considered a realistic extrapolation tool. For some examples of these models, see Caswell (2001) and Spencer and Ferson (1998). Population-based models have been discussed in greater detail in Chapter 4.

6.3.2 Seasonal and Successional Variations in Sensitivity of Communities

Temporal changes in environmental conditions due to diurnal or seasonal rhythms or long-term succession influence not only fate and bioavailability of toxicants, but also the physiological states of organisms, the age structure of populations, and the composition of assemblages that may be subject to chemical stress. In laboratory microcosms, the timing of toxicant addition has been shown to affect community sensitivity. Responses appear to be strongly influenced by developmental stage, species density, and differences in water chemistry at the time of addition (Kindig et al. 1983; Swartzman et al. 1990; Taub et al. 1991). This corresponds with the theory of Odum (1981) that the response of a biotic community to perturbations varies with the stage of its development (i.e., the stage in ecological succession).

During succession, biological assemblages may become better adapted to resist the nominal stresses that occur in the particular kind of environment in which the ecosystem is developing. In the trajectories of succession, where species richness and biomass increase, quick-response mechanisms and feedback loops may also increase within the system. In other words, the increased functional complexity of more mature systems does enhance the resistance of the system to small or short-term perturbations, as compared with the situation in early or pioneer stages of ecological succession (Odum 1981). This phenomenon was demonstrated experimentally in aquatic microbenthic biofilms. Microbenthic biofilms are consortia of autotrophic and heterotrophic organisms imbedded in a matrix of polymers and particles. The stage of biofilm maturation strongly influenced sensitivity to zinc and cadmium exposure (Ivorra 2000). More pronounced effects of the metal on the young than on the more mature biofilms were observed. As biofilms develop, internal cycling of materials increases and dependence on external conditions is reduced. In the more complex mature biofilms, effects of metal are diminished or delayed by structural and chemical barriers created by the presence of more taxa, mucus, and particles trapped, as well as by an altered pH regime with increased biomass and photosynthetic activity (Ivorra 2000).

When evaluating the influence of the time of year on responses of aquatic communities to chemical stress, it is convenient to make a distinction between the ecological threshold concentrations and the magnitude of effects that occur above these threshold concentrations. A lake enclosure study exploring the effects of a single application of pentachlorophenol (0, 4, 10, 24, 36, 54, 81, and 121 μg/L) to planktonic communities in spring, summer, autumn, and winter indicated that threshold concentrations for effects on planktonic communities ($NOEC_{community}$) varied little with season (24 to direct toxic 36 μg/L), whereas indirect effects (increase in the alga *Cryptomonas*) occurred at lower treatment concentrations in autumn (54 μg/L and higher) and in none of the treatments in winter (Willis et al. 2004). The variations in plankton community response to pentachlorophenol were apparent in the following order of decreasing sensitivity; autumn > winter–spring > summer. The overall small variations in community response observed were considered to be caused by physicochemical conditions and physiological states such as pH rather than major differences in species assemblages (Willis et al. 2004). Similarly, there was no evidence of a major difference in the direct effects of spring or late-summer application of lambda-cyhalothrin

(0, 10, 25, 50, 100, and 250 ng/L) to ditch enclosures. Multivariate analysis revealed that threshold concentrations for the total invertebrate community were close to 10 ng/L for both application periods, although at the lowest treatment concentration (10 ng/L), effects were observed primarily in 1 species — the phantom midge *Chaoborus obscuripes* (van Wijngaarden et al. 2005b). In addition, major differences in indirect effects at higher concentrations could not be observed between spring and late summer. However, recovery of sensitive macroinvertebrate populations (e.g., *Chaoborus obscuripes*) was slower after the late-summer application (van Wijngaarden et al. 2006).

Studies where timing of application has been shown to be important include those with carbaryl and atrazine. In experimental ponds, carbaryl addition well above the threshold concentration for direct toxic effects at different stages in the seasonal cycle induced distinct dynamics in indirect effects, as well as distinct recovery patterns in zooplankton communities. It was suggested that, in particular, temperature, competitive interactions, and population trends were significant factors influencing interactions between species and recovery of the zooplankton (Hanazato and Yasuno 1990). In outdoor microcosm studies, a chronic exposure to 10 µg/L atrazine on Lake Geneva plankton communities disrupted the phytoplankton community directly or indirectly in different periods of the year, but the magnitude and nature of the treatment-related effects observed differed between seasons. Different populations of algae in the phytoplankton community were inhibited, stimulated, or apparently unaffected by atrazine. The clear-water phase (June) was the period when the algal communities were the most sensitive to restructuring by atrazine, whereas they were the least sensitive during the spring blooms (March to April; Bérard et al. 1999). Inhibition of photosynthesis by atrazine is comparable to limiting conditions of light, and the toxicity of photosystem II inhibition is lower for plants that are adapted to low light conditions (Guasch and Sabater 1998). So, seasonal changes in light conditions may, at least in part, explain the observed seasonal differences in sensitivity of phytoplankton to atrazine. It may also be that the collective sensitivities of the various taxa comprising the community at any one time may be more or less sensitive than at another time.

The "Community-Level Aquatic Systems Studies: Interpretation Studies" (CLASSIC) guidance document, which deals with the interpretation of results of microcosm and mesocosm tests in the risk assessment procedure of pesticides, recommends that "regulatory" model ecosystem experiments be conducted in spring to midsummer (Giddings et al. 2002). On the basis of the limited number of model ecosystem experiments described above, it seems that threshold concentrations for effects observed in early-season studies are reasonably predictive for threshold concentrations later in the season. Above these threshold concentrations, however, the intensity and duration of the responses (direct and indirect effects) may vary during different periods of the year. Consequently, the extrapolation of NOEC$_{community}$ values from one season to another seems to be possible with lower uncertainty than hazard estimates of higher concentrations in which both direct and indirect effects are involved.

Models that can be used to address extrapolation of community-level threshold concentrations and effect concentrations from one season to another are physical models (e.g., enclosure studies performed in the same system in different periods of the year) or computer models that simulate the dynamics of realistic food webs and incorporate seasonal differences in sensitivity of functional groups. We could not find examples of such food-web model simulations in the literature, but in principle, food-web models such as IFEM (Bartell et al. 1988), AQUATOX (Park 1999), and C-COSM (Traas et al. 2004) can be adapted to do this. Pastorok et al. (2003) tabulated the application of various types of ecosystem models used to assess effects of toxic chemicals.

6.3.3 RECOVERY

An advantage of field studies is that information can be obtained about recovery of disturbed populations and ecosystem functions. Recovery time is a highly attractive measurement endpoint in ecological and ecotoxicological studies because it is comprehensible to regulators, managers, and the general public. The phenomenon of recovery is particularly important in cases where exposure to a stressor declines due to physical (e.g., hydrological), physicochemical (e.g., hydrolysis), or biological (e.g., bacterial breakdown) processes that result in the disappearance of the stressor.

When defining recovery, a distinction between actual and potential recovery can be made. Actual (or ecological) recovery implies the return of the perturbed measurement endpoint (e.g., species composition, population density, or dissolved oxygen concentration) to the window of natural variability in the ecosystem of concern, or to the level that is not significantly different from that in control or reference systems of a microcosm or mesocosm study. This does not mean, however, that we should consider endpoints as being recovered if the statistical difference primarily disappears due to an increase in variability in control test systems. Potential (or ecotoxicological) recovery is defined as the potential for recovery to occur following the disappearance of the stressor to a concentration at which it no longer has adverse toxic effects on the measurement endpoints of interest (Brock and Budde 1994; Maltby et al. 2001).

At the ecosystem level, recovery is related to "ecological resilience." According to Gunderson (2000), ecological resilience is 1) the time required for an ecosystem to return to an equilibrium or steady state following a perturbation or 2) the magnitude of disturbance that can be absorbed by the ecosystem before the system redefines its structure. In this second definition, the concept of ecological resilience presumes the existence of multiple stability domains. Ecological resilience thus refers to the width or limit of a stability domain and is defined by the magnitude of disturbance that a system can absorb before it changes stable states. For example, in shallow aquatic ecosystems different stable states may comprise communities dominated by submersed macrophytes (clear water), by dense mats of free-floating plants (e.g., *Lemna* covers), or by phytoplankton blooms (turbid water) (Scheffer 1998).

It is argued that, in an ecotoxicological sense, returning to an original state is not a property of complex ecosystems and, consequently, that assumptions such as recovery can lead to unrealistic predictions. This opinion is in line with the community conditioning hypothesis (Matthews et al. 1996), which states that ecological communities tend to preserve information about every event in their history, including stress of chemicals. The historical information can be stored in a variety of layers, from the genetic and molecular to the pattern and dynamics of interspecies interactions (Landis et al. 1996). Considering the view that no 2 ecological structures will ever be the same, an important question is, then, whether the effects of chemical stress can be distinguished from the effects of other natural stress factors. When the preserved information of the chemical perturbation is of a smaller "scale" than that caused by other natural stressors, its ecological significance is difficult to interpret. For this reason, we consider recovery a concept that cannot be ignored in ecological risk assessment, particularly when addressing the dynamics of populations.

When studying population responses to chemical stress, it may be convenient to make a distinction between internal and external recovery. Internal recovery depends on surviving individuals in the stressed ecosystem or on a reservoir of resting propagules (e.g., seeds and ephippia) not affected by the stressor. In contrast, external recovery depends on the immigration of individuals from neighboring ecosystems by active or passive dispersal.

In cases where the toxicant degrades rapidly and/or its bioavailability decreases below a critical threshold concentration, the recovery rate of affected populations depends, for a large part, on the life-cycle characteristics of the affected species of concern. Important life-cycle properties are the number of generations per year and related reproductive strategies (r-K), the presence of relatively insensitive (dormant) life stages, and the capacity of organisms to actively migrate from 1 site to another. In general, recovery of affected populations from chemical stress may be rapid if the following conditions apply:

- The substance is not persistent, the exposure regime is short term or pulsed, and the time between pulses is long enough for recovery.
- The physicochemical environment and ecologically important food-web interactions are not altered by the stressor, or are quickly restored.
- The generation time of the populations affected is short.
- There is a ready supply of propagules of eliminated populations through active immigration by mobile organisms or through passive immigration by, for example, wind and water transport.

The relationship between life-cycle characteristics and recovery of species is illustrated in Figure 6.5. This figure presents the long-term response of 4 aquatic invertebrates in experimental ditches after a single application of chlorpyrifos (van den Brink et al. 1996). In this study, at the highest treatment concentrations, the cladoceran *Simocephalus vetulus* (Figure 6.5C) recovered rapidly, as a result of its short generation time and the possession of a relatively insensitive life stage in the form of winter eggs (ephippia). Insect species with several emergence periods per year, such as *Cloeon dipterum* (Figure 6.5A), also recovered rapidly compared with insect species with only 1 or 2 generations per year (e.g., *Caenis horaria*; Figure 6.5B). Insects usually do not have aquatic life stages that are insensitive to chemical stress, but the winged adult stage offers the possibility to recolonize isolated aquatic systems. Where a species cannot easily reach an isolated system and does not have an insensitive aquatic life stage, there is a greater probability that the species will disappear from the isolated system for a longer period as a result of chemical

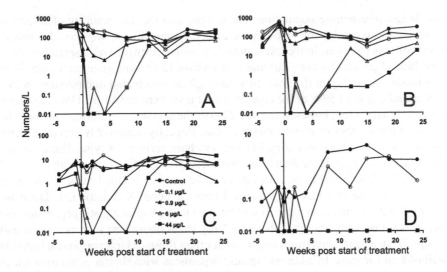

FIGURE 6.5 Dynamics of numbers of the aquatic stages of 2 ephemeropteran insects *Cloeon dipterum* (A) and *Caenis horaria* (B), and of the cladoceran *Simocephalus vetulus* (C) and the amphipod *Gammarus pulex* (D). *Source*: Redrawn from data from van den Brink et al. (1996).

stress. For example, *Gammarus pulex* (Figure 6.5D) became extinct in the isolated experimental ditches treated with greater concentrations of chlorpyrifos, and recovery only occurred after this species was reintroduced (not shown in the figure). In nonisolated water courses, however, *Gammarus pulex* may, after local elimination by toxicants, show a relatively rapid recovery as a result of successful recolonization by active swimming from upstream or downstream refugia (Liess 1998).

The main strategies that organisms apply to survive unfavorable periods are dormancy (escape in time) and dispersal (escape in space). In ephemeral freshwater habitats in particular, these adaptations may affect the ecological impact of toxicants (Lahr 2000). In temporary ponds in the Sahel (Africa) stressed by nonpersistent insecticides, the average time to recovery for adults of several backswimmer species (*Anisops*) was short, approximately 2 weeks. The recovery mechanism in this winged insect species was aerial recolonization. In contrast, fairy shrimps (*Streptocephalus sudanicus*) did not reappear until the next rainy season, almost 1 year later. Fairy shrimps produce 1 generation per season, which releases its resting eggs in the sediment. The eggs presumably need at least 1 period of desiccation before they will hatch. According to Lahr (2000), reserves of cysts in ephemeral habitats may be depleted during the early rainy season and before maturity of fairy shrimps is reached. In such cases, internal recovery will not occur.

Van der Geest (2001) related the available toxicity data obtained in laboratory single-species tests to the recovery times of the corresponding test organisms. He noted that, currently, little attention is paid to aquatic insect species with long life cycles (and consequently long recovery times) such as several species of mayflies, caddisflies, and stoneflies. For the reasons mentioned above, it seems logical to collect data on the life-cycle characteristics of the species from the ecosystem at risk. When invertebrate populations are at risk, and the acceptability of effects is based on the potential for recovery of affected populations, we should consider whether the populations in the test system (e.g., microcosm) are representative of the univoltine (1 generation per year) and multivoltine (2 or more generations per year) populations occurring in the field. In a macrophyte-dominated ditch system in The Netherlands, at least 59% of the aquatic insect taxa were univoltine (Van der Geest 2001). It is surprising, however, that for 27% of the species, no information could be found on the number of generations per year. Recently, reasonably comprehensive data became available concerning life history characteristics of typical aquatic species (see, for example, Heneghan et al. 1999; Usseglio-Polatera et al. 2000; Liess and Von der Ohe 2005; Poff et al. 2006) that can be used to aid the characterization of recovery potential of aquatic organisms. From the PondFX Aquatic Life Database (PondFX n.d.), some broad patterns with respect to the size of aquatic organisms and their generation time emerge (Table 6.6). Very small organisms such as rotifers and copepods have generation times much shorter than those of larger organisms such as mollusks and insects. In addition, aquatic organisms with shorter generation times are reported to recover faster than organisms with longer generation times (Niemi et al. 1990; Barnthouse 2004). Further developments of this database are provided by Freshwater Life (n.d.).

TABLE 6.6
Generation times for various groups of aquatic organisms

Taxon	Generation time in days, mean (range)
Phytoplankton	1
Lemna	3
Rotifera	8 (6 to 35)
Cladocera	14
Copepoda	61 (14 to 73)
Oligochaeta	105 (51 to 730)
Amphipoda	73 (105 to 250)
Ostracoda	121 (51 to 362)
Gastropoda	513 (105 to ?)
Bivalvia	256 (105 to ?)
Coleoptera	(209 to ?)
Diptera	(81 to 503)
Ephemeroptera	(81 to 730)
Hemiptera	(81 to 503)
Trichoptera	(162 to 1264)
Fish (short life cycle)	181
Fish (long life cycle)	1673

Source: As derived from the PondFX Aquatic Life Database (http://www.ent.orst.edu/PondFX) and Barnthouse (2004).

It follows from the above that the time of the year in which the exposure to a chemical stressor takes place may determine the extent of biological recovery. Pulsed exposures can have greater impacts on recovery patterns if they occur during critical life stages (particularly in the case of univoltine insects) or if they occur in autumn, when lower dispersal activities and lack of winter reproduction may delay recovery until the following spring. In experimental ponds and enclosures, it appeared that application of chemicals at different times of the year may induce different recovery patterns of the same invertebrate species (Hanazato and Yasuno 1990; Willis et al. 2004; van Wijngaarden et al. 2006). In most field studies, the time of occurrence of a specific short-term disturbance changed the recovery time of invertebrate populations by less than 1 year (Niemi et al. 1990).

The number of generations per year of certain invertebrate species may vary with latitude and, consequently, with temperature and the length of the growing season (Niemi et al. 1990). For example, in colder regions, the same insect species may be univoltine, whereas in warmer regions this species may have more generations per year. Some countries may have a sufficiently wide temperature range to have a range of single and multiple generations for species such as mayflies (Huryn 1996). Consequently, when recovery is taken into account in the assessment of acceptable concentrations, differences between latitudes may be of importance when extrapolating data from temperate to colder regions.

Estimates of recovery potential can be made through either empirical or modeling studies. Suitable empirical studies comprise microcosm and/or mesocosm tests in which the immigration of affected populations may be manipulated, for example, by reintroducing organisms or by preventing external recovery, such as by covering the test systems with nets (see Sherratt et al. 1999). It should be noted, however, that most microcosm and/or mesocosm experiments on the impact of chemical stress focus on the responses of dominant populations, which are often characterized by a relatively short life cycle (e.g., algae and invertebrates). Microcosm and mesocosm studies are generally less suitable to study the recovery of populations of larger organisms with a long life span (such as vertebrates). In addition, the duration of many published microcosm and mesocosm studies is too short to be able to derive the recovery period of sensitive populations. Another point of attention in the interpretation and extrapolation of responses observed in microcosm/mesocosm studies is that most experiments utilize isolated test systems. This means that eliminated populations with a limited dispersal capacity cannot rapidly recolonize these test systems. For these organisms, observations on their recovery in isolated microcosms and mesocosms should be considered as a worst-case scenario.

It appears from the above that microcosm and/or mesocosm tests are limited by the constraints of experimentation, in that usually only a limited number of recovery scenarios can be investigated. Consequently, modeling approaches may provide an alternative tool for investigating likely recovery rates under a range of conditions. Generic models, like the logistic growth mode (for example, see Barnthouse 2004) and life history and individual-based (meta)population models, which also may be spatially explicit, provide mathematical frameworks that offer the opportunity to explore the recovery potential of individual populations. For an overview of these life history and individual-based models, see Bartell et al. (2003) and Pastorok et al. (2003).

A recently developed metapopulation model to extrapolate responses of aquatic invertebrates as observed in mesocosms to assess their recovery potential in the field is provided by Van den Brink et al. (2007). When the primary interest is in the recovery of processes and functional groups, food-web models are the required mathematical tools (for an example, see Traas et al. 2004). Two drawbacks of these models are that they require detailed information on the species and functional groups of concern and that they are very specific to the species and functional groups and sites for which they are developed.

Modeling approaches need to be tailored to the issue and level of detail that need to be addressed in the risk assessment. With a simple modeling framework, Barnthouse (2004) showed that it is possible, at least in principle, to quantify expected population recovery rates from basic life history considerations. A logistic population growth model was used to compare population recovery rates for different types of aquatic organisms and to evaluate the influence of life history, disturbance frequency, and immigration on the time required for populations to recover from simulated chemical exposures. Generation time was found to be the most important determinant of population recovery rates. Generation time is a function of developmental rate and longevity. Estimates of these 2 parameters are much more widely available than are estimates of age- or stage-specific survival or reproduction rates (Barnthouse 2004).

6.3.4 ADAPTATION TO CHEMICAL STRESS

Sensitivity of populations to toxic stress may be influenced by the exposure history of test specimens. Van den Brink et al. (1997) exposed 2 strains of the alga *Chlamydomonas reinhardtii* to a concentration range of the photosynthesis-inhibiting herbicide linuron. One strain originated from a linuron-free medium, and the other from a culture previously treated with a relatively high concentration of linuron (i.e., 150 µg/L). *Chlamydomonas reinhardtii* from the pre-exposed strain had significantly larger relative growth rates when exposed to linuron concentrations of 150 and 500 µg/L than *C. reinhardtii* from the nonexposed strain (Figure 6.6). In this case, it is not clear whether the differences in response to the toxicological stress were the result of a physiological acclimation of individuals of the algal population or whether the tolerance was caused by the selection of relatively insensitive genotypes.

An adaptation can be defined as a trait that enhances an individual's fitness to a certain stressor. Tolerance is the ability of individuals to withstand exposure to toxicant concentrations that would normally be expected to cause physiological inhibition or mortality. Tolerance can be acquired by an individual due to phenotypic plastic responses following exposure to sublethal concentrations (physiological acclimation) or by inheritance (genetically based tolerance). Genetic adaptation implies that a population has been selected for an increased mean tolerance through natural selection acting upon genetically based interindividual variation in tolerance (Hoffmann and Parsons 1991). In contrast to genetically based tolerance, physiological acclimation is not inherited by offspring or retained by individuals when maintained in uncontaminated habitats. Acclimation occurs when a prior exposure results in increased resistance during subsequent exposures due to the induction of specific detoxification

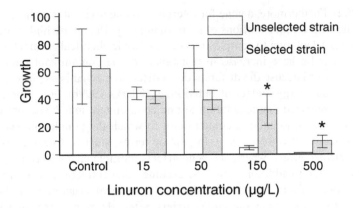

FIGURE 6.6 Relative growth of 2 strains of *Chlamydomonas reinhardtii*; the unselected strain originated from a linuron-free medium, and the selected strain from a culture previously treated with 150 μg/L linuron. *Note*: The 2 strains were exposed to 0, 15, 50, 150, and 500 μg/L linuron for 3 days. Significant differences between the strains (ANOVA, $p \le$ 0.05) are indicated by an asterisk.*Source*: Figure redrawn from data of van den Brink et al. (1997).

mechanisms (Mulvey and Diamond 1991). In practice, however, it may not be easy to make a proper distinction between physiological adaptation and genetically based tolerance. For example, Lam (1996) reported differences in acute toxicity of cadmium for 2 populations of the gastropod *Brotia hainanensis*, and although these differences persisted into newly hatched animals of the F1 generation, the differences disappeared by the time the F1 animals were 7 days old. This suggests a short-term maternal effect or acclimation response rather than genetically based tolerance (Lam 1996).

In addition to the example presented in Figure 6.6, several examples of rapid adaptation to certain toxicants have been reported. Rapid microevolution of tolerance to toxicants has been demonstrated in a number of algae, vascular plants, and arthropods, genetically responding to the use of pesticides (Caprio and Tabashnik 1992; Kasai and Hanazato 1995b), and in pathogenic bacteria responding to antibiotics. Swift genetic adaptation was also found in cases of anthropogenic metal input. Experiments with the oligochaete *Limnodrilus hoffmeisteri* (Klerks and Levinton 1989) and the midge *Chironomus riparius* (Postma and Davids 1995) indicate that adaptation to metals can develop very rapidly within a few generations, with similar observations for field invertebrate communities (Courtney and Clements 2000) and snails pre-exposed to arsenic (Golding et al. 1997).

Species that evolve genetic tolerance are frequently opportunists with short generation times and large numbers of eggs and seeds (Luoma 1977). This may simply reflect a greater number of opportunities for beneficial mutations. Many species, however, apparently do not have the potential to rapidly develop pollution tolerance (Grant 2002). Some estimates suggest that only about 10% of plant species present on adjacent uncontaminated sites are able to develop metal-tolerant forms and thus colonize severely contaminated sites. In these tolerant species, genetic adaptation to toxicant stress usually involves a single gene rather than polygenic inheritance

(Grant 2002). Furthermore, a gene for tolerance to one toxicant is of little value in a polluted site where concentrations of more than one pollutant are high enough to be acutely toxic. Adaptation to complex mixtures of toxic chemicals is much less likely to occur, either because increase in resistance to one contaminant reduces resistance to others or because the different detoxification mechanisms place competing demands on the energy budget of individuals (Klerks 1999b). For some toxicants, however, it is reported that adaptation to one contaminant may be accompanied by increased resistance to another contaminant to which the population has not been exposed (so-called "co-tolerance"). In genetic terms, such co-tolerance is referred to as a pleiotropic effect, where one gene affects multiple characteristics (Millward and Klerks 2002). In addition, it may be assumed that cotolerance is more likely to occur for groups of toxicants that are similar in chemical structure or in their mode of action, are transported by the same carriers, affect the same target sites in organisms, and are degraded along similar routes (Blanck 2002).

In an adapted population, amongst other factors, the persistence of genetic adaptation to a toxicant depends on the persistence of the toxicant, the degree of genetic isolation of the population, and the fitness cost (if any) of the adaptation. Isolated populations with low dispersal ability tend to have low temporal variability in the adapted toxicant-tolerant trait. For nonisolated populations, however, it is reported that gene flow among populations will prevent complex fixation of an adaptive trait. For example, the large temporal variation in cadmium adaptation in chironomid populations from a polluted stretch of a lowland river was explained by the influence of gene flow from upstream nonpolluted areas (Groenendijk 1999). In general, tolerant individuals rapidly disappear when released in clean environments due to competition with nonadapted strains. For example, in *Chironomus riparius*, both larval mortality and larval development time increased compared to reference populations when cadmium-tolerant midges were cultured in a clean environment. This was explained as the "cost" of being tolerant (Postma et al. 1995b). Fitness costs associated with tolerance to cadmium were related with changes in mineral metabolism that caused an increased need for essential metals such as zinc (Postma et al. 1995a). It has been demonstrated for several aquatic and terrestrial species that, in the absence of exposure, toxicant-resistant genotypes have reduced fitness relative to nonresistant genotypes (van Straalen and Hoffmann 2000). In addition, increased tolerance due to physiological acclimation is likely to incur energetic costs to the individual.

It can be concluded from the above that adaptation to toxicant stress may be expected when the following occurs:

- Toxic stress is predominantly caused by a single compound.
- Exposure to the chemical is relatively high and long term or repeated in time.
- The genetic basis of adaptation involves one or a few genes.
- The species has a short generation time and a large number of offspring.
- The adapted population is isolated (limited gene flow).
- The costs associated with tolerance are less than the advantage of decreased competition with nonadapted individuals and/or less tolerant populations.

According to Weis (2002), however, it is unwise to relax standards on the basis that adaptation can be expected, for the following reasons:

- Resistance does not develop in all populations.
- Adapted populations often appear stressed by the contamination in spite of the adaptation.
- Adapted populations may have reduced genetic diversity.
- Adapted populations may have high tissue burdens of contaminants and could pass these up the food chain.
- The selection of resistance may be associated with fitness costs that can translate into future problems for the adapted population.

Prolonged or repeated exposure to a toxicant may affect the tolerance not only of individual populations but also of whole communities. Increased community tolerance as a result of prolonged or repeated exposure to toxicants has been observed in indoor multispecies tests and in outdoor model ecosystem experiments. For example, effects of a pulsed treatment with the photosynthesis-inhibiting herbicide linuron on oxygen metabolism in macrophyte-dominated experimental ditches were studied by Kersting and Van Wijngaarden (1999). The experimental ditches were treated 3 times with linuron at 4-week intervals. After each treatment, the test systems were kept static for 1 week, after which they were flushed with noncontaminated surface water until the next treatment. Gross primary production (GPP) had an EC50 of 21 µg/L during the static period after the first treatment. The EC50 for GPP increased to 36 µg/L after the second treatment and to 80 µg/L after the third treatment. Between treatments, GPP recovered to the level of the controls. The mechanism of decreased sensitivity after each pulse was interpreted as adaptation of the assemblage of primary producers, similar to pollution-induced community tolerance (PICT; Blanck 2002).

Measurement of PICT usually involves carrying out short-term (multispecies) toxicity tests on whole communities from clean and contaminated sites. Pollution tolerance is quantified by reduced sensitivity of the toxicant in these tests. The increased tolerance may result from replacement of sensitive species by less sensitive ones, development of heritable tolerance by one or more species, and/or short-term nonheritable acclimation. A significant increase in community tolerance compared to the baseline tolerance at reference sites suggests that the community has been adversely affected by toxicants. In this way, PICT can establish causal linkages between contaminants and effects in monitoring studies (Blanck 2002).

The time needed to develop PICT is, among other factors, dependent on exposure concentration and duration. For example, in enclosures exposed to a high copper concentration (0.24 µM), increased tolerance to copper was found in phytoplankton communities after 2 days, whereas in enclosures treated with a low concentration (0.016 µM), increased tolerance was found after 12 days (Gustavson and Wängberg 1995). In order to demonstrate PICT, the duration of the exposure to toxicants should be long enough to allow succession of the community of interest to a more tolerant state. This may vary with the generation times of the species that form the community under investigation. The fact that the generation time of phytoplankton species is low (approximately 1 day; Table 6.5) explains the fast response of the phytoplankton community exposed to copper in the study described above. Also, the temporal resolution of disappearance of PICT will depend on the succession rate of the community after exposure to the contaminant has disappeared. In a bacterial community, the major part of community tolerance (70% to 90%) was lost within a week when metal-tolerant communities were extracted and reinoculated in sterilized uncontaminated soil (Díaz-Ravina and Bååth 2001). In a field study of TBT tolerance in marine periphyton, the gradual change in tolerance was strongly coupled to changes of TBT concentrations in water over the boating season, with sampling intervals of about a month (Blanck and Dahl 1996).

PICT has received significant attention in the last decade, particularly when applied to communities of periphyton, phytoplankton, bacteria, nematodes, and insects. Overall, these organisms have short generation times that offer opportunities for rapid adaptations. To date, mainly metals, herbicides, and biocides have been evaluated in PICT studies (Blanck 2002).

The examples of dynamics in tolerance of individuals, populations, and communities to toxicant stress presented above clearly show that, in ecological risk assessment, it is important to account for possible physiological and genetic adaptation when the organisms were previously exposed to the stressor of interest. In part, this may explain the variability in toxicity for the same species reared in different laboratories, or sampled in different periods of the year or in different localities. This variability in toxic responses may complicate attempts to extrapolate effects of toxicants from laboratory toxicity tests to natural systems. In the fish species *Melanotaenia nigrans* (Gale et al. 2003), the gastropod *Potamopyrgus antipodarum* (Jensen et al. 2001), the crustaceans *Artemia* and *Daphnia magna* (Barata et al. 2002b; Forbes 1998), and the duckweed *Lemna gibba* (Mazzo et al. 1998), differences in tolerance to toxic substances among clones or subpopulations varied between two- and eightfold. These differences are relatively small, when taking into account that the intra- and interlaboratory variations in protocol tests with the same standard test species are approximately within a factor of 3 and 5, respectively (Rand and Petrocelli 1985; Ferretti et al. 2004).

6.4 SUMMARY AND OUTLOOK

6.4.1 IS EXTRAPOLATION ACROSS TIME-VARYING EXPOSURE REGIMES POSSIBLE?

Extrapolation across time-varying exposure regimes is possible for well-studied compounds with known modes of toxic action. However, it must be recognized that such extrapolation may not always be accompanied by a high degree of certainty.

The response of organisms to toxic chemicals depends not only on exposure concentration but also on exposure duration. The time needed to reach incipient toxicity is dependent on the type of chemical, ambient environmental conditions such as temperature, and the species and endpoints selected. Time-to-event models can be used to extrapolate observed responses in time. Individual organism characteristics and/or environmental circumstances (temperature) can be incorporated in these models as covariables (Section 6.2.1).

In characterizing effects of contaminants to aquatic organisms, the possibility of reversibility of the toxic effect may be important, particularly when evaluating the risks of pulsed and multiple exposures. Reversibility of the effects is reported to depend on the type of chemical, the exposure concentration, and the duration of the exposure event. Effects on organisms may be similar when exposed for a short time to a greater concentration or for a longer time to a smaller concentration. Such reciprocity relationships may be used to estimate responses to shorter environmental exposures where they are less than those used in acute single-species tests, and when dealing with compounds that are taken up slowly and that bind irreversibly (i.e., some kind of damage repair is necessary, and elimination of the compound is not sufficient to take away damage; Section 6.2.2).

In the mechanistic models used to predict effects of time-variable exposure to organisms, a distinction can be made between 1) 1-step models that consider the toxicokinetic terms "uptake," "elimination," and "critical body residues" and 2) 2-step models that besides toxicokinetics also address the toxicodynamic terms "injury" and "repair." A disadvantage of these models is that their parameterization is compound-and species-specific and hence requires many experimental data (Section 6.2.3).

In laboratory single-species tests, a significant relationship may exist between acute and chronic effects in some organisms and for some types of chemicals. However, there may be considerable uncertainty associated with predictions of chronic effects from acute effects. In part, this uncertainty can be explained by the fact that different responses may be observed in acute tests (e.g., lethality) and chronic tests (e.g., reproduction). For pesticides and aquatic organisms, an uncertainty factor larger than 100 may be necessary when predictions with a very high certainty are required (e.g., the 95th centile). Uncertainty factors up to 100 may suffice for 90th centile predictions, and UFs of 10 for 50th centile predictions. When compared with pesticides, the required uncertainty factors to predict chronic toxicity from acute toxicity data may be higher for metals and organometals and lower for other inorganics and organics.

For most pesticides evaluated, an uncertainty factor of 10 to 15 seems to suffice to extrapolate a median acute HC5 to a median chronic HC5, at least when based on toxicity data of sensitive taxonomic groups. In addition, it appears from model ecosystem experiments with pesticides that threshold concentrations for chronic exposures are approximately a factor of 10 lower than those for acute exposure. For a wider generalization, however, more data are required on compounds that differ in toxic mode of action (Section 6.2.4).

Possible delayed responses cannot be ignored when interpreting toxicity tests. These latent responses may become apparent in organisms after the exposure is completed or in conjunction with other stressors (e.g., food shortage and release of toxicants stored in body lipids). It may even be required to include the responses of the offspring in the observations, which is not often done in routine toxicity testing. At the community level, complex interactions within and between populations and nonlinear biological dynamics may create variable latency periods between the exposure event and indirect effects in particular. Latent direct effects of contaminants can be addressed by specially designed bioassays. Extrapolation of latent indirect effects can be addressed in model ecosystem experiments or by computer models that simulate food webs (Section 6.2.5).

6.4.2 CAN TEMPORAL VARIABILITY BE ADDRESSED WHEN EXTRAPOLATING TOXICITY DATA?

When exposure concentrations are known and when predictions are based on adequate information on the structure, function, and dynamics of the populations and communities at risk, then it may be possible to quantify variability associated with extrapolation between toxicity data.

Young stages of aquatic invertebrates and fish are often more sensitive to toxicants than older life stages. The greater sensitivity of the younger stages can be explained by greater body surface area per unit body mass, higher metabolic rates, and less advanced development of detoxification and excretion mechanisms. This suggests that the sensitivity of populations may vary throughout the year. Failure to consider endpoints above the level of individuals, however, may lead to an over- or underestimation of risk. Extrapolating time- and/or development-related effects of chemicals on individuals to populations can be achieved using quantitative models. Especially when the effects of toxic chemicals are age or stage dependent, life history models can be considered a realistic extrapolation tool (Section 6.3.1).

When evaluating the influence of the time of year on responses of aquatic communities to chemical stress, it is convenient to make a distinction in threshold concentrations of direct toxic effects, and in the magnitude of effects that occur above these threshold concentrations. Only a limited number of (model) ecosystem experiments are available that allow a comparison of responses due to treatment with the same chemical in the same type of test system at different periods of the year. These studies indicate that, in freshwater communities, threshold concentrations for direct toxic effects may vary little with the season (within a factor of 2). At higher exposure concentrations, however, the intensity and duration of the responses (direct and indirect effects) may vary considerably between different periods of the year (Section 6.3.2).

Recovery potential of affected population and ecosystem functions is important in cases where exposure to the toxicant is not constant due to fast dissipation processes. The recovery rate of affected populations is highly dependent on life-cycle characteristics of the affected species of concern (e.g., generation time, offspring number, presence of dormant life stages, and migration capacity). In the ecotoxicological literature, relatively little experimental information can be found on the recovery potential of species with a long and/or complex life cycle. In addition, for many aquatic species, basic information on life-cycle characteristics is not readily available. The number of generations per year of invertebrate species may vary with latitude. Estimates of recovery potential can be made through either empirical (e.g., mesocosm experiments) or modeling studies. Generic models, like the logistic growth model, and life history and individual-based (meta)population models, which may be spatially explicit, can be used to explore the recovery potential of individual populations. When researchers are interested in the recovery of ecosystem processes (e.g., primary productivity) and functional groups, food-web models are the required tools (Section 6.3.3).

Sensitivity of populations and communities to toxic stress may be influenced by exposure history. Tolerance can be acquired by an individual due to physiological acclimation or by genetic adaptation. At the community level, PICT may result from replacement of sensitive species by less sensitive ones, genetic adaptation of one or more species, and/or physiological adaptation of individuals. Rapid genetic adaptation to toxicant stress usually involves a single gene. Experiments have demonstrated that genetic adaptation to metals and pesticides can develop within a few generations, particularly when it concerns species with a short generation time and a large number of offspring. In general, tolerant individuals disappear rapidly when released in clean environments, due to competition with nonadapted strains. Adaptation to chemical stress may be one of the reasons for the observed variability in toxicity for the same species reared in different laboratories, or sampled during different periods of the year or at different localities. Reported differences in tolerance to toxic substances among clones or subpopulations of aquatic organisms varied between two- and eightfold (Section 6.3.4).

ACKNOWLEDGMENTS

We thank John Chapman, John Deneer, Chris Hickey, Thomas LaPoint, Steve Maund, Hans Sanderson, and Paul Sibley for their constructive suggestions to improve this chapter.

7 Spatial Extrapolation in Ecological Effect Assessment of Chemicals

Theo C. M. Brock, Lorraine Maltby,
Christopher W. Hickey, John Chapman,
and Keith R. Solomon

CONTENTS

7.1 INTRODUCTION

Natural populations and communities are characterized as being spatially heterogeneous (Connell and Sousa 1983; Sousa 1984). This may result in a high degree of variability in their response to chemical stress. It is, however, neither financially nor practically feasible to test a large number of chemicals on a large number of species and communities in different localities. Therefore, the spatial extrapolation of ecotoxicological effect data is an important issue in ecological risk assessment.

Ecologists have long recognized the considerable differences in biological diversity that exist between ecoregions. For example, a larger biological diversity is reported for the tropics compared to temperate and polar zones (Hawkins 2001). In addition, certain regions are characterized by a high degree of endemism amongst their fauna and flora (e.g., ANZECC ARMCANZ 2000; Walker et al. 2001). However, there are limited ecotoxicological data on nontemperate species in general, and on endemic species in particular. Consequently, ecotoxicological generalizations developed for temperate species and assemblages in North America and Europe may be challenged when extrapolated to other parts of the world with different climates (e.g., the tropics and Antarctica) and/or a different evolutionary history (e.g., Australia and New Zealand). Climatic factors in these widely differing environments may markedly affect both the chemical fate and ecological response rates.

More and more papers are becoming available in the ecological literature showing that chaotic dynamics in population and community responses may play an important role (May 1974a; Perry et al. 2000; King et al. 2004). Also, in recent years, ecotoxicology has moved from a static view to a dynamic view, in which the risks of chemicals to populations, communities, and ecosystems are considered in their temporal and spatial context within landscapes (Fahrig and Freemark 1995; Johnson 2002). At the landscape level, the effect of a toxic event can extend beyond the area of direct impact due to metapopulation dynamics, but may disappear faster due to external recovery processes. The aim of assessing risks of chemicals at the landscape level is to make inferences about the longer term dynamics of larger systems. This approach also implies a move from the ecological risk assessment of single chemicals to that of realistic combinations of multiple stressors.

In extrapolation across spatial scales, a variety of models may be employed, including physical models (such as microcosms and mesocosms), statistical models, and computer simulations (Johnson and Rodgers 2005). Recent technological innovations make spatially explicit risk assessment feasible at scales relevant to biological populations and ecosystem processes. New mapping technologies based on remote sensing provide the basis for detailed topographic models from which relevant environmental and ecological conditions and spatial process rates can be initially deduced or hypothesized (Power et al. 2005).

This chapter aims to provide insight into 1) our current knowledge of spatial variability of ecological responses to chemical stress, and 2) the tools that can be used to extrapolate across space. The focus of this chapter will be on aquatic organisms and ecosystems, and it is divided into 2 major sections:

- Spatial variability and ecotoxicological data extrapolation: This section describes the current knowledge and available extrapolation tools with respect to the effect assessment of the same type of stressor in test systems of different sizes, in different types of aquatic ecosystem within a region, and in comparable ecosystems in different geographical regions.
- Landscape ecotoxicology: This section describes current developments and extrapolation tools used in landscape ecotoxicology. Because ecological effect assessment of chemicals at the landscape level requires the integration of both spatial and temporal aspects, this section in particular builds further on the data presented in Chapter 6 on temporal extrapolation in ecological effect assessment of chemicals.

7.2 SPATIAL VARIABILITY AND ECOTOXICOLOGICAL DATA EXTRAPOLATION

7.2.1 INTRODUCTION

Ecological risk assessment is well advanced in developed regions of the world (e.g., North America and Western Europe), but it is less advanced in many developing nations (Calow 1998b). Furthermore, whereas most of the data and science underpinning ecological risk assessment are based on North American and European ecosystems, much of the world's biodiversity is concentrated in the tropics. Tropical forests, for example, are thought to contain more than half the world's biodiversity, and almost a quarter of the world's freshwater fish species are found in the Amazon (World Conservation Monitoring Centre 1992). It is not as though environmental contaminants only pose a problem in northern temperate regions; on the contrary, many of the current environmental issues affect non–northern hemisphere, nontemperate ecosystems (e.g., Henriques et al. 1997; Lacher and Goldstein 1997; Dudgeon 2000; Leonard et al. 2000). The dominance of ecotoxicological information on northern temperate species and ecosystems raises a number of important questions relating to the spatial extrapolation of ecotoxicological data, including the following:

- Should ecological risk assessments be based on toxicity data for indigenous species only, or can existing toxicity data for nonindigenous species be used?
- To what extent can ecotoxicological data be extrapolated from one geographical region to another, and what are the constraints on this extrapolation?

Here we explore these questions by initially considering the geographical distribution of species, before reviewing current information on the spatial extrapolation of toxicity data. We consider the extent to which the sensitivity of temperate and tropical species to environmental contaminants differs, and then consider similarities in species sensitivity between hemispheres (northern hemisphere versus southern hemisphere) and within climatic regions (Palearctic versus Nearctic). We then go on to consider extrapolations between habitat types within a single geographical region (lotic versus lentic) before exploring the importance of geographical location and scale in experimental multispecies systems (micro- and/or mesocosms).

TABLE 7.1
Wallace's zoogeographical regions

Region	Location
Nearctic	North American continent south to the Tropic of Cancer
Paleartic	Europe, Asia north of the Himalayas, northern Arabia, and a narrow strip of coastal North Africa
Neotropical	South America, part of Mexico, and the West Indies.
Ethiopian (Afrotropical)	Africa south of the Atlas Mountains and Sahara Desert, Madagascar, and southern Arabia
Oriental	Asia south of the Himalayas: India, Indochina, South China, Malaysia, and the western islands of the Malaysian Archipelago
Australian	Eastern islands of the Malaysian Archipelago, Australia, Tasmania, New Zealand, Papua–New Guinea, and South Pacific oceanic islands.

Source: Lincoln et al. (1983).

7.2.2 GLOBAL DISTRIBUTION OF SPECIES

The global distribution of species is a consequence of evolutionary, geological, and ecological factors and processes, and has been influenced by human activity. The distribution of animals can be classified in terms of 6 zoogeographical regions (i.e., Nearctic, Neotropical, Palearctic, Oriental, Australian, and Ethiopian; Table 7.1) and, for plants, in terms of 6 floral regions (Holarctic, Neotropical, Paleotropical, Australian, Cape, and Antarctic) (Cox and Moore 1993). The major difference between the faunal and floral regions lies in the recognition by plant geographers of 1) an Antarctic floral region, 2) the location of the boundary between Southeast Asia and Australia, and 3) the recognition of a Cape floral region (Takhtajan 1986).

Whereas some groups of organisms have a cosmopolitan distribution (the mussel genus *Mytilus*, the rodent genus *Myomorpha*, and *Homo sapiens*), many are endemic to a continent (e.g., Australian monotremes), island (e.g., the New Zealand kiwi), or single location (e.g., Lake Tanganyika cichlids). Endemism is low in Nearctic and Palearctic regions compared to other zoogeographical regions. For example, 88% of Australian reptiles, 90% of Neotropical freshwater fish, and 98% of Madagascan amphibians are endemic. Furthermore, of the approximately 8500 freshwater fish species worldwide, approximately 21% are endemic to the Amazon and 12% are endemic to Lake Malawi. In contrast, only about 10% of Nearctic birds and 30% of Nearctic freshwater fish are endemic (World Conservation Monitoring Centre 1992).

In general, biodiversity increases toward the tropics (Gaston et al. 1995); however, global diversity patterns can differ between taxa. For instance, a recent study of the geographical distribution of stream insects concluded that whereas Ephemeroptera were most diverse in the Ethiopian region, Plecoptera were most diverse in the Nearctic. Plecoptera genera richness was the greatest at 40° latitude, whereas Ephemeroptera genera richness peaked at 10° and 30° to 40° latitude (Vinson and Hawkins 2003). Differences in the global distribution patterns of species will

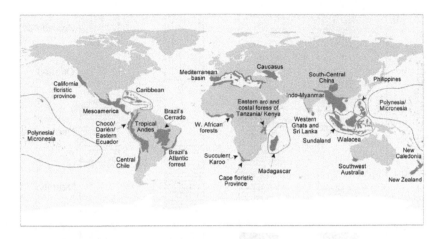

FIGURE 7.1 Global distribution of 25 biodiversity hotspots. *Source*: Identified by Myers et al. (2000).

result in spatial differences in community composition and hence potentially to spatial differences in sensitivity to environmental contaminants.

Tropical regions not only are the most biodiverse and have the highest levels of endemism, but they also contain many of the most vulnerable ecosystems. Myers et al. (2000) identified 25 biodiversity hotspots, areas that featured exceptional concentrations of endemic species and that were experiencing exceptional loss of habitat (Figure 7.1). Based on this analysis, the 5 main hotspots — Tropical Andes, Sundaland, Madagascar, Brazil's Atlantic forest, and the Caribbean — are in tropical regions (Neotropical, Ethiopian, or Oriental) and contain 20% of all plant and 16% of all vertebrate species worldwide. It should be noted that this prioritization scheme was based on vascular plants and vertebrates (excluding fish), a constraint imposed by data availability, and hence the areas identified may not be the major "hotspots" for other taxonomic groups. In fact, the highest levels of endemism for freshwater fish species are found in the Amazon River (circa 1800 species), Lake Malawi (circa 1000 species), and the Zaire River (circa 500 species), none of which are contained within the biodiversity hotspots identified by Myers et al. (2000); all 3 are, however, captured by the Global 200 priority scheme.

Global 200 identified 233 ecoregions whose biodiversity and representativeness (i.e., representative of distinct habitat types) are outstanding on a global scale (Olson and Dinerstein 1998). More than half (58%) of the ecoregions identified are terrestrial, 16% are freshwater, and 26% are marine; the dominance of terrestrial ecoregions is due in part to the higher endemism of terrestrial biota and the lack of biogeographic information for freshwater and marine organisms. Of the 36 freshwater ecoregions identified by Global 200, approximately half are Neotropical or Oriental (Figure 7.2) and include the major rivers of the world, such as the Amazon, Zaire, Mekong, Colorado, and Yangtze, and lakes, such as the rift valley lakes, Baikal, and the Great Basin lakes, as well as many smaller freshwater ecosystems.

According to Olson and Dinerstein (1998), most temperate freshwater biota are threatened by invasion of exotics, pollution, dams, and habitat degradation.

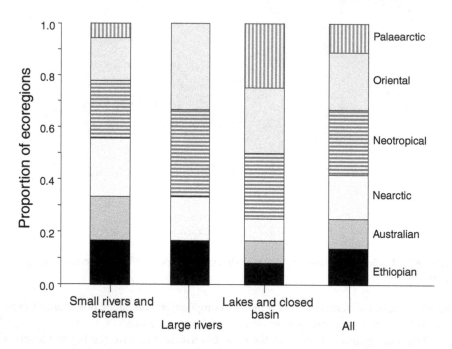

FIGURE 7.2 Geographical distribution of priority freshwater ecoregions. *Source*: Redrawn from data from Olson and Dinerstein (1998).

However, these threats are not restricted to temperate freshwaters. Habitat destruction in Asia (Dudgeon 2000), introduction of exotics in Africa (Lowe-McConnell 1994), and long-range transport of pollutants to polar regions have all had major impacts on freshwater ecosystems. In addition, environmental contaminants including those from mining activities (Mol et al. 2001; David 2003), petroleum industries, and agriculture (Abdullah et al. 1997) are causing problems in many developing regions.

7.2.3 SPATIAL EXTRAPOLATION OF TOXICITY DATA BETWEEN SPECIES

7.2.3.1 Temperate versus Tropical

As discussed in Chapter 6, numerous studies have demonstrated that toxicity generally increases with increasing temperature (Mayer and Ellersieck 1986; Donker et al. 1998; Boone and Bridges 1999). For example, a study of temperature–toxicity relationships of Australian species found that the toxicity of chlorpyrifos and endosulfan to fish and invertebrates generally increased with increasing temperature, but the trend with phenol was more complex (Patra 2000). The positive correlation between temperature and toxicity has led to the assumption that organisms from warm climates will be more sensitive than those from cold climates (Castillo et al. 1997; Peters et al. 1997). However, several authors have noted that fish endemic to coldwater habitats were slightly more sensitive than fish from temperate and subtropical habitats (LeBlanc 1984; Mayer and Ellersieck 1986). Moreover, Dyer et al. (1997) concluded that temperate fish were more sensitive than tropical fish, although this

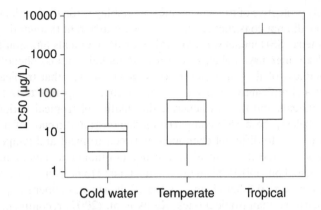

FIGURE 7.3 Box plot of LC50 values for coldwater, temperate fish species exposed to DDT. *Source*: Redrawn from data of Dyer et al. (1997).

difference was driven by only 1 of the 6 chemicals investigated, DDT (Figure 7.3). Interestingly, DDT (Mayer and Ellersieck 1986) and some of the pyrethroids (NRCC 1987) are among the few chemicals for which there is a negative correlation between water temperature and aquatic toxicity. Brix et al. (2001) also found a latitudinal gradient in copper sensitivity in fish, with tropical fish being less sensitive than temperate warmwater fish, who were, in turn, less sensitive than temperate coldwater fish.

Very few studies have compared the sensitivity of temperate and tropical invertebrates to environmental contaminants. Maltby et al. (2005) compared species sensitivity distributions for temperate and tropical arthropods exposed to the insecticides chlorpyrifos, fenithrothion, and carbofuran. In contrast to the fish studies discussed above, they reported a tendency for HC5 values to be lower for tropical arthropods (Figure 7.4), although this difference was not statistically significant.

Kwok et al. (2007) used species sensitivity distributions, constructed with acute toxicity data of freshwater animal species, to determine whether temperate data sets

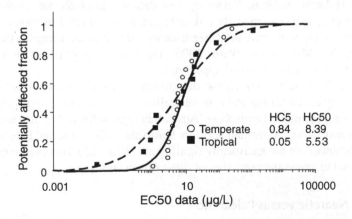

FIGURE 7.4 Species sensitivity distribution for tropical or temperate arthropods exposed to fenitrothion in acute toxicity tests. *Source*: Redrawn from data of Maltby et al. (2005).

are adequately protective of tropical species assemblages for 18 chemical substances. However, they did not construct separate SSDs for different taxonomic groups. The best-fit parametric SSD model was only valid for 10 chemicals (ammonia, cadmium, chromium, lead, mercury, nickel, zinc, lindane, malathion, and phenol) with a satisfactory goodness of fit. Furthermore, they demonstrated that tropical tests were conducted at a significantly higher temperature than temperate tests for 13 of the 18 tested chemicals and that, in general, the quality of tropical toxicity data was lower. Nevertheless, these SSD comparisons demonstrated trends of differences in species sensitivities to different chemicals between tropical and temperate aquatic organisms. For 6 of the 18 chemicals examined (ammonia, arsenic, zinc, chlorpyrifos, chlordane, and phenol), tropical organisms tended to be more sensitive than temperate ones. For several other chemicals tested, however, temperate species tended to be more sensitive than tropical ones. Kwok et al. (2007) recommended using an assessment factor of 10 if the water quality standard is primarily based on temperate species and a priori knowledge on the sensitivity of tropical species is very limited or not available. An extrapolation factor of 1 would protect 90% of species for more than 60% of chemicals.

Based on the above results, there is no evidence to support the contention that tropical fish are inherently more sensitive to environmental contaminants than temperate fish and only limited evidence to suggest that tropical invertebrates may be more sensitive than temperate invertebrates to selected chemicals. However, these observations are based on a very limited number of studies, and there are insufficient high-quality data available for the tropics to rigorously compare the sensitivity of temperate and tropical species.

7.2.3.2 Northern Hemisphere versus Southern Hemisphere

The majority of toxicity test data are generated using species from the northern hemisphere (i.e., Holarctic). For example, 9 of the 12 freshwater fish species used in the ecological risk assessment of atrazine (Solomon et al. 1996) and 27 of the 40 freshwater fish species used in the risk assessment of copper (Brix et al. 2001) are from Holarctic habitats. Relatively few data are available for southern hemisphere species, and consequently risk assessments conducted to protect southern hemisphere ecosystems have to utilize toxicity data obtained using northern hemisphere species (Muschal and Warne 2003). Does this matter? Based on the limited data currently available, it would appear not.

Hose and van den Brink (2004) demonstrated that there was no difference in the sensitivity of Australian and non-Australian fish species to endosulfan. Furthermore, although southern hemisphere (Australian region) freshwater fish are, on average, less sensitive to copper than northern hemisphere (Nearctic and Palearctic) fish, these differences are not statistically significant (Figure 7.5). However, more studies are required to assess the generality of these observations.

7.2.3.3 Nearctic versus Palearctic

As noted above, most toxicity data are available for Nearctic and Palearctic (in particular, western Palearctic) species. Although there are similarities in the fauna and

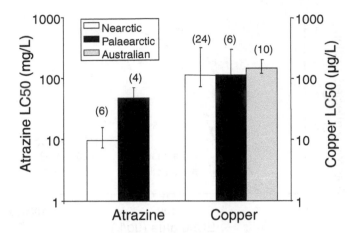

FIGURE 7.5 Median LC50 values, and interquartile ranges, for freshwater fish exposed to atrazine or copper in acute toxicity tests. *Note*: Fish are classified according to biogeographical region (i.e., Nearctic, Palearctic, and Australian), and the number of species in each category is given in parentheses. *Source*: Atrazine LC50 data are from Solomon et al. (1996), and copper data are from Brix et al. (2001).

flora of these 2 regions, there are also differences in biodiversity and community composition. For example, the freshwater fish faunas of the Nearctic and Palearctic both include pike (Esocidae), salmonids (Salmonidae), smelts (Osmeridae), sticklebacks (Gasterosteidae), and sturgeons (Acipenseridae). Both regions also have catfish, but in the Nearctic they belong to the family Ictaluridae, whereas in the Palearctic they belong to the family Siluridae. Oberdorff et al. (1997) compared the biodiversity of freshwater fish assemblages of 132 Holarctic rivers (91 western Palearctic and 41 Nearctic) and concluded that the western Palearctic rivers had a lower species diversity than the Nearctic rivers.

Few studies have compared the relative sensitivity of Nearctic and Palearctic species. Reanalyzing toxicity data for atrazine (Solomon et al. 1996) and copper (Brix et al. 2001) reveals no difference in the median sensitivity of Palearctic or Nearctic fish species to copper but a greater sensitivity of Nearctic fish to atrazine (Figure 7.5). Maltby et al. (2005) compared the sensitivity distributions of Nearctic and Palearctic arthropods to 4 insecticides (chlorpyrifos, fenithrothion, diazinon, and lindane). There was a tendency for the HC5 values generated using Palearctic species to be less than those generated using Nearctic species (e.g., Figure 7.6), but these differences were not statistically different.

These data indicate that the sensitivities of Nearctic and Palearctic species may vary but that these differences may be compound and taxon specific; Nearctic fish were more sensitive to atrazine, but Palearctic arthropods were more sensitive to fenitrothion. However, these observations are based on data for a very limited number of compounds and a small number of species and should therefore be interpreted with caution. Further research is required to determine the true extent of similarities or differences in the relative sensitivity of species from these 2 biogeographical regions.

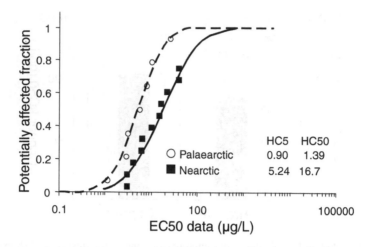

FIGURE 7.6 Species sensitivity distributions for Palearctic and Nearctic freshwater arthropods exposed to fenitrothion in acute toxicity tests. *Source*: Redrawn from data of Maltby et al. (2005).

7.2.3.4 Lentic versus Lotic Habitats

Freshwater ecosystems range in size from large lakes and rivers to small streams and ponds. All can be exposed to environmental contaminants, but risk assessment tends to focus on static water bodies, the rationale being that these provide a realistic worse case. But do they?

Relatively few studies have systematically compared the composition of freshwater communities, and even fewer have compared their relative sensitivity to environmental contaminants. A recent study, comparing the biodiversity of wetland macrophytes and aquatic macroinvertebrates of different freshwater ecosystems (rivers, streams, ditches, and ponds) within a lowland agricultural landscape, concluded that species richness was greatest for rivers and least for ditches (Williams et al. 2003). However, this depends on the spatial resolution. Although ponds had a lower average diversity than rivers, their community composition was more variable, and hence they supported the greatest number of species at a regional scale. Furthermore, although there was considerable overlap in the invertebrate assemblages of ditches with those of streams and ponds, there was little overlap between the invertebrate assemblages of ponds and those of rivers and streams. In contrast to the patterns in invertebrate communities, there was considerable similarity in the wetland plant assemblages of all water-body types.

Freshwater invertebrates in the orders Ephemeroptera, Plecoptera, and Trichoptera, the so-called "EPT taxa," are generally considered to be sensitive to environmental contaminants, whereas Diptera and oligochaetes are considered tolerant, an assumption underpinning several biotic indices of water quality (Norris and Georges 1993). This has led to the assertion that systems with a high proportion of EPT taxa are more likely to respond to chemical perturbation than those with a low proportion of EPT taxa (Versteeg et al. 1999). As EPT taxa are more common in flowing waters, it may be expected that lotic assemblages will be more sensitive to environmental

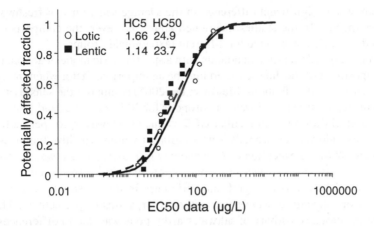

FIGURE 7.7 Species sensitivity distributions for lotic and lentic arthropods exposed to lindane in acute toxicity tests. *Source*: Redrawn from data of Maltby et al. (2005).

contaminants than lentic assemblages. Maltby et al. (2005) assessed the influence of habitat (lentic versus lotic) on the species sensitivity distributions of arthropods to 8 insecticides (carbaryl, chlorpyrifos, diazinon, fenitrothion, lambda-cyhalothrin, lindane, parathion-ethyl, and permethrin). There was no consistent pattern in the relative sensitivity of lentic or lotic species and no evidence of a significant difference in HC5 values among or within compounds. For example, single-species acute toxicity data were collated for freshwater arthropods exposed to lindane. Of the 39 species in the data set, 16 were found only in lentic habitats, and 9 only in lotic habitats; the remaining 14 species occurred in both lentic and lotic habitats and were therefore excluded from the analysis. There was no difference in species sensitivity distributions even though 47% of the lentic species were Diptera and 67% of the lotic species were EPT taxa (Figure 7.7). However, field and mesocosm studies of metal exposures for both North American and New Zealand lotic invertebrate communities have shown particularly high sensitivity for mayfly species and higher tolerance for caddisflies and stoneflies (Hickey and Clements 1998; Carlisle and Clements 1999; Clements et al. 2000). These studies suggest that general application of EPT indices in metal-polluted environments may not provide sensitive measures of community impacts.

7.2.3.5 Saltwater versus Freshwater Species

Toxicity data for saltwater organisms are often insufficient to assess risks. Freshwater toxicity data are usually more plentiful, and their use may provide a suitable surrogate for saltwater data. Wheeler et al. (2002b) used species sensitivity distributions to determine if freshwater data sets are adequately protective of saltwater species assemblages for 21 chemical substances. For ammonia and metal compounds, freshwater organisms tended to be more sensitive than saltwater species, whereas the opposite was true for pesticides and narcotic compounds (Wheeler et al. 2002b). De Zwart (2002), who compared 160 compounds, including 92 pesticides, concluded that there was no significant difference in the average sensitivity of freshwater and saltwater species. In the studies mentioned above, however, the comparisons were not based on SSDs constructed for different taxonomical groups.

It can be argued that for compounds with a specific toxic mode of action, the species compositions of the data sets used may be an important factor when interpreting species sensitivity distributions. Maltby et al. (2005) compared the sensitivity distributions of freshwater and saltwater arthropods for 10 insecticides and found no significant overall difference in median HC5 estimates. However, for permethrin and chlorpyrifos, saltwater arthropods were significantly more sensitive than freshwater arthropods. With the exception of *Chironomus salinarius*, all saltwater arthropods in these data sets were crustaceans, whereas the majority of freshwater arthropods were insects. There was no significant difference in the sensitivity distribution of freshwater or saltwater crustaceans to either chlorpyrifos or permethrin. Thus, the apparent increased sensitivity of saltwater arthropods was due to differences in the taxonomic composition of the data sets being compared, rather than a fundamental difference in the response of freshwater and saltwater taxa to insecticides (Maltby et al. 2005). These data suggest that in aquatic effect assessments of compounds with a specific toxic mode of action, the species sensitivity distribution approach can be used when the SSDs are constructed with freshwater and saltwater species of the most sensitive taxonomic group.

7.2.4 Size and Complexity of Test Systems and Community Responses

Experimental aquatic ecosystems have become widely used tools in ecotoxicology because they allow for a greater degree of control, replication, and repeatability than is achievable in natural ecosystems. The test systems in use vary from small indoor microcosms to large and complex outdoor experimental ecosystems. However, natural freshwater systems may also vary considerably in size and ecological complexity. Before addressing the spatial extrapolation of results of model ecosystem experiments that were conducted on different localities, the possible influence of the size and ecological complexity of test systems on responses to chemical stress will be discussed.

Differences in model ecosystem size and complexity are reported to have a profound effect on the enclosed community (Petersen and Hastings 2001). Important aspects of ecosystem and community functions may be controlled by keystone organisms too large or mobile to be confined in experiments that are smaller than the ecosystem of concern, such as large predatory fish. In addition, problems in interpreting micro- and/or meso-cosm experiments may be caused by inadequate or erroneous scaling of sediment–water interactions and potential artifacts associated with containerization (wall effects and water renewal times; Stephenson et al. 1986; Schindler 1998). Furthermore, large freshwater ecosystems are usually characterized by a diversity of habitats differing in abiotic and biotic properties (the pelagic or the littoral zone of lakes), whereas most artificial aquatic ecosystems usually simulate only one of these habitats.

Belanger (1997) analyzed data from more than 150 studies using model stream ecosystems ranging in size from 0.2 to 540 m long and from 0.05 to 4.3 m wide, and with a volume of 1.5 to 8×10^5 L. He concluded that although larger systems could be sampled more intensively and were more likely to contain fish, there was no relationship between test system size and the species richness of invertebrate, algal, or protozoan assemblages. Few studies have compared assemblages in model streams and natural streams, but those that have indicate that assemblages in model streams are representative of the natural streams from which they are derived (Belanger et al. 1995; Wong et al. 2004). Model ecosystems that simulate lentic aquatic ecosystems, however, usually contain species characteristic for deeper parts of freshwater ponds and often lack the species assemblages typical for littoral zones (Williams et al. 2002). Lentic model ecosystems used to assess the impact of chemicals can be conveniently divided into plankton- and macrophyte-dominated systems. Compared with macrophyte-dominated systems, plankton-dominated communities are usually characterized by a higher proportion of short-lived species (phytoplankton and zooplankton), lower biomass but higher turnover rates, and a less diverse macroinvertebrate community (Brock and Budde 1994).

The spatial scale and complexity of experimental tests systems may significantly affect the relationship between the fate and associated effects of contaminants. For example, in experiments introducing methoxychlor at the same initial concentration to enclosures of 3 sizes, Solomon et al. (1989) reported that reductions in zooplankton abundance in the smallest enclosures were numerically less, and their recovery rates faster, relative to the larger enclosures. Also, dissipation of methoxychlor from the water column was faster in the smallest enclosures, which was partially attributable to differences in (wall) surface area–volume ratios between different test systems (Solomon et al. 1989). Perez et al. (1991) examined the fate and ecological effects of the pesticide kepone in aquatic microcosms ranging in volume from 9 to 140 L and in which water depth was held constant. In the absence of kepone, the test systems showed size-dependent differences in the timing and magnitude of the phytoplankton bloom. Kepone reduced grazing pressure by zooplankton, resulting in an increase in phytoplankton densities. The time course of kepone in the overlying water was not affected by microcosm size, but the concentrations found in the sediment were dependent on the size of the test system (Perez et al. 1991). Other experimental studies with hydrophobic compounds reported that high densities of aquatic vascular plants may enhance dissipation rates from water and reduce concentrations in sediments (Crum and Brock 1994; Leistra et al. 2003). According to Johnson and Rodgers (2005), there is a need for more experimental studies specifically designed to manipulate scale as a controlled treatment variable. Furthermore, they stated that, in scaling, the exposure requires as much attention as the responses observed.

These examples described above indicate that certain properties of model ecosystems may affect the fate and exposure concentrations of a chemical and, consequently, the treatment-related effects observed. However, an important question remains: "Can responses of model ecosystem experiments be extrapolated between different types of experimental ecosystems if exposure concentrations are similar?"

It appears from several model ecosystem experiments with insecticides, where exposure concentrations are similar, that threshold concentrations for effects may be very similar between different types of test systems, at least when they contain representatives of sensitive taxonomic groups (in this case, arthropod populations). For example, both the observed half-life of chlorpyrifos of approximately 1 day and the $NOEC_{community}$ value of 0.1 µg/L observed in 18-L indoor plankton-dominated microcosms (van Wijngaarden et al. 2005a) were very similar to those observed in large 60 m³ outdoor macrophyte-dominated experimental ditches (van den Brink et al. 1996). The indoor microcosms were constructed with water and sediment material from the experimental ditches, and the most sensitive measurement endpoints comprised responses of microcrustaceans. Again, in these experimental ditches, populations of microcrustaceans were among the most sensitive measurement endpoints, along with several insect and macrocrustacean populations (van den Brink et al. 1996). Similarly, there was no evidence of a major difference in threshold concentration of chlorpyrifos in the 18-L indoor plankton-dominated systems that differed in ambient water temperature and nutrient concentrations. In both types of test systems that simulated temperate and Mediterranean conditions, an $NOEC_{community}$ of 0.1 µg/L was observed after a single application of chlorpyrifos, and again in both types of test systems microcrustaceans comprised the most sensitive measurement endpoint (van Wijngaarden et al. 2005a). A ditch enclosure experiment exploring effects of lambda-cyhalothrin application to macrophyte-dominated and plankton-dominated communities revealed similar dissipation rates of this compound, as well as similar threshold concentrations of direct toxic effects (close to 10 ng/L; Roessink et al. 2005).

In the examples described above, larger differences between types of test systems were observed at exposure concentrations well above the threshold concentration for direct toxic effects. For example, in the indoor plankton-dominated microcosms treated with chlorpyrifos (van Wijngaarden et al. 2005b), some clear indirect effects were observed after treatment with 1 µg/L (algal blooms and an increase in Rotifera), whereas no indirect responses could be detected in the more complex outdoor experimental ditches after treatment with 0.9 µg/L and 6 µg/L (van den Brink et al. 1996). Apparently, when evaluating indirect responses of chemical stress, laboratory microcosms sometimes show exaggerated responses to high concentrations of toxicants. In more structurally complex outdoor test systems, a greater number of feedback mechanisms may be available that dampen the indirect effects. Also, Roessink et al. (2005) concluded that, at higher concentrations of lambda-cyhalothrin, the magnitude and duration of effects differed between plankton-dominated enclosures (with a community characterized by short-lived organisms) and structurally more complex macrophyte-dominated enclosures. These observations are very much in line with those of Section 6.3.2 on seasonal variation in sensitivity of communities. The extrapolation of $NOEC_{community}$ values from one system to another seems to be possible with lower uncertainty than the extrapolation of hazard estimates of higher concentrations in which both direct and indirect effects, and recovery processes, are involved.

Computer models that simulate the dynamics of food webs might be used to extrapolate results of model ecosystem experiments to systems differing in size and ecological complexity. In principle, food-web models such as IFEM (Bartell et al. 1988), AQUATOX (Park 1999), and C-COSM (Traas 2004) can be adapted to do this, at least when detailed information on the ecology and ecotoxicology of the species and functional groups of concern is available.

7.2.5 GEOGRAPHICAL EXTRAPOLATION OF MODEL ECOSYSTEM EXPERIMENTS

Given the enormous natural variability in the structure and function of freshwater communities, it is reasonable to question the geographical extrapolation of model ecosystem experiments. Because most model ecosystems enclose parts of — or have been seeded with components of — natural communities, the geographical location of micro- and mesocosms will determine their species composition and hence potentially their sensitivity. From the data presented in the previous section (7.2.4), there is little doubt that structurally different ecosystems respond differently to toxic stress when exposed to concentrations well above threshold concentrations for direct toxic effects. The question is how unique such test systems are with respect to their threshold concentrations for direct toxic effects. The threshold concentration for direct effects is defined here as the highest concentration tested that causes no, or only a minor, effect on the most sensitive measurement endpoint. For pragmatic reasons, either effect class 1 or effect class 2 responses (see Chapter 1, Table 1.3) might be used as threshold concentrations. When answering the question raised above, it is very important to compare model ecosystem experiments characterized by similar exposure regimes. For this purpose, compounds were selected for which at least 5 $NOEC_{ecosystem}$ values could be derived from adequately performed model ecosystem experiments and that could be related to a specific exposure regime (short-term or long-term exposure; see Tables 7.2 to 7.4). These compounds

---

TABLE 7.2
Threshold concentrations for direct toxic effects of the most sensitive endpoints (responses of populations of crustaceans and insects) in model ecosystem experiments that studied the ecological impact of short term exposure to the insecticide chlorpyrifos

Application regime	Class 1	Class 2	Threshold concentration	Type of test system	Location	Reference
6 hour pulse	0.1 µg/L	—	0.1 µg/L	Experimental streams	Australia	Pusey et al. (1994)
Single	0.1 µg/L	0.3 µg/L	0.17 µg/L	Outdoor microcosms	Kansas, United States	Biever et al. (1994)
Single	0.1 µg/L	—	0.1 µg/L	Experimental ditches	The Netherlands	van den Brink et al. (1996)
Single	0.1 µg/L	—	0.1 µg/L	Lab microcosms, temperate	The Netherlands	van Wijngaarden et al. (2005b)
Single	0.1 µg/L	—	0.1µg/L	Lab microcosms, Mediterranean	The Netherlands	van Wijngaarden et al. (2005b)
Single	—	0.5 µg/L	0.5 µg/L	Lab microcosms	Minnesota, United States	Stay et al. (1985)

Note: For explanation of effect classes, see Table 1.3 of Chapter 1.

TABLE 7.3
Threshold concentrations for direct toxic effects of the most sensitive endpoints (community metabolism and responses of populations of primary producers) in model ecosystem experiments that studied the ecological impact of long-term exposure to the herbicide atrazine

Application regime	Class 1	Class 2	Threshold concentration	Type of test system	Reference
Single	—	5 μg/L	5 μg/L	Recirculating lab streams	Gruessner and Watzin (1996)
Constant	14 μg/L	25 μg/L	18.7 μg/L	Flow through lab streams	Nyström et al. (2000)
Single	—	2 μg/L	2 μg/L	Mesocosms (lentic)	Seguin et al. (2001)
Single	—	10 μg/L	10 μg/L	Lentic lab microcosms	Johnson (1986)
Single	5 μg/L	50 μg/L	15.8 μg/L	Lentic lab microcosms	Brockway et al. (1984)
Single	20 μg/L	—	20 μg/L	Lentic lab microcosms	Stay et al. (1985, 1989)
Repeated	5 μg/L	—	5 μg/L	Lentic lab microcosms	van den Brink et al. (1995)
Repeated	5 μg/L	10 μg/L	7.1 μg/L	Lentic field enclosures	Jüttner et al. (1995)
Repeated	10 μg/L	—	10 μg/L	Lentic lab microcosms	Pratt et al. (1988)

Note: For explanation of effect classes, see Table 1.3 of Chapter 1. Because atrazine is relatively persistent in the water of lentic test systems and recirculating experimental streams, these studies could also be used to assess effects of long term exposure.

comprise the surfactants dodecyl trimethyl ammonium chloride (C_{12}TMAC) and linear alkylbenzene sulfonate (LAS); the metal copper; and the pesticides atrazine, chlorpyrifos, and lambda-cyhalothrin.

Dyer and Belanger (1999) demonstrated that experimental stream assemblages had a similar sensitivity to municipal effluent as natural stream communities in the same region (Ohio, United States). There is also evidence that threshold concentrations for the surfactant C_{12}TMAC do not differ much between model stream experiments. Less than a twofold difference in $NOEC_{ecosystem}$ values (range 180 to 300 μg/L) is reported for long-term exposure to C_{12}TMAC derived from 5 artificial stream studies performed in different parts of the United States (Versteeg et al. 1999). A critical literature review of 6 experimental stream studies with LAS, however, revealed a fifteenfold difference in normalized $NOEC_{eco}$ values in the range of 74 to 1113 μg/L (Belanger et al. 2002). In addition, 7 lentic model ecosystem studies with LAS are described in this review paper. These studies, however, are more difficult to interpret due to differences in exposure regime employed and measurement endpoints selected (Belanger et al. 2002).

TABLE 7.4
Summary of threshold concentrations (geometric mean, range, 95% confidence limits, and spread) for 6 toxicants

Compound	Exposure regime	Geometric mean	Range (min–max ratio)	95% confidence limits (spread)	Reference
$C_{12}TMAC$	Long term	231 µg/L (n = 5)	180 to 300 µg/L (1.7)	192 to 277 µg/L (1.4)	Versteeg et al. (1999)
LAS	Long term	365 µg/L (n = 6)	74 to 1113 µg/L (15)	157 to 849 µg/L (5.4)	Belanger et al. (2002)
Cu	Long term	3.8 µg/L (n = 7)	2.1 to 6.3 µg/L (3)	2.9 to 5.3 µg/L (1.8)	Versteeg et al. (1999)
Atrazine	Long term	8.4 µg/L (n = 9)	2 to 20 µg/L (10)	5.3 to 13.4 µg/L (2.5)	Table 7.3
Chlorpyrifos	Short term	0.14 µg/L (n = 6)	0.1 to 0.5 µg/L (5)	0.08 to 0.24 µg/L (2.9)	Table 7.2
Lambda cyhalothrin	Short-term repeated pulses	6.9 ng/L (n = 6)	2.7 to 10.0 ng/L	4.3 to 11.2 ng/L (2.6)	Brock et al. (2006)

[a] The toxicants are the surfactants dodecyl trimethyl ammonium chloride ($C_{12}TMAC$); linear alkylbenzene sulfonate (LAS); the metal copper (Cu); and the pesticides atrazine, chlorpyrifos, and lambda cyhalothrin as observed in aquatic model ecosystem experiments.

A threefold difference in threshold concentrations is reported for long-term exposure to copper (adjusted to 50 mg/L $CaCO_3$ hardness) derived from 1 lentic mesocosm and 6 artificial stream studies conducted in the United States and Europe (Versteeg et al. 1999). The $NOEC_{ecosystem}$ values of these studies with copper were in the range of 2.1 to 6.6 µg/L.

Recently 2 reviews of model ecosystem experiments with herbicides (Brock et al. 2000a) and insecticides were performed (Brock et al. 2000b; van Wijngaarden et al. 2005b). The following criteria were applied in the selection of the studies:

- The test system represents a realistic freshwater community, and populations of various trophic levels are present.
- The description of the experimental setup is adequate and unambiguous.
- The exposure concentrations are well described.
- The investigated endpoints are sensitive to the substance, and effects can reasonably be expected to be related to the mode of action of the pesticide.
- The effects are statistically significant and show an unambiguous concentration–effect relationship, or the observed effects are in agreement with a dose–effect relationship from additional studies.

The responses observed for the most sensitive measurement endpoints (univariate or multivariate) at each exposure concentration were assigned to the 5 effects classes described in Table 1.3 (Chapter 1). Effect class 1 (effects could not be demonstrated) and effect class 2 (slight effects, usually observed on a single sampling date immediately post application only) were used to derive threshold concentrations for direct toxic effects. For the insecticide chlorpyrifos, 6 model ecosystem experiments were available (1 lotic and 5 lentic studies) from which reliable ecological threshold concentrations for short-term exposure could be derived (Table 7.2). These threshold concentrations ranged from 0.1 to 0.5 µg/L (a fivefold difference). For the herbicide atrazine, no less than 9 suitable model ecosystem experiments (2 lotic and 7 lentic) were available that allowed comparison of threshold concentrations for long-term exposure (Table 7.3). These threshold concentrations differed tenfold and ranged from 2 to 20 µg/L. Brock et al. (2006) summarized the results of several lentic model ecosystem experiments performed with the nonpersistent pyrethroid, lambda-cyhalothrin. All studies were characterized by multiple applications (repeated pulse exposures), and the effect classes 1 to 2 threshold concentrations ranged from 2.7 to 10 ng/L.

The variability in observed threshold concentrations for effects was relatively small (min–max ratio < 5) for the compounds C_{12}TMAC, copper, chlorpyrifos, and lambda-cyhalothrin, and relatively large for atrazine (10) and LAS (15). In part, this can be explained by experimental limitations such as the ecological complexity of the test systems used and the range in exposure concentrations and measurement endpoints selected. Other confounding factors may be related to seasonal and successional variations in sensitivity of communities and ambient environmental factors such as temperature and light conditions (see Section 6.3.2), as well as to the exposure history of the populations that inhabit the test systems (see Section 6.3.4). It can be argued that outliers should not unduly influence the conclusions. For this reason, analogous to the method used by Blanck et al. (2003), we calculated the 95% confidence limits and took the ratio of the upper and lower confidence limits (the spread) as a measure of the variability (Table 7.4). These ratios (spreads) can be considered as measures of uncertainty in the geographical extrapolation of threshold concentrations for effects. It can be concluded from the data presented in Table 7.4 that the geographical uncertainty factor (spread) for threshold concentrations is in the range of 1.4 to 5.4 for the compounds and model ecosystem experiments selected.

Our observations are similar to those of Blanck et al. (2003), who studied the variability in zinc tolerance in periphyton communities sampled from 15 European river stretches using the PICT concept. Due to differences in water chemistry, (history of) metal pollution, species composition, and other biotope characteristics, the regional uncertainty factor for Zn was estimated to range from 1.7 to 4.3, and the interregional uncertainty factor from 2.4 to 8.6, when extrapolating periphyton tolerances from river to river (Blanck et al. 2003).

7.3 LANDSCAPE ECOTOXICOLOGY

7.3.1 INTRODUCTION

The spatial distribution of pollutants in ecosystems and landscapes tends to be patchy or aggregated. In most areas, exposure to pollutants occurs at relatively low concentrations over extended periods of time; however, a limited number of areas may contain high concentrations of pollutants that may be persistent or occur for short periods of time (Widianarko 1997). In addition, the physicochemical properties that control exposure and organisms are not uniformly distributed in landscapes. In well-defined landscape units like undisturbed watersheds, species are more or less predictably structured according to gradients in current velocity and substrate composition of sediments. In riverine systems, resource requirements, such as different types of organic matter, predict the shifts in the relative abundance of feeding guilds over the longitudinal profile of a river, as described by the river continuum concept (Vannote et al. 1980). The ways in which tolerances and requirements of species interact and match the conditions and resources provided by certain habitats are addressed in the niche-assembly concept. Actual ecological communities, however, are governed by both niche-assembly and dispersal-assembly rules, along with ecological drift (Hubbell 2001). An example of how the distribution of fish species in a watershed was affected by toxicants is provided by Napier (1992) and Chapman et al. (1993). Downstream of an acid rock drainage-affected stream in inland Australia (due to mining activities), the tributaries in a forested catchment appeared to support a variety of small native fish species that were less common elsewhere. Apparently, the main stream, subject to chemical stress, was not able to act as a conduit for invasive introduced species such as trout or mosquitofish. Due to acidic water (pH 2.7) and very high concentrations of copper, zinc, cadmium, and lead, the only form of macroscopic life in the stream for the first 12 to 16 km was a metal-tolerant alga *Hormidium* sp. There was a steady increase in the mean number of macroinvertebrate taxa with distance downstream, aided by freshwater inputs from tributaries. By 32 km downstream of the abandoned mine, concentrations had fallen below Australian water quality criteria current at the time, yet the macroinvertebrate community still lacked some sensitive species of Ephemeroptera and Mollusca (Napier 1992; Chapman et al. 1993).

Examination of the effects of toxic chemicals at scales larger than those usually considered in environmental toxicology has led to the establishment of a new approach in ecotoxicology (i.e., landscape ecotoxicology). This approach has its focus on the spatiotemporal configuration of populations, communities, and ecosystems in the landscape as affected by the spatiotemporal distribution of toxic chemicals in the subjected landscape. Typically, in landscape ecotoxicology the problem is one of scaling up: moving from short-term studies of small areas (e.g., performed in the laboratory or in specific outdoor experimental ecosystems) to make inferences about the longer term dynamics of larger systems (Johnson 2002). According to Spromberg et al. (1998), the temporal and spatial heterogeneity of organisms and environmental toxicants is purposefully minimized in laboratory and field experiments, whether they are single-species tests or elaborate mesocosm tests. The reason for minimizing this variability is to generate sufficient statistical power to conduct meaningful statistical analyses with practical numbers of samples and replicates. This, however, excludes the heterogeneity of both abiotic and biotic parameters, a fundamental property of real ecological systems.

In a chemically disturbed landscape unit, exposure concentrations, as well as population and community responses, may show clear spatial patterns. Assessing risks at a landscape scale (e.g., watershed), however, requires the development of a conceptual model for the identification of chemical stressors against a background of other physicochemical and habitat factors that may shape biological communities. Such a model can only be developed after characterization and condition assessment of the landscape unit of interest. The characterization of the landscape unit will assist in problem definition and in identifying suitable targets for quantification of adverse biological effects. Geographical Information System (GIS) software is an important tool to conduct spatially explicit risk assessments for environmental problems ranging from small to large landscape units, to regions.

7.3.2 GIS AND RISK MAPS

A spatially explicit environmental risk assessment is one in which the estimates of risk differ for different sites in well-defined landscape units and may be used to refine the generic, usually conservative, first-tier risk assessment. The approaches usually rely on GIS software packages. GIS software packages are collections of tools representing specific assumptions about spatial correlations in the landscape unit under investigation. For example, a spatial analysis of the Mississippi cotton landscape in the United States was used to estimate realistic exposure concentrations in aquatic ecosystems of pyrethroid insecticides used on cotton. Image-processing techniques and GIS were applied to investigate the number and size of the water bodies in the landscape and their proximity to the agricultural fields with cotton. Results showed that these techniques can be used cost-effectively to characterize the agricultural landscape, and provide verifiable data to refine conservative model assumptions to assess exposure concentrations (Hendley et al. 2001). These approaches might also be used to provide a more realistic evaluation of the potential for effects. To accomplish this, however, information needs to be incorporated on the spatiotemporal variability in the species composition of the aquatic habitats at risk, and their sensitivity to these pyrethroid insecticides.

The cotton landscape example presented above reveals that landscape-level risk assessment can be conducted by investigating the influence of the surrounding landscape on the emission of insecticides to the water bodies of concern in order to characterize more realistically actual exposure concentrations. This relatively simple approach addresses variability within the landscape, but pays less attention to the interactions between water bodies. A more complex approach is to assess the fate and effects of a chemical (or combination of stressors) for the entire watershed and to consider this watershed as a true continuum. The latter approach may include all water bodies within a watershed and addresses their interdependence, for example, by studying the flow of water, chemicals, matter, and organisms between these systems. An example of such a watershed approach is the study of Pandovani et al. (2004). They used a landscape-level approach to assess aquatic exposure via spray drift of chlorpyrifos-methyl in the watershed of the Simeto River in Sicily (Italy).

GIS may also be used to assess the site-specific bioavailability of toxicants and their ecological risks. For example, Prusha and Clements (2004) related metal concentrations in the lotic insect *Arctopsyche grandis* to physicochemical characteristics measured in 16 streams. GIS was used to calculate landscape attributes in these streams where DOC concentrations were measured. Multiple linear regression showed that the percentage of forested area explained 47% and 59% of the variation in maximum and mean DOC concentrations, respectively. The maximum concentration of DOC was negatively correlated with the concentrations of Zn and Cd in *Arctopsyche*. The results indicate that the percentage of forested area within a watershed can be used to describe DOC concentrations, which in turn influence metal bioavailability.

Landscape analysis allows the development of "risk maps" that show the spatial distribution of concentrations of toxic chemicals. This information can be combined with information on the spatial distribution of ecological features, allowing potential "hotspots" to be identified (see Figure 5.3 in Chapter 5 as an example). Risk maps may have applications in the development of spatially differentiated management strategies. The main reason why risk maps are useful is that they present complex spatial information in a manner that is easily interpreted. However, according to Woodbury (2003), risk maps can also be misleading because they may suggest that there is more information than actually exists. To judge whether a map of predicted values is appropriate, it may help to indicate on the map where data were collected and to show data only for the area that might influence the analysis.

GIS and risk maps are useful tools, but they are not standard methods or a panacea to conduct spatially explicit risk assessments. Woodbury (2003) described the following do's and don'ts that help to improve the credibility of spatially explicit environmental risk assessments.

Do's

- Account for errors in spatial data.
- Recognize that interpolation and other steps in an analysis represent models with specific assumptions.
- Use available explanatory data first before performing spatial interpolation.
- Use related data that are sampled more densely to improve interpolated estimates.
- Use maps to communicate information about uncertainties in spatial data and uncertainties in spatially explicit analyses.

Don'ts

- Conflate data and models.
- Ignore physical barriers or other reasons why spatial correlations may differ in different locations.
- Ignore issues of scaling and aggregation.
- Ignore the propagation of errors in GIS analyses.

7.3.3 Spatially Explicit and Metapopulation Modeling Approaches

Upscaling in environmental risk assessment concerns, amongst other things, the extrapolation of ecotoxicological data derived from laboratory tests and model ecosystem experiments to larger areas by means of spatially explicit modeling tools. Spatially explicit models increase ecological realism by incorporating landscape structure and habitat quality explicitly and by considering the size and the distance between the landscape patches of concern, and the properties of the landscape between these patches, which might affect dispersal or movement of organisms in the landscape. Variation in habitat quality can be incorporated using spatially referenced data sets or land use patterns, landscape characteristics, and stressor distributions. In this respect, spatially explicit population models linked to GIS are promising tools to estimate the potential risks of major chemical stressors and of multiple stressors at the landscape level, particularly for mobile wildlife species that occupy a large territory and different types of habitats in the landscape (Bartell et al. 2003; Pastorok et al. 2003; Topping et al. 2003). An overview of techniques and models that can be used in the ecological risk assessment of riverine landscapes is provided by Leuven and Poudevigne (2002).

Metapopulation models explicitly deal with environmental heterogeneity in the distribution of habitats and organisms in the landscape. According to Hanski and Gyllenburg (1993), a metapopulation is a "population of populations" of the same species connected through immigration and emigration. In discontinuous habitats, corridors can connect habitat patches and integrate them into networks where populations can sustain in metapopulations. The minimum viable population (MVP) size is a population size below which patch extinction occurs. The carrying capacity (CC) is that population size that can just be maintained without a tendency to increase or decrease. A subpopulation may serve as a sink if it is below the MVP and drains immigrants, or it may serve as a source for nearby patches by providing immigrants to them. The addition of a toxicant to a source patch will have a greater impact than the same toxicant addition to a sink patch.

In densely populated areas, or in areas characterized by intensive agriculture, many rare and/or protected species are currently restricted to small areas of marginal habitat quality. These small populations run a high risk of extinction due to stochastic events (Lande 1993). Moreover, with habitats of marginal quality, their long-term population growth rate will be small, making them even more vulnerable to stochastic events (Klok 2000). Small populations are threatened by a variety of factors that endanger their persistence. Schaffer (1981) recognized 4 stochastic extinction pressures:

- Demographic stochasticity
- Environmental stochasticity (habitat quality degradation)
- Genetic stochasticity (inbreeding)
- Natural catastrophes (e.g., lethal disease)

Reduction in habitat quality by persistent pollutants can decrease the survival and reproduction of individuals that predominantly dwell on or near the polluted site and, in this way, increase the extinction probability of populations. This can become effective through direct exposure or through transfer of the pollutants through the food web. In mobile species, this also depends on the relative proportion of contaminated sites in the total range of landscape elements that individuals use to forage within their territory, the availability of alternative food sources, and the extent of habitat fragmentation.

Populations of endangered species often consist of small numbers that may be trapped in the so-called "extinction vortex." This means that the combination of inbreeding, demographic stochasticity, and genetic drift leads to feedback loops that make small populations even smaller (Caughley 1994). According to Klok (2000), the minimum viable population size is species specific because the life history of a species can have a large impact on the outcome of the aforementioned stochastic events. Moreover, the MVP size will also depend on the quality of the habitat and, consequently, on the temporal and spatial distribution and availability of pollutants.

Metapopulation models have been used to examine the dynamics of populations resulting from chemical stress. To date, most ecotoxicological studies that have used metapopulation models addressed the dynamics and response of terrestrial populations of wildlife and arthropod species. For example, Sherratt and Jepson (1993) showed that the persistence of a phytophagous predator population in the landscape is enhanced if only a few fields are sprayed by a pesticide, the application rate of the pesticide is low, or the intrinsic toxicity of the pesticide is low. Another important finding was that pesticide application patterns can cause the prey insect population to reach higher densities than would occur otherwise. Dispersal rates of the predator and the prey are important factors determining the prey population densities. Mauer and Holt (1996) used several types of metapopulation models to investigate the importance of migration and other factors determining the impacts of pesticides. An increase in migration rate among patches decreased the persistence of the population, and an increase in the rate of reproduction increased the persistence of the population in the landscape. The more toxic the pesticide and/or when more fields were sprayed, the less persistent was the population.

Spromberg et al. (1998) used a toxicant-treated metapopulation model to explore the range of possible dynamics of populations in contaminated field sites. A single-species metapopulation model was developed, and the distribution of the chemical was assumed to be limited to one patch and contagious within that patch. Both persistent and degradable toxicants were modeled. Five principal conclusions resulted from the simulation studies:

- Mortality in one subpopulation has ecologically significant effects on nondosed subpopulations ("action at a distance").
- When uncontaminated sites are indirectly affected by the migration of biota from contaminated sites, these uncontaminated sites cannot be reference systems.
- The arrangement of the patches is critical to the dynamics of the systems and the overall impact of a toxicant.
- Due to the contagious distribution of the toxicant and the stochastic function describing exposure and effects, multiple discrete outcomes often are possible from the same initial conditions (e.g., ranging from extinction to the reaching of the carrying capacity for a patch).
- If sufficient cleanup is not possible, it may be necessary to isolate the contaminated patch, allowing the formerly connected patches to regain more typical population dynamics.

Recently, metapopulation models have been successfully applied to assess the risks of contaminants to aquatic populations. A metapopulation model to extrapolate responses of the aquatic isopod *Asellus aquaticus* as observed in insecticide-stressed mesocosms to assess its recovery potential in drainage ditches, streams, and ponds is provided by van den Brink et al. (2007). They estimated realistic pyrethroid concentrations in these different types of aquatic ecosystems by means of exposure models used in the European legislation procedure for pesticides. It appeared that the rate of recovery of *Asellus* in pyrethroid-stressed drainage ditches was faster in the field than in the isolated mesocosms. However, the rate of recovery in drainage ditches was calculated to be lower than that in streams and ponds (van den Brink et al. 2007). In another study, the effects of flounder foraging behavior and habitat preferences on exposure to polychlorinated biphenyls in sediments were assessed by Linkov et al. (2002) using a tractable individual-based metapopulation model. In this study, the use of a spatially and temporally explicit model reduced the estimate of risk by an order of magnitude as compared with a nonspatial model (Linkov et al. 2002).

7.3.4 PROSPECTIVE AND RETROSPECTIVE RISK ASSESSMENT AT THE LANDSCAPE LEVEL

Any ecological risk assessment at the landscape level has to start with the question, "What has to be protected?" This protection aim preferably needs to include a spatial component (e.g., protecting the aquatic biodiversity from pesticide stress in watercourses neighboring agricultural fields). It may also include a temporal component: for example, consider only effects on the densities of aquatic populations to be acceptable in drainage ditches neighboring agricultural fields that show full recovery within a certain time period (e.g., 8 weeks) but do not allow these effects in main watercourses connected to these ditches (see Section 1.3.1 in Chapter 1 for a discussion on this topic).

All approaches to assessing ecological risks of toxicants at the landscape level heavily rely on the proper linking of exposure concentrations (or regimes) to ecotoxicological and ecological data. Relevant exposure concentrations in the landscape unit of concern can be obtained either by chemical monitoring (which is expensive), by applying fate models to derive PECs (characterized by uncertainty), or by a combination of monitoring and modeling. For example, in the probabilistic aquatic ecological risk assessment of atrazine in North American surface waters, both monitoring data and model predictions are used in the exposure assessment (Giddings et al. 2005).

It is clear that an accurate exposure prediction at the landscape level requires models calibrated and validated for the landscape unit of interest and that the input parameters used have a high precision and accuracy for the area of interest (see Section 1.7 in Chapter 1). However, in a prospective risk assessment for new chemicals not yet placed on the market, chemical monitoring data are not yet available, and exposure predictions at the landscape level may be characterized by a relatively high uncertainty because the scale and intensity of the use of these chemicals are not yet known. When performing prospective risk assessments at the landscape level, a cost-effective, generic method to get insight into possible exposure concentrations is the development of exposure scenarios. For example, within the European Union, harmonized approaches for conducting aquatic exposure assessments for agricultural pesticides have been developed. These are documented in the "FOCUS Surface Water Scenarios" report (FOCUS 2001). The realistic worst-case scenarios developed aim to predict exposure concentrations in ponds, ditches, and streams for different agricultural landscapes of Europe, taking into account agronomic and climatic conditions relevant to crops. Currently, 10 scenarios for the compartment surface water have been designed, which collectively represent agriculture in the European Union and which are used for the European registration procedure of pesticides (FOCUS 2001). When applying the exposure scenario approach at the landscape level, the fate model-specific driving climatic, soil (and slope), and ecosystem properties needs to be identified for the areas of interest using a sensitivity analysis (see, for example, FOCUS 2005; European Food Safety Authority 2006). These model specific driving properties are of high numerical relevance for the model output. A similar approach has been used in the identification of 11 agroecological regions in the United States that are used in modeling exposures to pesticides (Mangels 2001; Giddings et al. 2005).

After having collected data on (the dynamics of) exposure concentrations in the landscape unit of concern (by either monitoring or prediction), the second step in the risk assessment is the linking of exposure and effects. Lack of a clear conceptual basis for the interface between the exposure and effect assessment may lead to a low overall scientific quality of the risk assessment. This interface is defined by Boesten et al. (2007) as the type of concentration that gives the best correlation to ecotoxicological effects and is called the ecotoxicologically relevant concentration (ERC). The ecotoxicological considerations determining the ERC may include the following: 1) In which environmental compartment do the organisms at risk live (e.g., water or sediment)? 2) What is the mode of action of the toxicant? 3) What is bioavailable for the organism (see Chapter 2, Section 2.4)? 4) What is the influence of the exposure pattern (e.g., short peaks or constant concentration over long periods) on the type and degree of the effects? And 5) which information is available on the "time to effect" to determine whether short-term or long-term exposures are relevant (see Chapter 6, Section 6.2)? For instance, for a lentic aquatic insect living associated with macrophytes, the ERC could be the maximum concentration over time of the dissolved fraction for a fast-acting insecticide or some time-weighted average (TWA) concentration for a slow-acting fungicide (e.g., the 7-day or 21-day TWA). For sediment-dwelling insects that live predominantly in the top centimeters of the sediment, the ERC could be the maximum over time of the pore water concentration of the insecticide in the top 2 cm of the sediment. Note that at the ecosystem and landscape levels, the ERC may be different for different populations, life stages of species, and so on. After the ERCs for the relevant populations have been selected, the collected exposure data can be linked to the relevant ecotoxicological data. Key is that the same type of ERC should be used consistently in the exposure estimates related to the field and related to the ecotoxicological experiments used to predict the risks. To do this, matrix and media extrapolation (see Chapter 2), and models to predict effects of time-variable exposure (Chapter 6, Section 6.1.3) and mixtures (Chapter 5), may be required.

In translating measured and/or predicted exposure data of individual chemicals into ecological risks, a simple approach is to compare the exposure data with established water quality standards based on laboratory toxicity tests (e.g., the standard test species–UF approach; the HC5 derived from SSDs). This, however, does not provide any insight into the concentration–response relationships at the ecosystem level. A more complex approach, which may give insight into ecological responses that include direct and indirect effects and recovery, is the use of results of model ecosystem experiments performed with the toxicants of concern, or with other toxicants with a similar mode of action. A case-based reasoning (CBR) approach can be used that predicts the effects of a given concentration of a toxicant based on the outcome of already performed micro- and/or mesocosm experiments with toxicants with a similar mode of action, and that expresses the exposure concentration as fractions of toxic units. Case-based reasoning is a problem-solving paradigm that is able to utilize the specific knowledge of previously experienced, concrete problem situations (cases) for solving new problems. A very important feature of CBR is its ability to learn. By adding present experience into the case base, improved predictions can be made in the future (Aamodt and Plaza 1994). Recently, a case-based reasoning methodology, PERPEST, for the prediction of pesticide effects on freshwater ecosystems has been developed (van den Brink et al. 2002c, 2006b). Currently, the PERPEST database (PERPEST n.d.) contains more than 208 herbicide and 213 insecticide cases derived from more than 104 micro- and/or mesocosm experiments. Currently, the database has been extended with available data for fungicides. The PERPEST model searches for analogous situations in the database based on relevant environmental fate characteristics of the compound, exposure concentration, and type of ecosystem to be evaluated. A prediction is provided by using weighted averages of the effects reported in the most relevant literature references. PERPEST results in a prediction showing the probability of no, slight, or clear effects on the various grouped endpoints (van den Brink et al. 2002c). Figure 7.8 presents an example of how the PERPEST model can be used to evaluate the risks of the peak concentrations of chemical-monitoring data of the herbicide diuron in several waterways in the Bommelerwaard area of The Netherlands.

Another approach in translating measured and/or predicted exposure data, and other relevant information at the landscape level, into ecological risks is the use of metapopulation-modeling approaches (e.g., the INTASS methodology described by Emlen et al. 2006; studies referred to in Section 7.3.3) and food-web models (see Chapter 4, Section 4.5.4). For an overview of landscape models that can be used in risk assessment, see Mackay and Pastorok (2002). In prospective risk assessment at the landscape and regional level (e.g., for pesticide registration), a promising approach might be to develop harmonized, generic risk assessment approaches by developing realistic worst-case scenarios that combine fate- and effect-modeling approaches. Such a combined scenario approach allows one to extrapolate experimentally derived fate and effect endpoints to the landscape of interest (see, for example, Probst et al. 2005).

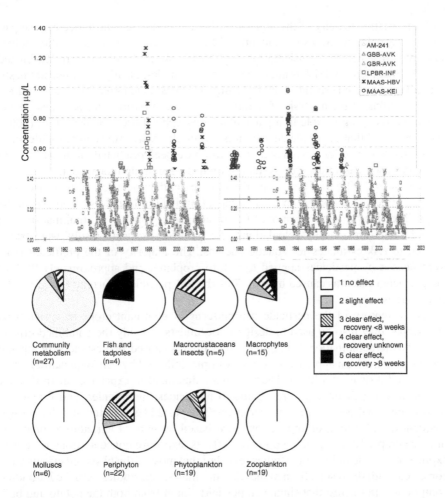

FIGURE 7.8 Dynamics in exposure concentrations of the herbicide diuron in 6 waterways as measured in the Bommelerwaard area, The Netherlands, from 1990 to 2002 (data from Steen 2002), and the probability of effects of the peak concentrations observed (1.25 µg diuron/L) as predicted by PERPEST.

7.3.5 Predicting the Impact of Multiple Stressors

There are different, complementary approaches to assessing ecological risks of multiple stressors at the landscape level. The reductionist approach aims at identifying risks to populations and ecosystems of concern on the basis of accumulated data on simple stressor–effect relationships, and by identifying the main stressors of concern. This can be achieved by combining field observations and experimentation. Experimentation has the advantage of providing evidence of causation.

Belden et al. (2007) used a 3-step approach to evaluate the relative toxicity and the occurrence pattern of pesticide mixtures in streams draining agricultural watersheds. First, a landscape of interest was identified as the corn–soybean crop setting in the United States. Second, the relative toxicity of mixtures was compared with the relative toxicity of the highest individual pesticide. Third, occurrence patterns of pesticide mixtures were identified for use in follow-up mechanistic studies. The study revealed that consideration of pesticide mixtures increased the estimated risk, but that increased risk is not large for most samples. Nearly all important pesticide mixtures were herbicides. So, focusing on a "pesticide usage landscape" allows simplifications of the analysis of potential pesticide mixtures and provides results that are directly applicable to a particular geographical area. Belden et al. (2007) also concluded that the pesticide mixtures within the "corn–soybean pesticide landscape" tended to be less complex than would have been predicted on the number of pesticides used within the crop system or on uncensored monitoring data.

De Zwart (2005) used a novel method to predict the effects of multiple stressors: caused by pesticides based on a GIS map of agricultural land use, comprising 51 crops. Through the application of SSDs for aquatic organisms, in combination with rules for mixture–toxicity calculations, the modeled exposure results were transformed to risk estimates for aquatic species. The majority of the predicted risks were caused by pesticides applied to potato cropland, and approximately 95% of the predicted risk was caused by only 7 of the 261 pesticides currently used in The Netherlands.

An inventory of the pesticide use patterns in dominant crops in agricultural landscapes may also be used to design experiments to study the combined effects of a realistic combination of multiple stress effects caused by pesticides in aquatic ecosystems. Such an approach was, for example, followed by van Wijngaarden et al. (2004) and Arts et al. (2006). These authors simulated in experimental freshwater ecosystems a realistic exposure regime to herbicides, insecticides, and fungicides used in bulb and potato crops, respectively. The pesticides selected and the dosage, frequency, and timing of application were based on normal agricultural practices in these crops. To interpret the observed effects, exposure concentrations were also expressed in toxic units. Both studies revealed that most of the observed effects were consistent with the results from higher tier and mesocosm studies with the individual compounds. In the test that simulated pesticide input from both the potato and bulb crops, the most severe effects on the aquatic community were caused by a pyrethroid insecticide. Multiple and repeated pesticide stresses played a small role within the applied pesticide package, because of the rapid dissipation of most substances. Both studies suggest that risk assessments based on the individual compounds would in these cases have been sufficiently protective for their uses in a crop protection program (van Wijngaarden et al. 2004; Arts et al. 2006).

If, through a proper landscape analysis, a realistic combination of multiple stressors is identified, food-web models may be used to predict and extrapolate their ecological effects to relevant ecosystems of the landscape unit of concern. An overview of models that can be used for the integrated assessment of eutrophication and organic contaminants in aquatic ecosystems is provided by Koelmans et al. (2001). Examples of aquatic food-web models that can be used or adapted to predict effects of multiple stressors are IFEM (Bartell et al. 1988), AQUATOX (Park 1999), and C-COSM (Traas 2004).

The examples described above mainly concern a reductionist approach in which, for example, "pesticide landscape" information is combined with experimentation and/or modeling. The alternative holistic approach takes the whole landscape or watershed into account and assumes that the impairment of interrelated ecosystems by realistic combinations of multiple stressors can be studied using indicators without knowing all the details of the internal structure and functioning of the systems at risk (Leuven and Poudevigne 2002). For example, an emphasis on sustaining water resources has led to the development of ecosystem indicators, useful in determining the "state of the watershed" (Richter et al. 2003). The biotic indices (such as the index of biotic integrity, the total number of invertebrate or EPT species per area, or the presence or absence of key indicator species) may also have their use in toxicological determinations of contaminant effects on the total system. This holistic approach heavily relies on field monitoring data and field gradient analysis approaches. What is key in monitoring programs and field gradient studies is the understanding of effects at a spatial scale that matches the resource: for aquatic invertebrates, this may be on the scale of a stream or shoreline reach (tens of meters), whereas for migratory fish like salmon, it may be on the scale of the watershed or landscape (hundreds of kilometers; Naiman and Turner 2000). In addition, linkage between ecological and stressor measures requires not only knowledge of the cause–effect relationships but also an understanding of ecological response times. Although selection of appropriate endpoints is dependent on the intended purpose of the ecosystem measures (Table 7.5), assessment indicators for effects may often represent intrinsically important species (e.g., major fisheries). However, a major consideration for selection of species monitoring implementation is the response time of the measure. The response time of many ecological measures may be long

TABLE 7.5
Generic considerations for selecting ecosystem measures based on their purposes

Purposes for ecosystem measures	Key characteristic	When to use
Intrinsic importance	Measure is the endpoint itself.	Population levels of economic species.
Early warning indicator	Rapid identification of potential effects.	Use when endpoint is slow to respond or has delayed effect; minimal time lag in response to stressor and rapid response rate; low signal to noise ratio of the measure and low discrimination; screening tool; and false positives acceptable.
Sensitive indicator	Reliability in predicting response.	Use when endpoint is relatively insensitive; high stressor specificity and high signal to noise ratio; minimizes false positives.
Process or functional indicator	Endpoint is a process.	Use for monitoring chemical or physical processes; complements structural measures.

Source: From Harwell et al. (1999).

(e.g., salmon population recovery), whereas others may be acceptably rapid (e.g., those of invertebrate populations) relative to exposure to or the removal of stressors. A long time lag for biological response will favor an emphasis on monitoring stressor measures or populations with rapid responses to the known stressor.

An example of a holistic approach to assess and predict the occurrence of chemical stressors in aquatic ecosystems is the "species at risk" (SPEAR) concept (Liess and Von der Ohe 2005). The SPEAR concept emphasizes the importance of considering ecological traits and recolonization processes on the landscape level for ecotoxicological risk assessment. Species can be grouped according to their vulnerability to toxicants by using the following ecological traits: sensitivity to toxicants, generation time, migration ability, and the presence of aquatic stages of insects during periods of high exposure. Liess and Von der Ohe (2005) also showed that the abundance of SPEAR species in pesticide-stressed streams is increased greatly when forested stream sections are present in upstream reaches. This positive influence of forested stream sections compensated for the negative effect of temporally high pesticide concentrations through recolonization.

7.3.6 FIELD MONITORING AND VERIFICATION OF EXTRAPOLATION TOOLS

Not only may long-term monitoring programs ensure that unexpected ecological impacts do not occur, but also the data may be used to validate and/or calibrate the extrapolation tools used in the risk assessment of chemicals (Figure 7.9). The selection of suitable measurement endpoints is an important step when setting up monitoring studies to assess ecosystem and watershed scale effects of chemical stressors. However, interpretation of monitoring data is often complex. Although monitoring programs can be considered a very useful reality check on exposure and effect predictions, uncertainties associated with monitoring data concern sampling constraints, representativeness of the monitoring sites, and causality between stressors and responses of ecological indicators.

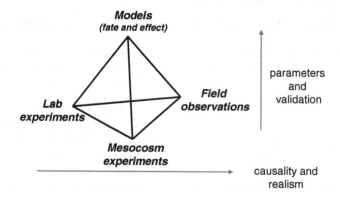

FIGURE 7.9 The integrative research approach to assess and predict stress and disturbance at the ecosystem and landscape level and to validate and calibrate the extrapolation tools developed. *Source*: Redrawn after Brock (1998).

Minimizing sampling constraints information on the spatial and temporal resolution of the monitoring data is required, especially when contamination is due to nonpersistent and/or water-soluble chemicals like pesticides. For example, monitoring programs that rely on spot sampling in rivers may fail to pick up contamination resulting from short-lived storm runoff events (Schulz 2004). Monitoring sites should of course be representative for the systems and ecological risk assessments of concern. Causality requires not only an overview of all potential stressors that may have impacted the measurement endpoints of concern, but also basic ecotoxicological information on the individual stressors. Using the hazard quotient approach, the list of potential chemicals that actually cause ecological risks can be narrowed down. A proper interpretation of chemical and ecological monitoring data for predicting site-specific ecological effects of contaminants usually requires multiple lines of evidence. Using more than one indicator increases the discriminatory power of identifying impaired habitats and reduces the possibility of "false negatives" (Menzies et al. 1996; Hall and Giddings 2000; Naiman and Turner 2000).

In the published literature, several monitoring studies are described that successfully linked dynamics in chemical exposure to catchment characteristics and responses of organisms. For example, reported aqueous-phase insecticide concentrations are negatively correlated with the catchment size, and relatively high concentrations (> 10 μg/L) were predominantly found in smaller scale catchments (< 100 km^2). In addition, several studies revealed a clear relationship between quantified, nonexperimental exposure and observed effects in situ on abundance, drift, community structure, and/or other ecological indicators that are in line with results of laboratory toxicity tests and micro- and/or mesocosm experiments (Schulz 2004). Leonard et al. (1999, 2000) found that populations of 5 common benthic macroinvertebrate taxa (mayflies and caddisflies) in the Namoi River, Australia, were reduced downstream of irrigated cotton-growing areas during the spraying season. Using multivariate analysis, this seasonal reduction was related to aqueous concentrations of endosulfan (determined from passive samplers) and concentrations in sediment. The passive samplers confirmed that endosulfan was entering the river through surface runoff attached to suspended particles during storm events. The β-endosulfan isomer remained strongly adsorbed to the larger (> 63 μm) particles, but the α-isomer readily desorbed and degraded in the water column to form endosulfan-sulfate. Toxicity tests with a local mayfly confirmed that the endosulfan-sulfate would have been the most likely cause of the decrease in population densities in macroinvertebrate taxa observed (Leonard et al. 2001).

Several field gradient studies revealed predictive relationships between chemical stressors and biotic responses. For example, exposure to mixtures of heavy metals has shown concentration–response relationships with macroinvertebrate community indices (Hickey and Clements 1998; Clements et al. 2000). This indicates that response thresholds for ecological indicators (e.g., species diversity and EPT taxa) may be available for use in toxic impacts, providing the nature of the toxicant is known (e.g., metals, organics, and ammonia). The field studies have shown strong concentration–response relationships with stressors, with metal response related to the cumulative criterion unit (CCU) value for the sum of the metals present (Hickey and Clements 1998; Clements et al. 2000). The CCU concept was discussed in greater detail in Chapter 5.

Other field gradient approaches have been successfully utilized for a range of stressors exposed to complex effluent gradients. Such studies include comparisons of community structures upstream and downstream of contaminant discharges, between reference and polluted sites, and along concentration gradients of an environmental stressor. These include river invertebrate community impacts from inorganic suspended solids (Quinn et al. 1992), sewage lagoon effluents (Quinn and Hickey 1993), responses of freshwater mussels to metals (Hickey et al. 1995a), and estuarine studies on responses of sediment-dwelling shellfish to organic contaminants (Hickey et al. 1995b). Such gradient approaches are also useful in determining hydraulic habitat preferences for invertebrates (Quinn and Hickey 1994) and attached algal biofilms (Biggs and Hickey 1994). Although such studies require careful consideration in the selection of sites and design, the analysis can provide strong evidence for effect thresholds in multistressor environments. Commonly the statistical analysis may not be able to link effects with a single variable in complex effluent because high correlation frequently occurs between constituents in comparably sourced effluents, and effects must be related to either surrogate measures or those with the highest correlation. Further studies may be required in such situations if management options require a clear identification of the major stressor.

The monitoring studies mentioned above can be considered as valuable retrospective tools to verify the field relevance of risk assessment procedures currently in use, including the extrapolation tools used to predict chemical stress at the ecosystem and landscape levels.

7.4 SUMMARY AND OUTLOOK

7.4.1 Can Toxicity Data Be Extrapolated Spatially?

The analyses conducted to date suggest that, although the composition of freshwater communities varies across biogeographical regions, climatic zones, and habitat types, the distribution of species sensitivities does not vary markedly. Tropical freshwater fish species are not generally more sensitive to environmental contaminants than temperate fish species. When compared with temperate species, there is a trend of a slight increase in the sensitivity of tropical invertebrates to a few selected chemicals only, whereas for a few other chemicals the reverse was observed. Overall, the quality of tropical toxicity data was lower than that of temperate ones. There is no evidence to suggest that northern hemisphere and southern hemisphere freshwater species differ systematically in their sensitivity to pesticides or metals. Within temperate regions, there is evidence of differences in sensitivity between Palearctic and Nearctic species, but this is taxon specific and based on very few observations. There is no evidence that lotic arthropod assemblages are generally more sensitive than lentic arthropod assemblages to insecticides, even though they contain a higher proportion of EPT taxa. No significant overall difference was found in median HC5 estimates for 10 insecticides when comparing the SSDs of freshwater and saltwater arthropods, at least when based on the most sensitive taxonomic group (e.g., crustaceans; Section 7.2.3).

There is no question that model ecosystems differing in size and complexity will also respond differently to high levels of chemical stress. In part, this is caused by differences in fate and dynamics in exposure concentrations. Under conditions where exposure concentrations are similar, however, it seems that threshold concentrations for effects are more or less similar between different types of test systems, at least when they contain enough representatives of sensitive taxonomic groups. The extrapolation of $NOEC_{eco}$ values from one system to another seems to be possible with lower uncertainty than hazard estimates of higher concentrations in which both direct and indirect effects and recovery processes are involved (Section 7.2.4).

In aquatic risk assessment, an important question at stake is how unique model ecosystem experiments are in their threshold concentrations for direct toxic effects. When comparing chemicals for which at least 5 aquatic $NOEC_{eco}$ values are available, it appears that geographical extrapolation of model ecosystem experiments is possible. The proposed geographical uncertainty factor (spread) for threshold concentrations was in the range of 1.4 to 5.4 (Section 7.2.5).

Therefore, there is no evidence to support the contention that the ecological risk assessments must necessarily be based on indigenous species and communities, at least when researchers are interested in threshold concentrations of effects. However, these conclusions are based on very limited data sets, and further research is required to establish the extent to which they can be generalized across geographical locations, taxonomic groups, and compounds.

7.4.2   Is There a Future for Landscape and Watershed Ecotoxicology?

Probably yes, and its role will certainly increase when implementing strategies and environmental laws to manage and assess risks of perturbations (including chemical stress) at the watershed level, as required by the European Water Framework Directive. Managing the impact of chemical stressors in watersheds and landscapes is not an easy task because, at this scale, multiple-stress impacts cannot be ignored. The discipline of landscape ecotoxicology is still in its infancy, and widely applicable tools and techniques to assess multiple-stress impacts at the landscape level are scarce. Nevertheless, several promising tools are available for implementing spatiotemporal considerations into the risk assessment process.

Available tools for implementing spatiotemporal considerations into the risk assessment process comprise metapopulation models and spatially explicit models. When linking these models to GIS, the ecological realism of model predictions is increased. In addition, GIS is an excellent means of compiling and sorting data, visualizing spatial relationships, and establishing sampling programs. To date, metapopulation models and spatially explicit models have been used predominantly to assess risks of contaminants and habitat quality on terrestrial wildlife species. There is, however, no reason to believe that these models, when adapted, cannot be used to assess ecological risks for aquatic populations. Output of these models seems to be valuable, but further testing and uncertainty analysis are certainly needed (Sections 7.3.2 and 7.3.3).

When studying the ecological status of watersheds and landscapes, monitoring techniques and indicators are widely used tools. The diagnostic indicators comprise indicator species and community metric approaches. To improve the interpretation of monitoring programs and "ecosystem health" indicators to assess risks of chemicals, the existing information in the region of concern should be appropriately mined, and data gaps filled. This can be done by constructing databases with information on life-cycle characteristics of species, their occurrence and mobility in the landscape, and their sensitivity to the chemicals of concern. In the prediction of site-specific ecological impacts of chemical stressors, it is important to use more than one indicator to increase the discriminatory power of identifying impaired sites and to reduce the possibility of false negatives. The most powerful applications of landscape ecotoxicology are when experimental studies, chemical and ecological monitoring, and model simulations are integrated in a weight-of-evidence approach (Sections 7.3.4 to 7.3.6).

ACKNOWLEDGMENTS

We thank Thomas LaPoint, Steve Maund, Paul Sibley, and Udo Hommen for their constructive suggestions to improve earlier drafts of this chapter.

8 Conclusions

Keith R. Solomon, Theo C. M. Brock,
Dick De Zwart, Scott D. Dyer, Leo Posthuma,
Sean M. Richards, Hans Sanderson, Paul K. Sibley,
and Paul J. van den Brink

CONTENTS

Extrapolations are integral to all forms of risk assessment and to all levels of refinement in the risk assessment process. Short of testing all combinations and permutations in physical experiments, it is not possible to conduct risk assessments and make risk management decisions without using extrapolations to one degree or another.

Extrapolation is commonly used in human health and ecotoxicological risk assessment, either in a formal process or, more commonly, through the use of uncertainty factors derived from empirical experience or observation. This book has attempted to characterize the extrapolation process as applied or potentially applied in ecotoxicology to allow the user to better understand the process. Because extrapolations are strongly dependent on the availability of knowledge and information, not all extrapolations will be appropriate for all situations. This book describes the techniques that are available for a range of extrapolation problems. It also offers insight into the practicability of use of those methods for different types of assessment problems. It presents guidance as to the techniques for extrapolation that are most appropriate to the situation. It is also recognized that extrapolations will never be completely precise — toxicity data from which many are derived will always show small variance. Thus, in recommending research directions, we have focused on needs related to extrapolations with large uncertainty. We believe that these recommendations will be helpful in developing future extrapolation procedures that will reduce uncertainty and improve ecotoxicological risk assessment.

Extrapolation is used in predictive risk assessment procedures, such as the setting of criteria or guidelines, as well as retrospective risk assessments, where organisms have already been exposed to a stressor. Both criteria setting and risk assessments are conducted via tiers, which become more sophisticated and refined as one utilizes the higher tiers. As a result, extrapolation procedures become more refined, more complex, more demanding of data, and, hopefully, more realistic. This tiered approach is presented in all chapters and in the technical guidance and is key to balancing the correct degree of complexity in the assessment to the goals and objectives of the risk assessment.

Exposure and effects data are the keys to risk assessment, and, in the same way, extrapolation is utilized in the characterization of both exposures and effects. This book was organized around this theme, with exposures discussed in Chapter 2 and effects in the chapters that follow.

8.1   MATRIX EXTRAPOLATIONS

Matrix extrapolations are critical in understanding the bioavailability and fate of toxic substances and are driven by interactions between the matrix and the toxicant. These affect the uptake of substances by biota, and thus increase or decrease the body dose and resultant responses. There are large differences in the methods used for extrapolating matrix effects between organic and inorganic substances. Organic compounds normally follow the rules of equilibrium partitioning between the large-molecule organic constituents of the matrix and the lipid content of the exposed organisms. These processes can be relatively easily modeled because the input data, such as partition coefficients, fugacities, and proportions of partitioning compartments, are available. For inorganic toxicants, mainly heavy metals, speciation governs availability. In the water compartment, metal speciation can be addressed in a mechanistic way, and models are available to calculate the proportion of the metal species capable of entering organisms. The input to those calculations requires a quantification of various water chemistry parameters such as pH, hardness, DOC, and so on. For the soil and sediment compartments, the bioavailable fraction of the metals is, in general, empirically related to soil and sediment characteristics such as pH, cation exchange capacity, and calcium content. In general, there is less knowledge of soil processes that govern bioavailability than is the case for water, and this has been identified as a research need.

Three general levels of complexity are recognized in matrix and media extrapolation. The simplest approach assumes that all toxicants are completely available to be taken up by the biota. In this case, no extrapolation is required. The second level of complexity requires the calculation of bioavailable fractions of toxicants, whereas at the highest level of complexity the influence of physiological responses to toxicant uptake is considered.

8.2   EXTRAPOLATION OF EFFECTS

8.2.1   (Q)SARs

The prediction of toxicity through the use of quantitative structure-activity relationships ([Q]SAR) is the most basic of effect extrapolations and is applied when no toxicity or response data have been measured for the substances in question. (Q)SAR extrapolation relies on relationships between the physicochemical properties of a substance and its biological properties. Widely used in the initial screening of industrial and other commercial substances for potential toxicity, (Q)SAR predictions are often subjected to large uncertainty factors because of a perceived lack of confidence in the accuracy of the predictions. This lack of confidence is not deserved in all situations. Where the mechanism of action of the substance is via baseline toxicity or narcosis, (Q)SAR models can predict responses with reasonable accuracy. However, where other mechanisms of toxicity are involved, such as those mediated by interactions with a specific receptor, (Q)SARs may provide inaccurate extrapolations, unless the training sets on which they are based are extensive and the toxic mode of action is specific to certain functional groups or moieties on the molecule. (Q)SARs for substances with novel modes of action and for which training sets are not available will be the least accurate.

Despite some drawbacks, (Q)SARs can be useful in dealing with large groups of substances when no data are available as well as for setting priorities for further characterization and in suggesting hypotheses for further experimental evaluation. Future directions in (Q)SAR include the

- addition of more effect data for more substances to the training data sets to increase accuracy.
- integration of (Q)SARs used in drug design and pharmacology to incorporate mechanisms of actions that are conserved across taxa.
- development of specific (Q)SARs to address measures of effect and responses that are unique to ecotoxicology.

8.2.2 EXTRAPOLATION ACROSS LEVELS OF BIOLOGICAL ORGANIZATION

Levels of biological organization (e.g., cell, organ, individual, population, and community) are important for extrapolation, because more extrapolation data are available for lower levels of organization. Extrapolation across levels of organization, however, presents significant challenges, especially where these levels are far apart. Extrapolations over short biological distances have been conducted with some success. This is especially the case where the bioindicator is a biochemical or physiological response (biomarker), the mechanisms of action of which are well understood, and when the response is substance specific. Where greater biological distances occur, or the mechanisms of action are nonspecific and poorly understood, extrapolation has been less successful. Thus, extrapolation of population-level effects from biochemical responses in single organisms is generally not possible and will present interesting challenges in future research.

Responses from single-species laboratory tests have provided useful information for extrapolating effects to populations under natural conditions; however, these data are largely derived from aquatic species that are used in standardized tests. Other groups of organisms, such as terrestrial arthropods, are poorly represented in routine testing, and their biology is often not as well understood. Population and food-web models have been useful in extrapolating from laboratory data to the field but may require too many parameters to be widely applied outside of a few well-studied species. Generic population or food-web models applied to classes of organisms with similar life cycle characteristics may be useful.

Extrapolation between structure and function is an area in ecotoxicology that is not well understood. At the ecosystem level, current protective regulations often rely on the "functional redundancy" concept, applied in conjunction with results from species-poor model systems with high resilience. Evidence from recent studies in ecology suggests that functional redundancy exists within natural communities, but the relative contribution of different species to specific functions is far from equal, and the disappearance of certain species, such as dominant species or keystone species, can have large effects on overall system function and the occurrence of other species.

Extrapolation between levels of biological organization is predicated on the correct choice of parameters to measure. There is always a danger in basing approaches on untested assumptions, such as assuming species equivalence when applying the species sensitivity distribution (SSD) model. Lack of knowledge of biotic interactions is also important: for example, simply because A influences B does not mean that B will always respond to a change in A. Although we may have a good understanding of toxicological effects, such as mode of action, and this is important for understanding individual and ecological responses, there is a greater need for understanding the ecological consequences of the responses caused by substances in the environment.

Future research directions include

- increasing the amount of toxicity data for terrestrial organisms to expand the range of organisms tested.
- the development and use of generic food-web models for classes of organisms with similar life histories and characteristics.
- research to better categorize the functional role of organisms in ecosystems so that these may be better integrated into risk assessments.

8.2.3 Mixture Extrapolation

In the environment, organisms are seldom exposed to single substances; in fact, mixtures are the rule rather than the exception. Mixture extrapolation is necessary for risk assessment purposes, but techniques for mixture extrapolations are not strongly validated, except for those for some classes of substances in single species. There is an almost complete lack of data on mixture effects at the community and ecosystem levels. Nonetheless, the available data indicate strongly that mixture extrapolation should be seriously considered as an alternative to no mixture extrapolation.

Our current understanding of mixture extrapolation is based on simple pharmacodynamic concepts of noninteractive joint action, such as simple similar action and simple independent action, with the associated extrapolation models concentration addition and response addition. These models are used for various types of extrapolations. Although mode of action is important when considering possible mixture interactions and extrapolations, the concept of the ecological mode of action needs to be expanded, as was also concluded for extrapolation across levels of biological organization. Mixture extrapolation should consider environmental (matrix)–chemical interactions, toxicokinetic and toxicodynamic interactions in organisms, and ecological interactions between organisms and populations.

A tiered system for mixture extrapolation is proposed. The lowest tier is based on extrapolation using toxicological point-estimate information such as EC50 values. This translates into the use of toxic units, toxic equivalencies, and similar techniques. The use of the entire concentration–response relationships of the separate compounds is recommended for Tier-2, in conjunction with the use of either concentration or response addition as a modeling approach. In Tier-3, a mixed-model approach can be considered, to more specifically address considerations on toxic modes of action. In the latter case, the approach may be extended to allow incorporation of the responses of different ecological receptors (Tier-4). Research needs have been clearly identified in community-level mixture assessments.

Suggested research directions include the

- integration of ecological mode of action in the assessment of mixture toxicity.
- incorporation of matrix–chemical interactions, toxicokinetic and toxicodynamic interactions in organisms, and ecological interactions between organisms and populations into the assessment of the effect of mixtures in the ecosystem.

8.2.4 TEMPORAL EXTRAPOLATION

The response of organisms in the environment to toxic substances depends on exposure concentration and exposure duration. The time needed to reach incipient toxicity is dependent on the type of chemical; ambient environmental conditions, such as temperature; species tested; and endpoints selected. Extrapolation across time-varying exposure regimes is possible for well-studied compounds with known modes of toxic action. Several key properties of organisms that are generally not well characterized in most toxicity tests can influence temporal extrapolations. Key amongst these is the possibility of reversibility of the toxic effects, which can be important when evaluating the risks of pulsed and multiple exposures. Reversibility of the effects depends on the type of chemical, exposure concentration, and duration of the exposure event. Critical body residue, in combination with toxicokinetics, has been used to extrapolate from constant to pulsed exposure and for exposures of varying lengths. Recent developments of new models to address these extrapolations, such as ACE (Ellersieck et al. 2003), are promising.

Most toxicity data for ecological effects originate from acute laboratory tests on single species. Although, in some cases, a predictable relationship may exist between acute and chronic effects in some organisms and for some types of chemicals, there is still some uncertainty associated with predictions of chronic effects from acute effects. In part this uncertainty can be explained by the fact that different responses, such as lethality, may be observed in acute and chronic tests where growth and development are normally assessed. Generalizations in these extrapolations have been incorporated into regression models that are available for general use. However, for some substances, uncertainty factors larger than 100 may be required for lower tier extrapolations from acute to chronic conditions. Based on analyses of SSDs fitted to acute and chronic toxicity data for pesticides tested in aquatic organisms, an uncertainty factor of 10 appears to suffice to extrapolate a median acute HC5 to a median chronic HC5, at least when based on toxicity data of sensitive taxonomic groups. In addition, it appears from model ecosystem experiments with pesticides that threshold concentrations for chronic exposures are approximately a factor of 10 lower than those for acute exposures. For a wider generalization, however, more data are required on compounds that differ in toxic modes of action.

Possible delayed responses cannot be ignored when interpreting and extrapolating from toxicity tests. These latent responses may become apparent in organisms after the exposure is completed or in conjunction with other stressors. It may even be required to include the responses of the offspring in the observations, which is not often done in routine toxicity testing. Complex interactions within and between populations and nonlinear biological dynamics may create variable latency periods between the exposure event and indirect effects in particular, further complicating extrapolations in communities.

Life history models may be useful for considering age- or stage-specific variability in sensitivity in the extrapolation of response; however, it should not be assumed that smaller or younger stages are always the most sensitive. The few studies that have addressed the impact of time of year on responses of aquatic communities to a stressor indicate that, in freshwater communities, threshold concentrations for direct toxic effects vary within a factor of 2 among seasons — well within the normal range of variation observed in laboratory toxicity tests. However, at greater exposures, the intensity and duration of direct and indirect responses may vary considerably between different periods of the year because of the influence of climatic and seasonal factors on recovery.

Where exposures are short and infrequent, the recovery potential of affected populations and ecosystem functions is important in extrapolating potential responses. Rate of recovery is highly dependent on the life-cycle characteristics of the affected species. In the ecotoxicological literature, relatively little experimental information can be found on the recovery potential of species with a long and/or complex life cycle. In addition, for many aquatic species, basic information on life-cycle characteristics is not readily available. To further complicate matters, the number of generations per year of invertebrate species may vary with latitude.

Tolerance can be acquired by an individual due to physiological acclimation or by genetic selection. At the community level, pollution-induced tolerance may result from the replacement of sensitive species by less sensitive ones, genetic adaptation of one or more species, and/or physiological adaptation of individuals. Rapid genetic adaptation to toxicant stress usually involves a single gene. Field observations and experiments have demonstrated that genetic adaptation to metals and pesticides can develop within a few generations, particularly when the species in question have short generation times and large numbers of offspring.

Research directions include

- the incorporation of mode and mechanism of action information, particularly recovery, in extrapolation of effects for time-varying exposures.

- developing a better understanding of the mechanisms of latent responses to toxic substances so that these can be predicted and integrated into risk assessment extrapolations.
- more consideration of recovery potential of species and communities in the risk assessment process.

8.2.5 Spatial Extrapolation

Because most of the routine toxicity testing of environmentally important substances is carried out using species from temperate regions, spatial extrapolation is an important process in extending risk assessments to tropical or polar regions. Although the composition of freshwater communities varies across biogeographical regions, climatic zones, and habitat types, studies have shown that the distribution of species sensitivities does not vary markedly as long as the comparison is confined within similar taxa. Tropical freshwater fish species are not generally more sensitive to environmental contaminants than temperate fish species. There is no evidence to suggest that northern hemisphere and southern hemisphere freshwater species differ systematically in their sensitivity to pesticides or metals. Within temperate regions, there is evidence of differences in sensitivity between Palearctic and Nearctic species, but this is taxon specific and based on very few observations. There is no evidence that lotic arthropod assemblages are generally more sensitive than lentic arthropod assemblages to insecticides, even though they exhibit differences in taxonomic makeup. At this time, there is no evidence to support the contention that the ecological risk assessments must necessarily be based on indigenous species and communities, at least when interested in threshold concentrations of effects within taxa. However, these conclusions are based on very limited data sets, and further research is required to establish the extent to which they can be generalized across geographical locations, taxonomic groups, and compounds.

Size of test system causes significant differences in fate and, therefore, exposure concentrations. However, where exposure regimens are similar, threshold concentrations for similar effects in different types of test systems also are similar, at least when they contain enough representatives of sensitive taxonomic groups. The extrapolation of $NOEC_{ecosystemvalues}$ from one system to another is possible with lower uncertainty than in hazard estimates of higher concentrations in which both direct and indirect effects and recovery processes are involved.

Managing the impact of chemical stressors in watersheds and landscapes is not an easy task because multiple-stress impacts cannot be ignored at this scale. The discipline of landscape ecotoxicology is still in its infancy, and widely applicable tools and techniques to assess multiple-stress impacts at the landscape level are scarce. Nevertheless, recent advances in geomatics, along with the introduction of several promising tools, are greatly improving the capacity to incorporate spatiotemporal information into the risk assessment process. These tools include geographic information systems (GIS), remote sensing, and global positioning systems (GPS), which provide a basic approach for compiling and sorting data, visualizing spatial relationships, and establishing sampling programs.

To date, metapopulation models and spatially explicit models have been used predominantly to assess risks of contaminants and habitat quality on terrestrial wildlife species. There is, however, no reason to believe that these models, when adapted, cannot be used to assess ecological risks for aquatic populations. Output of these models is valuable, but further testing and uncertainty analyses are needed.

Diagnostic indicators such as indicator species and community metric approaches are useful in extrapolation between smaller test units to landscapes and between landscapes themselves. The use of these indicators in extrapolation can be improved by constructing databases with information on the life-cycle characteristics of species, their occurrence and mobility in the landscape, and their sensitivity to the chemicals of concern. In the extrapolation of site-specific ecological impacts of chemical stressors, it is important to use more than one indicator to increase the discriminatory power of identifying impaired sites and to reduce the possibility of false negatives (type 2 errors, in which responses are present but not observed).

Research directions include

- increased and better testing of species from different regions, such as the tropics and polar regions, to better characterize differences, if any, in organisms from these regions.
- better incorporation of geographical information in risk assessment for aquatic organisms.
- better use of life-cycle characteristics of species, such as range of distribution and mobility, in extrapolation of effects to the larger landscape.

8.2.6 Guidance on Extrapolation

There are many extrapolation methods, of different complexities, and with different purposes and suitabilities for prospective and retrospective risk assessments. A compilation of the methods is insufficient to guide the choice of procedures to use when assessors need to conduct risk assessments. Therefore, a practical and pragmatic guide to extrapolations and their everyday use is provided in the last chapter. It defines a general stepwise approach to identifying the types of extrapolation (matrix and media, (Q)SARs, mixtures, etc.) that are most relevant for an assessment problem, and it defines an overall approach to the assignment of tiers.

There may be differences between the Guidance chapter and current extrapolation schemes used in existing regulations. This relates to the fact that current regulations were not designed to result in a set of systematically and globally homogeneous approaches. The approach here was to derive guidance from some scientific principles for cases where guidance is lacking. The guidance was not meant to replace currently used approaches that serve their purposes well. The Guidance chapter was designed to serve as a stand-alone document, but it is obviously closely linked to the other chapters of this book as a source of additional information and theory.

8.2.7 Calibration and Validation

As in all other areas of science, validation of extrapolations for all possible situations is not possible. Calibration can be achieved and is, in fact, considered in a number of regression models used in extrapolation as well as in the traditional use of an uncertainty factor (UF), although these may be selected to be very conservative. As one moves from the lower tiers of extrapolation through the use of UFs to more data-rich approaches with increasing realism, the need for conservative assumptions decreases. However, with this comes an increased risk of a false negative — extrapolating no or low risk when risk actually exists or is significant. One of the major questions that may be asked in the context of extrapolation is the issue of validation; this question was one of the main drivers for collecting the suite of extrapolation methods described in the chapters of this book, along with considering prediction accuracy where possible.

Extrapolation results are, by definition, predictions on the performance of entities for which data are lacking. Both from a scientific perspective as well as from the perspective of practical decisions based on extrapolation, there is a need to consider not only the outcomes of an extrapolation per se but also the question of whether the outcome is supported by a certain degree of validity. All this relates to the issue generally referred to as "validation."

A validated model is generally seen in practice as a model that

- is suited to execute a task in practice and
- yields an outcome that is trustworthy as to representing the assessment endpoint.

In the previous chapters, and in the design of the tiered system described in Chapter 10 (Tables 10.2 to 10.4), it is shown that models differ as to their feasibility of use in practice. The more a model is based on simple and uniform theory (whether or not the theory fully covers the issue), the more it is used in daily practice. For example, empirical formulae that describe exposure as a function of matrix characteristics can be employed often, because one usually has information on the most relevant parameter values at the site of interest. In contrast, (semi)mechanistic modeling with the biotic ligand approach is probably better suited for some highly specific assessment problems, because it requires more specific data, or alternatively assumptions, before it can be applied. In the context of mixtures, a universal fallback option has always been the model of concentration addition, which is easy to use, especially when it is applied in the format of point estimates for toxicity, like through dimensionless toxic units. This shows that practical adoption of extrapolation models in part depends on practicability, not necessarily on validation. Suter et al. (2002) already noted that regulatory adoption of a model for practical use can be an argument to consider a model sufficiently validated, without necessarily meaning that a model is perfect.

Further, the validation of a model needs the definition of the criterion for establishing that a model has been validated. How well should a model predict effects precisely, and what are the bounds between which one calls a model (sufficiently) valid? It also needs the definition of the context against which a model is to be considered "valid." For example, "validation" of the SSD model has generally been based on whether the so-called "hazardous concentration" for 5% of the species (HC5) is a concentration that is conservative (sufficiently protective) compared to the no-effect concentration in multispecies mesocosm or field tests. In that sense, the model has performed well for both aquatic and terrestrial systems (e.g., Emans et al. 1993; Okkerman et al. 1993; Posthuma et al. 1998; Versteeg et al. 1999; van den Brink et al. 2002a; Maltby et al. 2005). No case study performed up till now has shown that (regulatory) unacceptable effects occur at concentrations lower than the risk limit, probably with the exception of the study of Mulder et al. (2004), in which the decline of Lepidoptera in the field was associated to changes in host plants responding to low-level exposures where each metal was below the individual HC5 for Dutch soils. However, validation studies conducted to date have not falsified the hypothesis that the SSD model is *always* a valid approach for an assessment. When a more refined question is asked — for example, "Does an SSD predict ecosystem performance at a known concentration that is higher than the HC5?" — then the performance of SSDs as an extrapolation approach appears dependent on the characteristics of the assessed system. SSDs appeared to perform relatively well in predicting acute responses in aquatic ecosystems, but in chronic exposures and in soil, the performance of the model is less satisfactory (van den Brink et al. 2002a).

The SSD example shows that one model, the SSD, can be considered sufficiently predictive for simple questions (e.g., "Is the HC5 protective of community responses?") but of more limited — or even insufficient — predictive capacity for specific assessment questions such as those regarding the effects of exposure on biodiversity. This shows that general conclusions on the "validity of a model" cannot be drawn. For the extrapolation methods for which validation studies have been done, the examples (and limitations) have been provided in the preceding chapters.

From the history of the use of extrapolation methods and validation, it is evident that the design of a method preceded the validation; at least in those cases there was an attempt at validation. However, at present, the issue of validation is not an integral part of the design of an extrapolation method. In software engineering, there is a specific design phase for this activity, often referred to as beta testing. In this design phase, the "bugs" that remain after moving from the training (coding) phase to the phase of practical use are removed, and software credibility is often considerably increased. Such a general approach is lacking in the case of the design of extrapolation methods. This may decrease the accuracy and adoption of extrapolation. For example, when the training data set (through which a method was developed) is not related to the extrapolation problem under investigation, this may go unnoticed when the risk assessor does not pose critical questions on the relevance of the training data set for the problem being investigated. For example, when using extrapolation formulae on the partition coefficient of metals in field soils, for example, as a function of pH and organic matter content, the set of field soils from which the formulae were derived needs to be considered. This is a "within-method" type of scrutiny of the appropriateness of the approach. It is recommended to undertake such critical investigations into the methods used: on which principles are they founded, what data were used as a training set, and are principles and data appropriate for the problem? If not, which type of bias (e.g., magnitude, or direction) might be expected?

With this in mind, a number of new directions are suggested:

- Greater attention should be paid to the design of extrapolation methods with the objective of validation, that is, describing for which situations the approach appears accurate, to what extent accuracy can be expected, critical tests of the extrapolation method to problems outside the scope of the training set, and so on.

- Greater attention should be paid to developing standard exposure scenarios, particularly for areas outside the temperate regions. These scenarios could be tested in physical models (microcosms or mesocosms), whereby lower tier methods are scrutinized by comparison of predicted effects to observed effects in higher tier experimental approaches. They could also be tested by ecoepidemiological approaches, although that would require additional attention for the problem of disentangling effects of toxic substances and natural variability and the effects of other stressors.

Finally, the authors of and other contributors to this book hope that it will be useful and that it will increase our understanding of the process of extrapolation in the ecological context as well as stimulate additional research into refining methods of extrapolation. We thank the American Chemistry Council for their support of this project.

9 Glossary

This glossary is provided to help readers with terminology and abbreviations that are commonly used in ecotoxicology and risk assessment.

abiotic: Not associated with living organisms.

absorption: 1) Penetration of a substance into an organism by various processes, some specialized, some involving expenditure of energy (active transport), some involving a carrier system, and others involving passive movement down an electrochemical gradient. In mammals, absorption is usually through the respiratory tract, or skin. 2) The process of 1 material (absorbent) being retained by another (absorbate). This may be the physical solution of a gas, liquid, or solid in a liquid; attachment of dissolved molecules of a gas, vapor, liquid, or dissolved substance to a solid surface by physical forces; or the like.

acute: Responses occurring within a short period in relation to the life span of the organism (usually 4 days for fish). It can be used to define either the exposure (acute test) or the response to an exposure (acute effect). For animals that are dosed orally, "acute" refers to instantaneous exposure. In humans, "acute" often refers to exposure in a 24-hour period.

acute toxicity: The harmful effects of a substance or mixture of substances occurring after a brief exposure, usually 48 to 96 hours. *See also* chronic toxicity.

adsorption: An increase in the concentration of a dissolved substance at the interface of a condensed and a liquid phase due to the operation of surface forces. With pesticides, it is normally the increase in the concentration of a pesticide at the interface of soil colloidal clay or organic matter. *Antonym*: desorption.

adverse effect: Change in the morphology, physiology, growth, development, reproduction, or life span of an organism, system, or subpopulation that results in impairment of the capacity to compensate for additional stress, or an increase in susceptibility to other influences.

analysis of effects: A phase in an ecological risk assessment in which the effect measures are characterized along with associated uncertainties.

analysis of exposure: A phase in an ecological risk assessment in which the spatial and temporal distribution and intensity exposures are characterized along with associated uncertainties.

antagonism: Antagonism arises when the combined effect of 2 or more substances is smaller than the combined individual effects of the substances.

assessment endpoint: An explicit expression of the environmental value to be protected. An assessment endpoint must include an entity and a specific property of that entity.

background concentration: The concentration of a substance in environmental media that are not contaminated by the sources being assessed or any other local sources. Background concentrations are due to regional contamination or natural occurrence.

battery toxicity testing: The parallel application of a range of different toxicity tests.

benthos: Nonplanktonic animals (not being suspended in water) associated with freshwater substrata (upper layer of the sediment in rivers and ponds) at the sediment–water interface.

bioaccumulation: The net accumulation of a substance by an organism due to uptake from all environmental media, including food that occurs because the rate of intake exceeds the organism's ability to remove the substance from the body.

bioassay: Old term for toxicity test.

bioavailability: The extent to which the form of a substance is susceptible to being taken up by an organism. A substance is said to be bioavailable if it is in a form that is readily taken up (e.g., dissolved) rather than a less available form (e.g., sorbed to solids or to dissolved organic matter).

bioconcentration: 1) The net accumulation of a substance by an organism due to uptake from the matrix, but excluding uptake from food. 2) The degree to which a substance will partition from the matrix into an organism.

bioconcentration factor (BCF): Ratio between the concentration of a substance in an organism or tissue and the concentration in the environmental matrix (usually water) at apparent equilibrium during the uptake phase.

biomagnification: An increase in the concentration of a substance in organisms higher in the food chain. The tissue concentration increases at each trophic level in the food web when there is efficient uptake and slow rate of elimination.

biomarker: Indicator signaling an event or condition in a biological system or sample and giving a measure of exposure, effect, or susceptibility. Such an indicator may be a measurable chemical, biochemical, physiological, behavioral, or other alteration within an organism.

biomass: Material produced by the growth of microorganisms, plants, or animals.

centile: Similar to a quantile, but with the proportion expressed as a percentage. The median is the 50th centile.

chronic: Responses occurring after an extended time relative to the life span of an organism (conventionally taken to include at least one-tenth of the life span). Long-term effects are related to changes in metabolism, growth, reproduction, and/or the ability to survive. Also: exposures greater than 3 dissipation half-lives of the substance in the organism.

chronic toxicity: The harmful effects of a substance or mixture of substances occurring after an extended exposure. *See also* acute toxicity.

community (biotic): A biotic community consists of all plants, animals, and microbes occupying the same area at the same time. However, the term is commonly used to refer to a subset of the community such as the fish community or the benthic macroinvertebrate community.

conceptual model: A representation of the hypothesized causal relationship between the source of contamination and the response of the endpoint entities.

contaminant: A substance that is present in the environment due to release from an anthropogenic source and is believed to be potentially harmful.

critical body burden: The concentration of a substance in an organism at the time of appearance of the symptoms or response.

critical toxicity values (CTVs): Endpoints used to describe effects of substances of organisms for the purposes of risk assessment.

cumulative distribution function (CDF): A function expressing the probability that a random variable is less than or equal to a certain value. The CDF is obtained by integration of the probability density function (PDF) for a continuous random variable, or summation of the PDF in the case of a discrete random variable.

cumulative effect: Overall change that occurs after repeated doses or exposures to a substance or physical stressor.

cumulative risk: Probability of any defined harmful effect occurring through a common toxic effect associated with concurrent exposure by all relevant pathways and routes of exposure to a group of chemicals that share a common mechanism of toxicity.

desorption: Decrease in the amount of adsorbed substance (e.g., pesticide) at the interphase of the soil colloids (clay or organic matter). *Antonym*: adsorption.

deterministic hazard (risk) assessment (DHA): An assessment of hazard where the hazard (incorrectly referred to as "risk") is estimated from a single datum for effect and for exposure, and uncertainty in these 2 measures is not characterized.

direct effect: An effect resulting from an agent acting on the assessment endpoint or other ecological component of interest itself, not through effects on other components of the ecosystem. *See also* indirect effect.

domain: Domain is the types and/or classes of compounds that may be studied by a (Q)SAR model.

ecological risk assessment (ERA): A process that evaluates the likelihood that adverse ecological effects may occur or are occurring as a result of exposure to 1 or more stressors.

ecosystem: A collection of populations (microorganisms, plants, and animals) that occur in the same place at the same time, forming a functional system. This collection potentially interacts with all biotic and abiotic entities in the system.

ecosystem function: A biologically based process such as carbon fixation, energy, or nutrient cycling taking place in an ecosystem.

ecosystem structure: The composition of the biological community in an ecosystem in terms of the number of species and the number of organisms. *See* food web.

ecotoxicity: The property of a substance to produce adverse effects in an ecosystem or more than 1 of its components.

ecotoxicology: The study of toxic effects of substances and physical agents in living organisms, especially on populations and communities within defined ecosystems; it includes transfer pathways of these agents and their interaction with the environment.

effect criterion: The type of effect or response observed in a toxicity test (e.g., immobility).

effect measure: A concentration at which a specific response is observed in a toxicity test, that is, the values derived from a toxicity test that characterizes the results of the test (e.g., LC50 or NOEC).

effective concentration *n* (ECn) or effective dose *n* (EDn): Concentration (dose) that causes the designated response (e.g., a behavioral trait) in *n*% of the population observed. The EC50 is the median effective concentration, whereas the ED50 is the median effective dose. The EC values and their 95% confidence limits are usually derived by statistical analysis of effects observed at several test concentrations, after a fixed period of exposure. The duration of exposure must be specified (e.g. 96-hour EC50). *See also* lethal concentration *n* (LCn) or lethal dose *n* (LDn).

endpoint entity: An organism, population, species, community, or ecosystem function that has been chosen for protection. The endpoint entity is 1 component of the definition of an assessment endpoint.

environmental quality criterion (EQC): The concentration of a potentially toxic substance that can be allowed in an environmental medium over a defined period. The term is used in this book as a general term, for which also "environmental quality objective" (EQO) and "environmental quality standard" (EQS) are used as synonyms in different contexts.

estimated environmental concentration (EEC): Predicted concentration of a substance within an environmental compartment based on estimates of quantities released, discharge patterns, and inherent disposition of the pesticide (fate and distribution) as well as the nature of the specific receiving ecosystems. Also known as expected environmental concentration (EEC).

expected environmental concentration (EEC): That concentration of a stressor substance that is expected to be found in the environment being assessed. *See* predicted environmental concentration.

exposure: The contact or co-occurrence of a stressor with a receptor organism, population, or community. For chemical agents and microorganisms, exposure is usually expressed in numerical terms of substance concentration, duration, and frequency. For physical agents such as radiation, exposure is expressed as intensity.

exposure characterization: The component of an ecological risk assessment that estimates the exposure resulting from a release or occurrence in a medium of a stressor. It includes estimation of transport, fate, and uptake.

exposure pathway: The physical route by which a stressor moves from a source to a biological receptor. A pathway may involve exchange among multiple media and may include transformation of the contaminant.

exposure route: The means by which a stressor enters an organism, such as inhalation, stomatal uptake, or ingestion.

extrapolation: 1) An estimation of a numerical value of an empirical (measured) function at a point outside the range of data that were used to calibrate the function. 2) The use of data derived from observations to estimate values for unobserved entities or conditions.

food web: Interrelationships between the individual populations of species related to the transfer of energy.

frequency distribution: The organization of data to show how often certain values or ranges of values occur.

Geographic Information System (GIS): A system that allows for the interrelation of quality data (as well as other information) from a diversity of sources based on multilayered geographical information-processing techniques.

hazard (toxic): The set of inherent properties of a stressor or mixture of stressors that makes it capable of causing adverse effects in humans or the environment when a particular intensity of exposure occurs. *See also* risk.

hazard assessment (HA): Comparison of the intrinsic ability to cause harm with expected environmental concentration. In Europe, it is typically a comparison of predicted environmental concentration (PEC) with predicted no-effect concentration (PNEC). It is normally based on a single value for effects and exposure. It is sometimes incorrectly referred to as risk assessment.

hazard quotient (HQ): A ratio between the exposure concentration or dose and the effect concentration or dose, where both are expressed in the same units. Incorrectly referred to as the risk quotient (RQ).

hazardous concentration p (HCp): Hazardous concentration for $p\%$ (e.g., 5%, or HC5) of the species.

indirect effect: An effect resulting from the action of an agent on some components of the ecosystem, which in turn affects the assessment endpoint or other ecological component of interest. Indirect effects of substance contaminants include reduced abundance due to adverse effects on food species or on plants that provide habitat structure. *See also* direct effect.

intervention value: A screening criterion (in The Netherlands) based on risks to human health and ecological receptors and processes. The ecotoxicological component of the intervention value is the hazardous concentration 50 (HC_{50}), the concentration at which 50% of species are assumed to be protected.

joint action: Two or more substances exerting their effects simultaneously. These effects may or may not be additive, synergistic, or antagonistic.

K_{OC}: *See* soil organic carbon partition coefficient.

K_{OW}: *See* octanol–water partition coefficient.

lethal concentration n (LCn) or lethal dose n (LDn): The concentration or dose of a substance in water that is estimated to be lethal to $n\%$ of the test organisms. The LC50 is the median lethal concentration; the LD50 is the median lethal dose. The LC values and their 95% confidence limits are usually derived by statistical analysis of mortalities in several test concentrations, after a fixed period of exposure. The duration of exposure must be specified (e.g., 96-hour LC50).

life-cycle assessment (LCA): A method for determining the relative environmental impacts of products and technologies based on the consequences of their life cycle, from extraction of raw materials to disposal of the product following use.

line of evidence: Two or more substances exerting their effects simultaneously. These effects may or may not be synergistic or antagonistic. Each line of evidence is qualitatively different from any others used in the risk characterization. In ecotoxicological assessments, the most commonly used lines of evidence are based on 1) biological surveys, 2) toxicity tests of contaminated media, and 3) toxicity tests of individual substances.

LOEC$_{ecosystem}$: The lowest concentration of a test substance to which an ecosystem is exposed that causes an observed and statistically significantly different effect on the most sensitive endpoint as compared with the controls.

log-normal distribution: A distribution that is classically bell-shaped and symmetrical only when the data are transformed to a logarithm. *See also* normal distribution.

lowest-observed-adverse-effect concentration (LOAEC): The lowest level of exposure to a substance in a test that causes statistically significant differences from the controls in a measured negative or adverse response. *See also* lowest-observed-effect concentration.

lowest-observed-effect concentration (LOEC): The lowest concentration of a test substance to which organisms are exposed that causes an observed and statistically significantly different effect (adverse or not) on the organism as compared with the controls. "Level" is sometimes incorrectly substituted for "concentration." *See also* lowest-observed-adverse-effect concentration.

maximum allowable toxicant concentration (MATC): Geometric mean of the lowest-observed-effect concentration (LOEC) and no-observed-effect concentration (NOEC).

measure of effect: *See* effect measure (equivalent to the earlier term "measurement endpoint").

measure of exposure: A measurable characteristic of a contaminant or other agent that is used to quantify exposure.

mechanism of action: The process by which a physiological alteration is induced. It is often used interchangeably with the term toxic mode of action but is usually more specific. For example, the mode of action of an agent on a population may be lethality, and its mechanism of action may be acute narcosis, cholinesterase inhibition, or uncoupling of oxidative phosphorylation.

mechanistic model: A mathematical or functional representation of some component of an ecosystem with parameters that can be adjusted to closely describe a set of empirical data.

median lethal concentration: A statistically or graphically estimated concentration of a stressor that is expected to be lethal to 50% of a group of organisms under specified conditions.

mode of action: Biochemical effect that occurs at the lowest dose or concentration or is the earliest among a number of biochemical effects that could, understandably, lead to adverse effects in the organism. A more precise term is the "primary mode of action" of a substance. However, there may also be other biochemical effects that occur later or at higher doses (i.e., secondary modes of action) that also may contribute to adverse effects.

model: A formal representation of some component of the world, or a mathematical function with parameters that can be adjusted so that the function closely describes a set of empirical data.

model uncertainty: The component of the uncertainty concerning an estimated value that is due to possible misspecification of a model used for the estimation. It may be due to the choice of the form of the model, its parameters, or its bounds.

Monte Carlo simulation: An iterative resampling technique frequently used in uncertainty analysis in risk assessments to estimate the distribution of a model's output parameter.

NOEC$_{community}$: The highest concentration of a test substance to which a community of organisms is exposed that does not cause any observed and statistically significant effects on the most sensitive endpoint in the community as compared with the controls.

NOEC$_{ecosystem}$: The highest concentration of a test substance to which an ecosystem is exposed that does not cause any observed and statistically significant effects on the most sensitive endpoint as compared with the controls.

no-observed-adverse-effect concentration (NOAEC): Greatest concentration or amount of a substance, found by experiment or observation, which causes no detectable adverse alteration of morphology, functional capacity, growth, development, or life span of the target organism under defined conditions of exposure.

no-observed-effect concentration (NOEC): Greatest concentration or amount of a substance, found by experiment or observation, that causes no alterations (adverse or otherwise) of morphology, functional capacity, growth, development, or life span of target organisms distinguishable from those observed in normal (control) organisms of the same species and strain under the same defined conditions of exposure. "NOEC" customarily refers to the most sensitive effect unless otherwise specified. No effect level (NEL) and no effect concentration (NEC) are used as equivalent terms. Sometimes "level" is incorrectly used in place of "concentration."

normal distribution: The classical statistical bell-shaped distribution that is symmetric and parametrically simple in that it can be fully characterized by 2 parameters: its mean and variance. The normal distribution is observed in situations where many independent additive effects are influencing the values of the variates. *See also* log-normal.

normalization: Alteration of a substance concentration or other property (usually by dividing by a factor) to reduce variance due to some characteristic of an organism or its environment (e.g., division of the body burden of a substance by the organism's lipid content to generate a lipid-normalized concentration).

octanol–water partition coefficient (K_{OW}): Partition coefficient for a substance in the 2-phase system octan-1-ol–water. The K_{OW} indicates the relative hydrophobicity of a substance and its potential for bioconcentration or bioaccumulation.

partition coefficient: Ratio of the concentrations of a substance in solution in 2 phases that are in equilibrium. *See also* soil organic carbon partition coefficient and octanol–water partition coefficient.

percentile: Incorrect term for centile.

pesticide: A substance (or device) used to kill, reduce, or mitigate the effect of one organism (the pest) on another (usually a plant, human, or domestic animal).

pKa: The negative of the base-10 logarithm of the acid dissociation equilibrium constant, Ka, of a compound. *Note*: The smaller the number, the more acidic the compound.

pKb: The negative of the base-10 logarithm of the basic reaction equilibrium constant of a compound. *Note*: The lower the number, the more basic (alkaline) the compound.

plant protection product (PPP): A pesticide used to protect plants from damage by pests.

pollution-induced community tolerance (PICT): Based on responses from short-term (multispecies) toxicity tests on whole communities from clean and contaminated sites. Pollution tolerance is quantified by reduced sensitivity of the toxicant in these tests.

population: An aggregate of interbreeding individuals of a species, occupying a specific location in space and time.

predicted environmental concentration (PEC): The concentration of a substance in the environment that is predicted or calculated from its properties, its use and discharge patterns, and the associated quantities.

predicted no-effect concentration (PNEC): The maximum exposure concentration that, on the basis of current knowledge, is likely to be tolerated by an organism without producing any adverse effect. Sometimes "level" is incorrectly used in place of "concentration."

probabilistic risk assessment (PRA): Risk assessment where the probability or likelihood of adverse effects is estimated from more than 1 datum and the uncertainty is characterized. *See also* deterministic hazard (risk) assessment.

probability: According to the frequentist view, probability is the frequency of an event in an infinite repetition of identical and independent trials. In the Bayesian view, probability is a measure for the degree of belief in possible values of a random variable. In both views, probability is a measure of uncertainty of some outcome of an experiment, extrapolation, or prediction.

probability density function (PDF): For a continuous random variable, the PDF expresses the probability that the random variable belongs to some very small interval. For a discrete random variable, the PDF expresses the probability that the random variable is equal to a specific (discrete) value.

problem formulation: The phase in an ecological risk assessment in which the goals of the assessment are defined and the methods for achieving those goals are specified.

quantile: The value of a random variable that corresponds to a specified proportion of the probability density function of that random variable. Quantiles can be determined from the inverse cumulative distribution function. The median is the 0.50th quantile. The quartiles are the 0.25th, 0.50th, and 0.75th quantiles. The centiles are the 0.01th, 0.02th, and so on, quantiles.

(quantitative) structure-activity relationship ([Q]SAR): The use of (quantitative) structure and activity relationships to estimate the toxic potency of a substance. *See also* structure-activity relationship.

random variable: A probabilistic (i.e., uncertain) quantity that may assume different possible values in either a continuous or a discrete way.

receptor: An organism, population, or community that is exposed to contaminants. Receptors may or may not be assessment endpoint entities. The term "receptor" is also used to refer to sites in proteins or membranes where a substance may bind and produce a biochemical or physiological effect.

recovery: The extent of return of a population, community, or ecosystem function to a condition that existed before being affected by a stressor. Due to the complex and dynamic nature of ecological systems, the attributes of a "recovered" system must be carefully defined.

refugia: Areas in a ecosystem that are not exposed to, or are exposed only to small concentrations (< NOEC) of, a stressor and from which affected areas can recover.

resiliency: The degree to which a population of an organism or a community can tolerate a perturbation without the structure or function being affected.

risk (toxic): The predicted or actual probability of occurrence of an adverse effect on humans or the environment as a result of exposure to a stressor or mixture of stressors. *See also* hazard.

risk assessment (RA): A process that entails some or all of the following elements: hazard identification, effects characterization, exposure characterization, and risk characterization. It is the identification and quantification of the risk resulting from a specific use or occurrence of a stressor, including the establishment of concentration– or dose–response relationships and target organisms.

risk characterization: A phase of ecological risk assessment that integrates the exposure and stressor response profiles to evaluate the likelihood of adverse ecological effects associated with exposure to the contaminants.

risk management: The process of deciding what regulatory or remedial actions to take, justifying the decision, and implementing the decision.

risk quotient (RQ): *See* hazard quotient.

safety factor: A factor applied to an observed or estimated toxic concentration or dose to arrive at a criterion or standard that is considered safe. The terms "safety factor" and "uncertainty factor" are often used synonymously, but uncertainty factor is preferred. *See also* uncertainty factor.

soil organic carbon partition coefficient (K_{OC}): Ratio of concentrations of a substance sorbed in the organic matter component of soil or sediment to that in the aqueous phase at equilibrium. The K_{OC} is calculated by dividing the K_d value by the fraction of organic carbon present in the soil or sediment.

soil partition coefficient (K_d): 1) Experimental ratio of the concentration of a substance in the soil to that in the aqueous (dissolved) phase at equilibrium. 2) Distribution coefficient reflecting the relative affinity of a substance for adsorption by soil solids and its potential for leaching through soil. The K_d is valid only for the specific concentration and solid–solution ratio of the test. *See also* K_{OC}.

speciation: Determination of the exact chemical form or compound in which an element occurs in a sample. For example, whether arsenic occurs in the form of trivalent or pentavalent ions or as part of an organic molecule, and the quantitative distribution of the different chemical forms that may coexist.

species sensitivity distribution (SSD): A probability density function (PDF) or cumulative distribution function (CDF) of the toxicity of a certain substance or mixture of substances to a set of species that may be defined as a taxon, assemblage, or community. Empirically, a PDF or CDF is estimated from a sample of toxicity data for the specified species set.

stressor: A chemical, physical, or biological agent that acts on and causes an adverse response in an organism.

structure activity relationship (SAR): A process whereby the effect (toxicity) of a substance is estimated from its physical and chemical properties.

surface water: All water naturally open to the atmosphere (rivers, lakes, reservoirs, streams, impoundments, seas, estuaries, etc.) and all springs, wells, or other collectors that are directly influenced by surface water.

susceptibility: The relative condition of an organism or other ecological component lacking the power to resist a particular stressor. It is inversely proportional to the magnitude of the exposure required to cause the response.

synergism: Toxicological interaction in which the combined effect of 2 or more substances is greater than the simple sum of the effects of each substance.

test endpoint: The responses measured in a bioassay or toxicity test. *See also* effect criterion.

toxic: Capable of causing injury or harm to an organism.

toxicity: 1) Capacity to cause injury to a living organism defined with reference to the quantity of substance administered or absorbed, the way in which the substance is administered and distributed in time (single or repeated doses), the type and severity of injury, the time needed to produce the injury, the nature of the organisms affected, and other relevant conditions. 2) Adverse effects of a substance on a living organism defined with reference to the quantity of substance administered or absorbed, the way in which the substance is administered (e.g., inhalation, ingestion, topical application, or injection) and distributed in time (single or repeated doses), the type and severity of injury, the time needed to produce the injury, the nature of the organisms affected, and other relevant conditions.

toxicity test: The determination of the effect of a substance on a group of selected organisms under defined conditions. A toxicity test usually measures either 1) the proportions of organisms affected (quantal) or 2) the degree of effect shown (graded or quantitative) after exposure to a range of stressor concentrations or a mixture of stressors.

toxic mode of action (TMoA): A phenomenological description of how an effect is induced. *See also* mechanism of action.

toxic unit (TU): The concentration of a substance expressed as a fraction or proportion of its effective concentration (measured in the same units). It may be calculated as follows: toxic unit = actual concentration of substance in solution/LC50. If this number is greater than 1.0, more than half of a group of exposed organisms will be killed by the substance. If it is less than 1.0, less than half of the organisms will be killed.

toxin: A substance produced by an organism that causes injury or harm to another.

training data set: A set of data used to develop a model or a data set used to develop a quantitative structure-activity relationship.

uncertainty: Imperfect knowledge concerning the present or future state of the system under consideration; a component of risk resulting from imperfect knowledge of the intensity of effect or of its spatial and temporal pattern of expression.

uncertainty factor (UF): A factor applied to an exposure or effect concentration or dose or hazard quotient to correct for unidentified sources of uncertainty. *See also* safety factor.

xenobiotic: Substance that is not natural to the organism in question. May be natural or anthropogenic in origins.

NOTE

This glossary was adapted from a number of sources, mainly Posthuma et al. (2002b) and Stephenson et al. (2006, with permission).

10 Guidance on the Application of Extrapolation Methods in Ecological Exposure and Effects Characterization of Chemicals

Leo Posthuma, Dick De Zwart,
Keith R. Solomon, and Theo C. M. Brock

CONTENTS

10.1   INTRODUCTION

Ecological risk assessments cannot be done without applying extrapolation methods. Sufficient data to execute such risk assessments are usually lacking. For example, there may be no toxicity data for the suspect substance, the tested species may differ from the species in the assessed ecosystem, exposure is to single substances in test systems but mixtures occur in the field, or risks are to be assessed for communities rather than for species. The lack of data is a consequence of practical and ethical considerations.

Risk is generally considered as a product of the probability of an adverse effect and the magnitude of that effect. In ecotoxicology, risk depends on the probability and intensity of exposure and the sensitivity of the exposed organisms, whereby the interpretation of risk can involve aspects of space and time and value judgments (e.g., believing that 1 species is more important than another). The sensitivity is often determined in laboratory toxicity tests, in which dose– or concentration–effect curves are established. Extrapolation methods exist for both components of risk and the additional aspects.

With respect to exposures, there are many possible toxicants. More than 100000 chemicals exist (Commission of the European Communities, 1990), the number of possible mixtures is almost infinite, and the exposure of biota to chemicals is influenced by matrix characteristics. With respect to sensitivity and responses of biota, there is great natural variability among species and ecosystem types and in the array of conditions in which they occur. It is evident that an array of extrapolation types is needed for risk assessment.

Reviews on the issue of extrapolation demonstrated that indeed a plethora of extrapolation methods exists. This is noted in earlier studies and overviews, such as the workshop on extrapolation organized by the OECD (OECD 1992) and the overview volumes of Suter on ecological risk assessment (Suter et al. 1993) and site-specific risk assessment (Suter et al. 2000). Over the last decades, the need for and use of extrapolation techniques have increased. This situation occurred concomitantly with a diversification of methods that are applied in practice. This book provides clear examples of this.

Extrapolation methods are used for various types of risk assessment. Methods may be used in the process of deriving environmental quality objectives, in the registration of new substances, and in the process of site-specific risk assessment. Suter (1993) called these approaches prospective (the former 2) and retrospective (the latter) risk assessments. The specific process in which extrapolation methods are used has implications for the concepts to be applied and the data to be used as input in extrapolation. Strictly described approaches are in place for the derivation of environmental quality criteria (EQCs) and the registration of pesticides and newly developed substances. The prescribed approaches for deriving EQCs can differ between jurisdictions. The approaches for retrospective investigations have more degrees of freedom. A characteristic of the latter approach is that the methods can make use of measured local exposure levels and can estimate local risk with known precision (or known uncertainty!). The latter is uncommon for EQCs.

The diversity of existing extrapolation techniques also relates to the types of extrapolation problems. Extrapolation can consist of range extrapolation, implying intra- or extrapolation using an available data set. It can also be a specific extrapolation from 1 data set to parameters in another realm (e.g., from total concentrations to bioavailable concentrations, or from species sensitivities to community-level responses).

Finally, the diversity of extrapolation techniques relates to the diversity of technical solutions that have been defined in the face of the various extrapolation problems. Methods may range from simple to complex, or from empirical–statistical methods that describe sets of observations (but do not aim to explain them) to mechanism-based approaches (in which a hypothesized mechanism was guiding in the derivation of the extrapolation method). In addition, they may range from those routinely accepted in formal risk assessment frameworks to unique problem-specific approaches, and from laboratory-based extrapolations consisting of 1 or various kinds of modeling to physical experiments that are set up to mimic the situation of concern (with the aim to reduce the need for extrapolation modeling).

When there is a practical need for a set of methods and when those methods are used in an unsystematic way, there is a need for developing guidance. Guidance generally is of help to practitioners who need to solve practical risk assessment problems. This guidance offers guidance on the use of the extrapolation methods that are compiled in this book. The guidance discusses issues that are relevant to making systematic use of the available methods, such as the concept of tiering, and the ordering of extrapolation methods according to their scientific principles and expected precision.

We are aware that various tiered systems are in use in different jurisdictions, and for different assessment problems. It is not our view that these methods should necessarily be changed on the basis of this overview. These systems have a formal status, and serve their purpose. Our purpose is to provide guidance for situations where there is not yet a prescribed approach, and to guide the way through the possible options.

10.2 NEED FOR AND CHARACTERISTICS OF TIERED SYSTEMS

Ecological risk assessments are driven not only by science but also by pragmatism. When many methods are available, ecological risk assessments driven only by science would likely result in the application of the best of available methods. This would result in more precise predictions and similarity of outcomes when different researchers conduct the same assessment. However, this would be a costly approach when applied for all problems. Pragmatism, cost, and efficacy are drivers for practical ecological risk assessments too. However, the application of practical arguments as sole drivers could result in too simplified and incorrect risk assessments and thus wrongly informed decisions. When both pragmatism and science drive the assessment, one can understand the development of tiered systems, with conservative, simple, fast, and less costly approaches in the lower tiers, and increasing complexity, time consumption, costs, and hopefully better predictions at higher tiers. Lower tier methods are usually referred to as screening-level or preliminary assessments, followed by a refined assessment or site-specific assessment in the next higher tiers.

The essential features of tiering are shown in Figure 10.1. The general objective of designing tiered systems is that an assessment problem should be solved by a limited and efficient use of resources, with enough information generated to make informed and responsible decisions. When a simple problem definition allows for extrapolation methods yielding gross insights, these may be sufficient for that simple problem. A key requisite for the proper functioning of a tiered system is that higher tier assessment results are generally more realistic and less conservative than lower tiers.

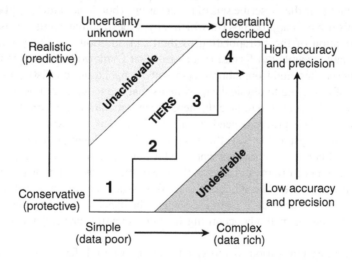

FIGURE 10.1 Tiers in the risk assessment process, showing the refining of the process through acquisition of additional data.

With no a priori idea on how to design a tiered approach, the application of tiered approaches could easily result in an array of different outcomes for the same assessment problem. The outcome would, for example, be dependent on the specific field of expertise of the assessor. In that particular field, this assessor could effectively and rapidly apply state-of-the-art, higher tier methods, whereas other assessors would start at the lowest tier. This way of organizing risk assessments would result in widely different approaches being applied to a single type of problem, and hence outcomes that differ in degree of specificity and certainty. Applying extrapolation methods without guidance on how to do that would eventually result in an adverse public perception on the quality and accuracy of the output of risk assessments. Credibility would be at stake. To avoid this potentially "damaging practice," tiered systems in ecological risk assessment should be designed according to some general rules. This means the following:

1) The system as a whole is internally consistent.
2) The system can address a logical array of increasingly more specific assessment questions.
3) The system is cost-effective.

These arguments together represent both the science-driven and the practice-driven views on tiering.

Thus, the concept of extrapolation in ecological risk assessments requires systematic evaluation of both the available scientific methods as well as the optimization of their practical usage.

10.3 CONCEPTUAL ISSUES IN EXTRAPOLATION

Various tiered risk assessment schemes have been published; for some examples, see Table 10.1. Tiered approaches are used to address risk problems for "classical" chemical substances, pharmaceuticals, pathogens, and so forth, for both ecological and human risk assessments.

Conceptually, the tiering of these risk assessment systems makes use of 3 axes, each representing an array of possibilities to go from simple to complex (Figure 10.2):

1) An axis for problem type, that is, the scientific problem to be solved by the extrapolation problem (e.g., from simple and generic assessment questions to more complex and specific ones)
2) An axis for data type (e.g., source of data on which extrapolations or decisions are based — e.g., laboratory data or field data)
3) An axis for the process of data handling (e.g., deterministic or probabilistic handling of the data)

The axes are chosen so that they are practicable in solving the assessment problem, that is, choices made on these axes should be made so that they reduce the uncertainties embedded in the available data as much as possible. For example, the

TABLE 10.1
Some examples of risk assessment systems with a tiered approach

Regulated issue	Reference
Water contamination: ARAMDG	SETAC (1994)
Water contamination: Water Environmental Research Foundation (WERF) approach and software	Parkhurst et al. (1996)
Bacterial pathogens in drinking water	Rusin et al. (1997)
New and existing chemical substances: EU	Vermeire et al. (1997)
New and existing chemical substances: USEPA	USEPA (1998)
Endocrine-disrupting chemicals: European Chemical Industry Council (CEFIC) proposal	Hutchinson et al. (2000)
Plant protection products: EU	Campbell et al. (1999)
Human pharmaceuticals: EU	Straub (2002)

European Union, in Directive 92/32/EC and EC Council Regulation (EC) 793/93, requires the risk assessment of chemicals. Under these regulations, the European Union System for the Evaluation of Substances (EUSES) software carries out tiered risk assessments of increasing complexity on the basis of increasing data requirements related to scientific problem type (Vermeire et al. 1997). In the risk assessment of plant protection products, the approaches taken in the risk assessment scheme are organized taking account of data type according to EU Council Directive 91/414 (European Union 1991); that is, laboratory data are used in the lower tiers, and data from field tests in the highest tier. By this method, reduction of uncertainties is sought not only by reduction of uncertainties in exposure extrapolation models but also by better mimicking testing under realistic (field) conditions in each step (Campbell et al. 1999). Finally, many risk assessments are developing ways to address the problem in a probabilistic way (e.g., Hart 2001; EUFRAM 2005).

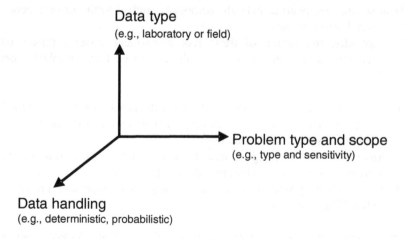

Data type
(e.g., laboratory or field)

Problem type and scope
(e.g., type and sensitivity)

Data handling
(e.g., deterministic, probabilistic)

FIGURE 10.2 Three axes along which extrapolation methods in tiered systems can be organized.

In the different risk assessment approaches that exist (e.g., Table 10.1), there are different ways to combine the 3 axes, in prescribing the approaches that are needed to address the specific problem, collect the data, and handle the data and models. Due to their different origins and the different problems that they address, the existing tiered systems are thus very different. One method is, however, not necessarily better or worse than the other. For this guidance, we have organized the methods according to the method characteristics themselves, not to existing tiered frameworks.

A set of "scopes" can be recognized when working with extrapolation. Basic in all possible extrapolations is the issue of range extrapolation. Do the data that are available on any extrapolation aspect cover the data range on which the assessment focuses?

Furthermore, 3 major types of extrapolation can be distinguished regarding the focus of the extrapolation:

- At the level of the matrix and media in which exposure takes place, extrapolation is needed to predict exposure in the matrices and media that have not been tested.
- At the level of substances and their toxicity, extrapolation is needed to predict the toxicity characteristics of substances that have not been tested.
- At the level of the exposed biota, extrapolation is needed to predict effects at the level of concern (e.g., populations and communities) from lower levels of biological organization (e.g., tests on individuals exposed in test systems).

Additionally, there are further issues to be addressed in executing a risk assessment because their influences shape the true responses to be expected in field conditions. These are as follows:

- Are there more substances of potential relevance, and, if so, can and should extrapolation take mixed exposure and interactive effects into account?
- Are there spatial and/or temporal distributions of the substances and of the biotic entity of concern that warrant spatially and/or temporally explicit risk assessment?

Although distinctions among scopes are an important lead in characterizing and choosing extrapolation methods, it should be clearly noted that 1 risk assessment may require the use of various techniques simultaneously or in sequence. For example, one always needs to consider both exposure and effect assessment, implying for example the need to use substance-to-substance extrapolation, matrix and media extrapolation, and extrapolation across levels of biological organization, with or without range extrapolation at any level, and with or without mixture issues or spatial or temporal issues being taken into account.

The sequential approaches should thereby follow the path from cause to effect. That is, a chemical mixture extrapolation should consider first the exposure issues of all relevant substances, then the effects issues per substance, and finally the effects of the mixture.

When the scope is known and the extrapolation methods are defined, the data for the risk assessment steps (and extrapolations) can be collected. For tiered systems, this implies that choices need to be made on the manner in which uncertainties are addressed and what to do when these are only addressed by simple methods. A distinction was already made between prospective use and retrospective use of extrapolation methods. In both cases, extrapolations are being applied, but the way in which existing methods are selected for an assessment problem can differ.

In the prospective context, a common and simple way to handle uncertainty is the use of uncertainty factors (UFs). These may suffice to derive a "safe" concentration of a substance associated to a predefined protection level to be used generically — that is, it is "safe" even in worst-case conditions. The greater the uncertainty in models or data for the extrapolations, the larger the overall UF in the lower tiers. The UF is applied to the risk assessment to account for unquantified uncertainties. In some cases, the factor depends on the amount of available data, or UFs per extrapolation are multiplied to provide the final factor (e.g., $10 \times 10 \times 10$ as the UF for 3 assessment steps yields an overall factor of 1000).

In a retrospective context (such as for contaminated environments), UFs are conceptually illogical for any step of the assessment. The retrospective use of risk assessment implies a wish to accurately quantify the site-specific risks of existing situations (to underpin decisions), rather than to offer a protective guideline per se. Whereas UFs may be a leading principle in prospective risk assessment or criteria setting in which conservatism is the preferred bias, accuracy in the end product is the principle for retrospective risk assessment. Confidence intervals are the logical way to show uncertainty in the retrospective risk assessment. Because deterministic approaches do not allow calculation of confidence intervals, this suggests the use of probabilistic approaches in higher tiers of risk assessments.

It is noteworthy that comparisons of existing assessment schemes reveal dissimilarities in the use of extrapolation methods and their input data between different jurisdictions and between prospective and retrospective assessment schemes. This is clearly apparent from, for example, a set of scientific comparisons of 5% hazardous concentration (HC5) values for different substances. Absolute HC5 values and their lower confidence values were different among the different statistical models that can be used to describe a species sensitivity distribution (SSD; Wheeler et al. 2002a). As different countries have made different choices in the prescribed modeling by SSDs (regarding data quality, preferred model, etc.), it is clear that different jurisdictions may have different environmental quality criteria for the same substance. Considering the science, the absolute values could be the same in view of the fact that the assessment problem, the available extrapolation methods, and the possible set of input data are (scientifically) similar across jurisdictions. When it is possible, however, to look at the confidence intervals, the numerical differences resulting from different details in method choice become smaller because confidence intervals show overlap.

A detailed review of all extrapolation methods is given in this book on extrapolation. The scientific and pragmatic views on extrapolation for separate processes in ecotoxicological risk assessment are presented in the previous chapters in detail. To enable use of all techniques in a productive way, this guidance provides guidance in using the listed extrapolation methods in a way that is founded both in science as well as on daily practices. A classification is provided to highlight the hidden features of the extrapolation techniques, so as to help in clarifying discussions on the issue of choice of extrapolation techniques. The purpose of the recommendations in this guidance is the promotion of discussion and thought on tackling extrapolation decisions rather than the setting of strict rules for extrapolation.

10.4 GUIDANCE FOR STEPWISE USE OF EXTRAPOLATION

10.4.1 PREPARING FOR AN ASSESSMENT

The basic features of ecological risk assessment schemes are very similar throughout the world. Usually, one focuses on effects (concentration or dose response information), exposure, and risk characterization. The following paragraphs summarize how extrapolation practices can be developed in such a way that a consistent pattern emerges.

The development of an extrapolation scheme for a specific problem may frequently need to start from a regulatory perspective, that is, the set of extrapolation methods is already clearly defined for a given assessment problem. Note that, for such existing approaches, the issue of a tiered system does apply as well, and that it may pertain to a slightly different perspective on tiering. In the current guidance, tiering is proposed for a set of extrapolation techniques pertaining to a single issue of the assessment problem. For example, various methods can be used to predict bioavailability from total ambient concentrations, and these methods can be tiered from simple and conservative to specific, mechanism-based, and more realistic approaches. In existing tiered systems for ecological risk assessment, the tiers are related to the overall specificity of the selected sets of approaches to address a problem.

10.4.2 EXTENDING ASSESSMENTS WITH EXTRAPOLATION GUIDANCE

In view of the diversity of available extrapolation options, risk assessment protocols would profit from the definition of clear guidance on the use of extrapolation methods. Unless specifically excluded, preferred procedures would include the following:

Step 1: Motivating the needs for extrapolation
Step 2: Identifying the issues that would require attention due to their influence on risk assessment results
Step 3: Identifying possible extrapolation methods
Step 4: Assigning available methods to tiers prior to working through an assessment
Step 5: Choosing a consistent set of extrapolation methods
Step 6: Collecting and judging the data that are needed
Step 7: Working with the set of extrapolation methods
Step 8: Interpreting assessment results

A standard protocol can be derived when there is a need for repeated or regulatory use.

These items will be addressed in subsequent paragraphs. Thereafter, various sets of technical questions are posed, so as to help decide on the need and options for extrapolation.

10.4.3 STEP 1: MOTIVATE THE NEED OF EXTRAPOLATION

The first issue to be tackled is to consider whether extrapolation is needed at all. General "scientific" questions can be posed (e.g., do we have appropriate data, or is it likely that we can easily find data, that are of direct help in decision making without extrapolation?). Note that posing these questions may not be needed in cases where risk assessment is executed within strictly prescribed rules. In that case, the whole process boils down to following the prescribed protocol.

Systematically, the possible subjects of extrapolation can be considered. In line with this book, such issues are as follows:

- Do we have original data on the substance or substance mixture under investigation?
- Are these data relevant regarding the matrix and medium of exposure, the ecological receptor (species, function, community, etc.), and the temporal/ spatial setting?

And, more specifically, these questions can be translated into questions such as these:

- Can (Q)SAR be of help to address lack of data on substance properties and toxicity?
- Can matrix and media extrapolation techniques be of help to address bio-availability differences among test media and the field?
- Can extrapolation of effect measures across levels of biological organization be of help?
- Is there a need for accounting for aggregation of risks because of mixtures?
- Are there special temporal or spatial conditions that magnify or reduce the probability of exposure or effects?

As a result, answering these questions yields a list of potentially relevant extrapolation items. Each of these items can be detailed into other questions, but at this stage, it is sufficient to consider whether there is, or seems to be, a need for extrapolation, and not how that need is fulfilled.

10.4.4 STEP 2: IDENTIFICATION OF ISSUES
WITH HIGH NUMERICAL RELEVANCE

The list of extrapolation items can be subjected to a test in terms of pragmatic issues. Choosing amongst all possible extrapolation methods may be needed due to resource limitations (pragmatism) or preferred methods and/or estimated lack of data or methods (science). This stage consists of a "preliminary sensitivity analysis" to identify the items of most concern, that is, those extrapolation subjects that are potentially most important in a numerical sense. All other extrapolations may be scientifically worthwhile, but either they are too expensive or they influence the numerical outcome of the risk assessment only in a small way. Techniques that can be applied are expert judgment, a limited literature search, and reconsideration of extrapolation practices regarding resource use. The result of this step is a preliminary ranking of extrapolation needs; focus is placed on the most sensitive aspects of the problem.

10.4.5 STEP 3: IDENTIFICATION OF POSSIBLE EXTRAPOLATION METHODS

The next step is the identification of extrapolation methods that are available, or methods that need to be derived on the basis of available knowledge.

In the first case, a literature review provides a large set of available and applied techniques, both at the conceptual level (e.g., how does it work, what does it do, how is the technique founded in science, and how well is a technique validated?) and in the technical details. When no validated technique is available, general scientific knowledge, such as deriving a novel type of specific extrapolation model from specific data, can be developed.

When a model-based extrapolation technique is unavailable, it may be possible to work along the axis of Figure 10.2 concerning the choice of data, that is, to choose for the use of physical models. With the term "physical model," it is meant that the model itself consists of creating an experimental or observational situation that mimics the situation of concern. This option is often applied in the registration of pesticides, where microcosm, mesocosm, and/or field experiments are used to characterize the impacts of pesticides on nontarget species. When the physical models do not, in some aspects, resemble the situation of concern, 1 or several extrapolation steps may be needed.

The result of this stage is a list of possible techniques, their operational details, and their degree of validation. This process may show a lack of existing extrapolation techniques, 1 unique method, or a number of different extrapolation techniques that can be used to address the same problem. For example, bioavailability extrapolations can be made on statistical-based techniques, describing phenomena without understanding their cause (such as "transfer functions" of the following type: exposure = f[pH, organic matter concentration, hardness, and so on]), and also on mechanism-based techniques such as the biotic ligand model.

As a result, the investigator has a provisional list of extrapolation methods, ranging from modeling approaches to experimentation (i.e., physical models of reality). If more than 1 method is available, the next step is to choose amongst them, or to estimate the level of specificity that is to be expected from applying the singular available model.

10.4.6 STEP 4: ASSIGNING METHODS TO TIERS

To fulfill the general requisite of tiering, higher tiers should yield generally more accurate, more precise, and in most cases also lowered estimates of risk. In the lowest tier, UFs are introduced to reflect that many issues that may modify (reduce) risk

TABLE 10.2
A proposal for conceptual tiering of extrapolation approaches in ecological risk assessment

Generalized tiers			
Characteristics	Major and minor tiers[a]	Extrapolation	Approach
None	0	None	None
Simple generic	1a	Deterministic	Uncertainty factor (UF) or simple model
	1b	Probabilistic	Data-dependent UF
Moderately simple generic	2a	Deterministic	Statistical model, deterministic use
	2b	Probabilistic	Statistical model, probabilistic use
Complex specific	3a	Deterministic	Complex statistical model, partially mechanistic orientation deterministic use
	3b	Probabilistic	Complex statistical model, partially mechanistic orientation deterministic use
Highly specific	4a	Deterministic	Mechanistic model, deterministic use
	4b	Probabilistic	Mechanistic model, probabilistic use
Special	Special	Deterministic	Specific approach (deterministic)
	Special	Probabilistic	Specific approach (probabilistic)

Note: Major tiers (0 to 4) and minor tiers (a and b) are discriminated. For further explanation, see the text.

are neglected. The higher the tier, the more specific the data that are used and the more the need for large UFs is reduced.

Even without specific extrapolation problems, it is possible to give a general overview of the ranking of extrapolation methods in this way. Such a general ranking can be used to rank a specific set of extrapolation techniques that address the same problem, or to give a gross identification of the level of operation of a sole technique. The general overview is based on scientific reasoning, and is shown in Table 10.2. Note that this implies that methods are organized according to a ranking of underlying principles (statistical methods are at a lower tier than mechanistic ones), but that various methods can be categorized at 2 tiers, depending on context. For example, an HC5 derived from an SSD is a fixed value, and is called deterministic by virtue of this, but the SSD itself is a probabilistic technique. Hence, practitioners of extrapolation can use the tables presented here as help to categorize the methods they could practically use for a problem they encounter. They should consider that it seems impossible to create overview,[*] tables of extrapolation methods that are completely without categorization problems. The cases of matrix and media extrapolation and

---

* That is, there is no simple decision tree that links a single, final set of extrapolation methods to a problem definition.

of mixture extrapolation are used to illustrate the organization of methods over tiers, and introduce the tier characterization terms.

The first general question to be answered is whether or not one will make use of data-driven extrapolation methods. Nonuse (Tier-0, "No extrapolation") can be chosen in case there is no need (Step 1, Step 2) or no option for extrapolation (Step 3). An interesting case in this respect is the discussion that is ongoing in the realm of risk assessment of mixtures. Especially here, the available ecotoxicity data on mixtures in many case studies show that all substances may contribute to toxicity, as is substantiated by the fact that mixture effects frequently occur at a mixture concentration that contains approximately 1 toxic unit of all substances (i.e., close to the prediction of the model of concentration additivity [CA]). Various existing assessment techniques, however, prescribe a limited use of CA to substances that share the same toxic mode of action (TMoA), and do not use mixture extrapolation for substances with dissimilar TMoAs. The latter may conceptually be justified (because the concept of CA is theoretically derived for substances with the same TMoA). Considering the outcomes of the majority of mixture studies, it may be numerically better to consider overall mixture risks than to refrain from using the CA as a mechanistically justifiable (sole) fallback option. In the case of mixtures, the competing models for extrapolation (in addition to CA, there are also the response addition [RA] and mixed-model approaches) provide numerically similar outcomes that are most often much closer to reality than no extrapolation (see Drescher and Bödeker 1995). In choosing between adopting or rejecting extrapolation, one has to answer the following question: which of the options introduces the most error into the outcome and thus into the decision to be taken? Such as in the mixture case, it might be justifiable to choose a slightly or conceptually "wrong" extrapolation method rather than "no extrapolation." In any case, any such decision should be clearly communicated.

The second general question to be answered is whether answering an assessment problem would profit from a probabilistic assessment over a deterministic one. In a probabilistic assessment, the aspect of variation in the data that are used in the extrapolation is propagated in the extrapolation outcome. That is, the result is a distribution of possible outcomes, with some answers being more probable than others. A probabilistic approach is often considered to be at a higher tier than a deterministic approach when the same modeling principles are addressed. This introduces a subtiering (the minor tiers a and b, for deterministic and probabilistic, respectively) when considering the overall tiers.

Finally, the overall major tiers (0 to 4) can be identified according to the characteristics given below. Note that further subtiering, as indicated in Table 10.3, is possible from Tier-2 onward. For example, in matrix and media extrapolation, one may apply statistical modeling to derive transfer functions that describe the entry of substances in organisms as a function of matrix characteristics. These models can be simple so as to address gross problems, such as the function

$$\text{Uptake} = \text{constant} + a \times \text{pH},$$

or it can be more complex, such as the function

$$\text{Uptake} = \text{constant} + a \times \text{pH} + b \times \text{Organic Matter} + c \times \text{Hardness} + d\ldots,$$

TABLE 10.3
Decision matrix for selecting extrapolation methods

Type of extrapolation	Problem definition	Problem subdefinition	Method	Chapter	Section
0: No extrapolation					
0: None	Prediction of actual exposure	Metals and organics in water, soil, and sediment	Assume completely available, no extrapolation	2: Matrix/media extrapolation	
	Prediction of mixture effects	No addition	Do not evaluate mixture toxicity	5: Mixture extrapolation	
	Geographical extrapolation of toxicity data	Toxicity differences between groups of species	Assume all groups of taxa are equally sensitive	7: Spatial extrapolation	
1: Simple generic					
1: Uncertainty factor (UF) or simple model extrapolation	Prediction of responses	Biomarker to individual	UF based on empirical studies	4: Across levels of biological organization	4.2.2
	Prediction of responses	Single-species test result to population	UF based on empirical studies		4.3.1
	Prediction of responses	Populations to community	UF based on empirical studies		4.4.1
	Prediction of mixture effects	Risk quotient concentration addition	TU, TEQ, TEF, and CCU addition	5: Mixture extrapolation	
	Time-varying exposure regimes	Empirical	UF	6: Temporal extrapolation	6.2.3
	Extrapolation of toxicity data between habitats	Toxicity differences between flowing and stagnant water organisms	UF based on empirical studies	7: Spatial extrapolation	7.2.4

Geographical extrapolation of toxicity data	Toxicity differences between fresh- and saltwater organisms		UF based on empirical studies	7: Spatial extrapolation	7.2.3
	Toxicity differences between Nearctic and Palearctic regions		UF based on empirical studies	7: Spatial extrapolation	7.2.3
	Toxicity differences between southern and northern hemispheres		UF based on empirical studies	7: Spatial extrapolation	7.2.3
	Toxicity differences between temperate and tropical regions		UF based on empirical studies	7: Spatial extrapolation	7.2.3
2: Extrapolation by statistical model	Prediction of chemical properties	(Q)SAR	Basic phys/chem $\rightarrow K_{OW}$, water sol, and so on; EPIWIN	3: (Q)SAR extrapolation	3.2.2
	Prediction of acute aquatic toxicity	Baseline toxicity	Taxa-specific linear regressions, ECOSAR		3.2.4
	Prediction of chronic aquatic toxicity	Baseline toxicity	Taxa-specific linear regressions, ECOSAR		
3: Extrapolation by semimechanistic models	Prediction of acute terrestrial toxicity	Baseline toxicity	Limited to a small number of taxa, linear regressions		3.3.3
	Time-varying exposure regimes	Incipient toxicity	ACR ratio, ACE	6: Temporal extrapolation	6.2.4
2: Moderately simple generic					
2: Extrapolation by statistical model	Prediction of actual exposure	Metals in soil and sediment	Empirical K_d–regression models	2: Matrix and/or media extrapolation	2.4.1
	Prediction of actual effect	Intrinsic differences in species sensitivity	SSD comparison		2.4.3

(Continued)

TABLE 10.3 (CONTINUED)
Decision matrix for selecting extrapolation methods

Type of extrapolation	Problem definition	Problem subdefinition	Method	Chapter	Section
	Prediction of acute aquatic toxicity	Specific toxicity	Multilinear, multivariate ordination; artificial neural networks	3: (Q)SAR extrapolation	3.2.4
	Prediction of chronic aquatic toxicity	Specific toxicity	Multilinear, multivariate ordination; artificial neural networks		
	Prediction of responses	Biomarker to individual	Statistical correlation	4: Across levels of biological organization	4.2.2
	Prediction of responses	Single-species test result to population	HERBEST and PERPEST approaches		4.3.1
	Prediction of responses	Populations to community	SSD		4.4.2
	Prediction of responses	Structure to function	Quantitative cause–effect relationships		4.5.3
	Prediction of mixture effects	SSD concentration addition	msPAF CA	5: Mixture extrapolation	—
		SSD response addition	msPAF RA		
	Reversibility of effects	Reversibility of effects	Empirical	6: Temporal extrapolation	6.2.2
	Latency of responses	Latent responses: direct and/or indirect effects	TMoA estimates, empirical	6: Temporal extrapolation	6.2.5
3: Extrapolation by semimechanistic model	Prediction of actual exposure	Organic toxicants in soil and sediment	Multiple-compartment equilibrium-partitioning modeling	2: Matrix/media extrapolation	2.4.1
		Organic toxicants in water	Single-compartment equilibrium-partitioning modeling		
		Prospective distribution modeling	Makay-type modeling		

	Prediction of chemical properties	(Q)SAR	Molecular descriptors	3: (Q)SAR extrapolation	3.2.2
4: Extrapolation by mechanistic model	Prediction of actual exposure	Metals in water	Speciation modeling	2: Matrix and/or media extrapolation	2.4.1
5: Specific approach (not 0, 1, 2, 3, or 4)	Measured mixture effects	Effluent bioassays	WET experimentation	5: Mixture extrapolation	—
3: Complex specific					
2: Extrapolation by statistical model	Prediction of chronic aquatic toxicity	Specific toxicity and/or TMoA	Receptor-specific (Q)SARs (comparative molecular field analysis [CoMFA])	3: (Q)SAR extrapolation	3.2.4
	Prediction of species sensitivity	QSSR	ICE		3.4.2
	Prediction of responses	Biomarker to individual	Quantitative cause–effect relationships	4: Extrapolation across levels of biological organization	4.2.2
	Prediction of mixture effects	SSD mixed model	msPAF mixed TMoA	5: Mixture extrapolation	—
	Time-varying exposure regimes	Pulsed exposure	RADAR	6: Temporal extrapolation	6.2.3
	Spatial variation in community responses	Comparing model ecosystem studies performed in different geographical regions	Statistical comparison	7: Spatial extrapolation	7.2.4
3: Extrapolation by semimechanistic models	Prediction of actual exposure	Organic toxicants in water	Multiple-compartment equilibrium-partitioning modeling	2: Matrix and/or media extrapolation	2.4.1
	Prediction of responses	Single-species test result to population	Simple pop models (exponential models)	4: Extrapolation across levels of biological organization	4.3.1

(Continued)

TABLE 10.3 (CONTINUED)
Decision matrix for selecting extrapolation methods

Type of extrapolation	Problem definition	Problem subdefinition	Method	Chapter	Section
4: Extrapolation by mechanistic model	Prediction of responses	Community to ecosystem and/or landscape	Simple food-web models	4: Extrapolation across levels of biological organization	4.5.4
	Time-varying exposure regimes	Mechanistic	DEBTox	6: Temporal extrapolation	6.2.3
5: Specific approach (not 0, 1, 2, 3, or 4)	Prediction of actual effect	Combined availability and/or sensitivity modeling	Indirect effects modeling	2: Matrix and/or media extrapolation	2.4.3
	Prediction of responses	Populations to community	Multispecies, microcosm, and mesocosm experiments	4: Extrapolation across levels of biological organization	4.4.1
	Prediction of responses	Community to ecosystem and/or landscape	Microcosm and mesocosm studies		4.5.1
	Measured mixture effects	Bioassays on environmental concentrates	pT measurements	5: Mixture extrapolation	5.7
4: Highly specific					
2: Extrapolation by statistical model	Prediction of mixture effects	SSD mixed-model split taxa	msPAF mixed TMoA–receptor	5: Mixture extrapolation	5.6.3
	Detection of differences in metapopulation responses in impacted landscapes	Toxicity as part of multiple stress	GIS modeling	7: Spatial extrapolation	7.3.2 and 7.3.5
3: Extrapolation by semimechanistic models	Prediction of actual exposure	Organic toxicants in water	Dynamic multiple-compartment modeling	2: Matrix and/or media extrapolation	2.4.1
	Detection of differences in metapopulation responses in impacted landscapes	Population dynamics	Metapopulation modeling	7: Spatial extrapolation	7.3.3

4: Extrapolation by mechanistic model	Prediction of actual effect	Metals in water	BLM modeling	2: Matrix and/or media extrapolation	2.4.1
	Prediction of chronic aquatic toxicity	Specific toxicity/TMoA and/or physiology	3-D modeling; PBTK	3: (Q)SAR extrapolation	3.2.5
	Prediction of responses	Biomarker to individual	DEB approaches	4: Extrapolation across levels of biological organization	4.2.2
	Prediction of responses	Single-species test result to population	Complex models (life stage, IBM, and energetic models)		4.3.2
	Prediction of responses	Community to ecosystem and/or landscape	Food-web models, metapopulation models		4.5.4
	Latency of responses	Latent responses: direct and/or indirect effects	CATS, AQUATOX, and CASM	6: Temporal extrapolation	6.2.5
5: Specific approach (not 0, 1, 2, 3, or 4)	Prediction of responses	Populations to community	Food-web models	4: Extrapolation across levels of biological organization	4.5.4
	Variable sensitivity within species	Seasonal, life history differences	Model systems	6: Temporal extrapolation	6.3.2
	Recovery of effects	Recovery experiments	Model systems		6.3.3
	Adaptation to exposure	Adaptation measurements	PICT		6.3.4
	Detection of differences in metapopulation responses in impacted landscapes	Ecological indicators	Monitoring	7: Spatial extrapolation	7.3.5 and 7.3.6
	Physical models	Mimic situation of concern in microcosms, mesocosms, or field tests		All	

Note: Overview of extrapolation techniques addressed elsewhere in this book, ordered according to the generalized tiering system of Table 10.2, and thereafter ordered according to the subject of extrapolation (matrix and media, mixtures, etc.). Major tiers (0 to 4) and minor tiers (a and b) are discriminated. For further explanation, see the text and the subject chapters of this book.

which takes into account all statistically significant variables in the "training set." Note further that the generalized tiering system may lead to 2 possible positions for a single approach — an extrapolation model can be of a purely statistical kind, but, upon closer inspection, it can pertain to a semimechanistic model too. For example, a statistical model for bioavailability may have been designed to be fitted to the data, but may in hindsight represent some major parameters of a mechanism-based model (e.g., the biotic ligand model). A specific tier may need to be reserved for very specific approaches that do not fit the outline provided here.

Conceptually, the generalized tiering concept can now be given as follows:

Tier-0: "No extrapolation," as treated above. There is

- no incentive to consider extrapolation at all,
- there are preliminary quantitative judgments that suggest a low numerical influence of the pertinent factor when subjected to extrapolation, or
- there is no technique for extrapolation, neither an existent technique nor a potential newly derived one.

Tier-1: A "simple generic approach." In this case, the uncertainties encountered in the assessment are captured by a UF, or by a simple generic model. When used in a deterministic way, the UF increases or decreases according to estimated uncertainties in the data. The following values for UFs are frequently encountered:

- 10: relatively large data set, close to the subject of study
- 100: less data, less linked to the subject of study
- 1000: little data, low relevance for the subject of study

In other cases, the UFs are multiplied, usually yielding low hazardous concentration values that are supposed to offer sufficient protection in worst-case conditions. In mixture assessments, the simple generic model consists of the use of point estimates from concentration–effect curves, in combination with the concentration addition model, to address mixture problems.

Data-dependent uncertainty factors might be identified as a probabilistic variant of a simple generic approach. For example, calculation of the lower confidence limit of the 95% confidence interval of the 5th centile of an SSD can be considered a statistically derived UF. The UF, when presented, would be different between substances, depending on the available data. Next to the use of SSD-derived UFs, whereby SSDs are considered Tier-2 methods (because they are of a statistical kind), other methods could be envisaged to derive Tier-1b — data-driven uncertainty factors.

Tier-2: The tier with "moderately simple generic approaches." In the second tier, instead of UFs, a statistical model is applied to handle the known data. The statistical model is thereby not derived on the basis of a mechanistic working hypothesis, but solely on the basis of the fact that the factor under investigation does numerically matter. In the case of mixtures, a generalized working hypothesis is adopted for all substances. Tier-2 extrapolations are numerical approximations of the true phenomena, not mechanistic ones. SSDs, for example, are statistical descriptions of the fact that species differ with respect to sensitivity; they do not explain why those species differ in sensitivity. In this case, the principle of parsimony (a pragmatic argument) is chosen as a major determinant for selecting a model. When the simpler model suffices as to the specific assessment problem, why use a more complex one? An example of this is the derivation of the above-mentioned transfer functions: why take the function with all statistically significant variables in cases where the last 4 variables together only contribute 5% to the total variation in the training set? Why not take, for example, only pH in the case of extrapolations of metal exposures?

Tier-3: The tier with "complex specific approaches." In the third tier, a mechanistic or semimechanistic model is adopted in addition to an empirical, statistical-modeling part. For example, the use of SSDs is a statistically oriented approach, but the use of this model can be refined by using mechanism-oriented insights (i.e., by considering separate species groups that are inherently more sensitive than others for the individual substance or the components of a mixture). In this way, the approach becomes more complex to handle, but likely also yields more accurate results than the "moderately simple generic" approach.

Tier-4: The "highly specific set of approaches." There are 2 kinds of approaches here, one based on the modeling axis of Figure 10.2 and the other on the choice of more appropriate data that are directly linked to the protection target.
  1) In the case of extrapolations based on modeling, for example, the simple or more complex empirical transfer functions of the lower tiers of exposure assessments are replaced by biotic ligand modeling (BLM). In BLM, the specific characteristics of the system are investigated (e.g., sorption sites on a gill, sorption sites in the matrix, and the concentrations of the target ion and competing ions), shaped into a model, and applied after measurement of the mechanistically justified parameters. Such modeling is better described as semimechanistic, because in contrast to physics, the biological systems in their natural or test environment cannot usually be fully described by mechanistic models. There are always additional parameters that are unknown or not recognized. Nonetheless, a population model does not address the statistical features of a population; it tries to capture the most relevant biological processes that determine population size, and is thus considered a semimechanistic model here.
  2) In collecting the most appropriate data, one can recognize the use of highly specific testing approaches, which can be called physical models of the real system of concern. For example, in the pesticide registration process, one can make use of (multispecies) microcosms,

mesocosms, or field tests. These tests are designed to mimic the situation of concern as closely as possible. The required measurements, for example, on species composition and food-web parameters, are directly made rather than modeled, and they are, therefore, highly specific. As a strategy for extrapolation in which the research aims to mimic the real situation as closely as possible by physical experiments, this should reduce the need for making sets of extrapolations because they mimic the situation of concern.

Note that the assigning of extrapolation methods to tiers can be ambiguous. For example, (Q)SARs are most often of a purely statistical kind (Tier-2), but also do touch upon mechanistically relevant molecular parameters (Tier-3), which shows that there is room for a choice of tiers in (Q)SAR extrapolation. Further, if 1 extrapolation model addresses both exposure and effects, it may be that exposure is addressed in a "simple generic way," and effects in a "moderately simple generic way," which poses a problem when one has to assign this method to the above set of tiers. These examples illustrate that the set of tiers proposed here should not be interpreted as clear-cut discrimination between extrapolation types, but as guidance toward providing insight into the consistency of what one is doing. The more a set of extrapolation methods can be assigned to the same level of scientific tiering, or to adjacent tiers, the more the extrapolation process as a whole can be considered scientifically consistent as to their common design.

At the end of this chapter (Section 10.4.7) the extrapolation techniques in this book have been arranged according to the system proposed here.

- In Table 10.3, the overview is organized according to the generalized tiers, and thereafter to the subjects of extrapolation, so that one can select conceptually consistent sets of extrapolation methods (e.g., for matrix and media extrapolation and mixture extrapolation).
- In Table 10.4, the same overview is presented; however, it is now organized from the perspective of the subjects of extrapolation (e.g., matrix and media extrapolations), so that one can inspect the presence of extrapolation methods, at different tiers, for a given extrapolation problem.

10.4.7 STEP 5: CHOOSING A CONSISTENT SET OF EXTRAPOLATION METHODS

Usually, a number of extrapolations are needed for a single assessment. In many cases, bioavailability is an issue of concern, as well as others such as mixture extrapolation and extrapolation from 1 level of organization to the other (e.g., species–community extrapolation). When the need for various extrapolations has been established, and the techniques listed, one can fill out the generalized tiered system with the selected methods that are conceptually consistent (e.g., statistics based or mechanism based), thereby addressing the assessment problem with a certain degree of specificity. Moreover, the system can be considered technically consistent, in that the efforts spent in each tier are roughly equivalent. For example, using transfer functions to control for bioavailability is an empirical statistics-based process,

TABLE 10.4
Overview of available extrapolation techniques

Tier	Type of extrapolation	Problem definition	Problem subdefinition	Method
2: Matrix and/or media extrapolation				
0: No extrapolation	0: None	Prediction of actual exposure	Metals and organics in water, soil, and sediment	Assume completely available, no extrapolation
2: Moderately simple generic	2: Extrapolation by statistical model	Prediction of actual exposure	Metals in soil and sediment	Empirical K_d-regression models
	3: Extrapolation by semimechanistic model	Prediction of actual effect	Intrinsic differences in species sensitivity	SSD comparison
		Prediction of actual exposure	Organic toxicants in soil and sediment	Multiple-compartment equilibrium–partitioning modeling
			Organic toxicants in water	Single-compartment equilibrium–partitioning modeling
				MacKay-type modeling
	4: Extrapolation by mechanistic model	Prediction of actual exposure	Prospective distribution modeling	Speciation modeling
			Metals in water	
3: Complex specific	3: Extrapolation by semimechanistic model	Prediction of actual exposure	Organic toxicants in water	Multiple-compartment equilibrium–partitioning modeling
	5: Specific approach (not 0, 1, 2, 3, or 4)	Prediction of actual effect	Combined availability and/or sensitivity modeling	Indirect effects modeling
4: Highly specific	3: Extrapolation by semimechanistic models	Prediction of actual exposure	Organic toxicants in water	Dynamic multiple compartment modeling

(Continued)

TABLE 10.4 (CONTINUED)
Overview of available extrapolation techniques

Tier	Type of extrapolation	Problem definition	Problem subdefinition	Method
	4: Extrapolation by mechanistic model	Prediction of actual effect	Metals in water	BLM modeling
3: (Q)SAR extrapolation				
1: Simple generic	2: Extrapolation by statistical model	Prediction of chemical properties	(Q)SAR	Basic phys/chem --> K_{ow}, water sol, and so on; EPIWIN
		Prediction of acute aquatic toxicity	Baseline toxicity	Taxa-specific linear regressions, ECOSAR
		Prediction of chronic aquatic toxicity	Baseline toxicity	Taxa-specific linear regressions, ECOSAR
		Prediction of acute terrestrial toxicity	Baseline toxicity	Limited to small number of taxa, linear regressions
2: Moderately simple generic	2: Extrapolation by statistical models	Prediction of acute aquatic toxicity	Specific toxicity	Multilinear, multivariate ordination; artificial neural networks
		Prediction of chronic aquatic toxicity	Specific toxicity	Multilinear, multivariate ordination; artificial neural networks
3: Complex specific	3: Extrapolation by semimechanistic models	Prediction of chemical properties	(Q)SAR	Molecular descriptors
	2: Extrapolation by statistical model	Prediction of chronic aquatic toxicity	Specific toxicity and/or TMoA	Receptor-specific (Q)SARs (comparative molecular field analysis [CoMFA])
		Prediction of species sensitivity	QSSR	ICE
4: Highly specific	4: Extrapolation by mechanistic model	Prediction of chronic aquatic toxicity	Specific toxicity and/or TMoA and/or Physiology	3-D modeling; PBTK

4: Extrapolation across levels of biological organization

1: Simple generic	1: UF extrapolation	Prediction of responses	Biomarker to individual	UF based on empirical studies
		Prediction of responses	Single-species test result to population	UF based on empirical studies
2: Moderately simple generic		Prediction of responses	Populations to community	UF based on empirical studies
	2: Extrapolation by statistical model	Prediction of responses	Biomarker to individual	Statistical correlation
		Prediction of responses	Single-species test result to population	HERBEST or PERPEST approaches SSD
		Prediction of responses	Populations to community	Quantitative cause–effect relationships
		Prediction of responses	Structure to function	
3: Complex specific	2: Extrapolation by statistical model	Prediction of responses	Biomarker to individual	Quantitative cause–effect relationships
	3: Extrapolation by semimechanistic models	Prediction of responses	Single-species test result to population	Simple pop models (exponential models)
	4: Extrapolation by mechanistic model	Prediction of responses	Community to ecosystem and/or landscape	Simple food-web models
	5: Specific approach (not 0, 1, 2, 3, or 4)	Prediction of responses	Populations to community	Multispecies, microcosm, and mesocosm experiments
		Prediction of responses	Community to ecosystem and/or landscape	Microcosm and mesocosm studies
4: Highly specific	4: Extrapolation by mechanistic model	Prediction of responses	Biomarker to individual	DEB approaches
		Prediction of responses	Single-species test result to population	Complex models (life stage, IBM, and energetic models)
		Prediction of responses	Community to ecosystem and/or landscape	Food-web models, metapopulation models
	5: Specific approach (not 0, 1, 2, 3, or 4)	Prediction of responses	Populations to community	Food-web models

(Continued)

TABLE 10.4 (CONTINUED)
Overview of available extrapolation techniques

Tier	Type of extrapolation	Problem definition	Problem subdefinition	Method
5: Mixture extrapolation				
0: No extrapolation	0: None	Prediction of mixture effects	No addition	Do not evaluate mixture toxicity
1: Simple generic	1: UF extrapolation	Prediction of mixture effects	Risk quotient concentration addition	TU, TEQ, TEF, and CCU addition
2: Moderately simple generic	2: Extrapolation by statistical model	Prediction of mixture effects	SSD concentration addition	msPAF CA
	5: Specific approach (not 0, 1, 2, 3, or 4)	Measured mixture effects	SSD response addition / Effluent bioassays	msPAF RA / WET experimentation
3: Complex specific	2: Extrapolation by statistical model	Prediction of mixture effects	SSD mixed model	msPAF mixed TMoA
	5: Specific approach (not 0, 1, 2, 3, or 4)	Measured mixture effects	Bioassays on environmental concentrates	pT measurements
4: Highly specific	2: Extrapolation by statistical model	Prediction of mixture effects	SSD mixed-model split taxa	msPAF mixed TMoA–Receptor
6: Temporal extrapolation				
1: Simple generic	1: UF extrapolation	Time-varying exposure regimes	Empirical	UF
	3: Extrapolation by semimechanistic models	Time-varying exposure regimes	Incipient toxicity	ACR ratio, ACE
2: Moderately simple generic	2: Extrapolation by statistical model	Reversibility of effects	Reversibility of effects	Empirical
		Latency of responses	Latent responses: direct/indirect effects	TMoA estimates

3: Complex specific	2: Extrapolation by statistical model	Time-varying exposure regimes	Pulsed exposure	RADAR
	3: Extrapolation by semimechanistic models	Variable sensitivity within species	Seasonal, life history differences	Life history population models
	4: Extrapolation by mechanistic model	Time-varying exposure regimes	Mechanistic	DEBTox
4: Highly specific	4: Extrapolation by mechanistic model	Latency of responses	Latent responses: direct and/or indirect effects	CATS, AQUATOX, and CASM
	5: Specific approach (not 0, 1, 2, 3, or 4)	Variable sensitivity within species	Seasonal, life history differences	Model systems
		Recovery of effects	Recovery experiments	Model systems
		Adaptation to exposure	Adaptation measurements	PICT

7: Spatial extrapolation

0: No extrapolation	0: None	Geographical extrapolation of toxicity data	Toxicity differences between groups of species	Assume all groups of taxa are equally sensitive
1: Simple generic	1: UF extrapolation	Extrapolation of toxicity data between habitats	Toxicity differences between flowing and stagnant water organisms	UF based on empirical studies
			Toxicity differences between fresh- and saltwater organisms	UF based on empirical studies
		Geographical extrapolation of toxicity data	Toxicity differences between North American and European polar regions	UF based on empirical studies
			Toxicity differences between southern and northern hemispheres	UF based on empirical studies
			Toxicity differences between temperate and tropical regions	UF based on empirical studies

(Continued)

TABLE 10.4 (CONTINUED)
Overview of available extrapolation techniques

Tier	Type of extrapolation	Problem definition	Problem subdefinition	Method
3: Complex specific	2: Extrapolation by statistical model	Spatial variation in community responses	Comparing model ecosystem studies performed in different geographical regions	Statistical comparison
	3: Extrapolation by semimechanistic models	Spatial variation in community responses	Comparing model ecosystem studies of different size and complexity	Food-web models, IFEM, AQUATOX, and C-COSM
4: Highly specific	2: Extrapolation by statistical model	Detection of differences in metapopulation responses in impacted landscapes	Toxicity as part of multiple stress	GIS modeling
	3: Extrapolation by semimechanistic models	Detection of differences in metapopulation responses in impacted landscapes	Population dynamics	Metapopulation modeling
	5: Specific approach (not 0, 1, 2, 3, or 4)	Detection of differences in metapopulation responses in impacted landscapes	Ecological indicators	Monitoring

Note: Table ordered according to the subject of extrapolation (matrix and media, mixtures, etc.), and thereafter to tiers. Major tiers (0 to 4) and subtiers (a and b) are discriminated. For further explanation, see the text.

conceptually matching to the statistics-based approach of SSDs. Utilizing these 2 approaches in 1 assessment is both scientifically and technically more consistent than using conceptually diverse approaches.

Note that, although the criterion of consistency may be a general guideline for designing a whole assessment scheme including extrapolations, it is not always necessary, possible, or appropriate. For example, it may be acceptable to use a statistics-based approach for bioavailability extrapolation (e.g., transfer functions) in combination with semimechanistic population models that are founded in the biology of the species. In many cases, the conceptually ideal combination of extrapolation approaches might be beyond reach, because of lack of techniques of similar kinds, or lack of data. In such cases, conceptually nonsimilar techniques can be used and are used in an assessment, as long as the basic feature of the tiering system is acknowledged. That is, a Tier-3 effects approach can be used together with a (conservative) Tier-1 exposure approach, because the net outcome will still be more conservative than using Tier-3 approaches for both exposure and effects. Verdonck et al. (2003) have, however, warned about complete misinterpretations that could follow from the use of different measures for the exposure and effects assessments (e.g., pore water exposure concentrations and total-concentration-based effects assessment). In any case, the different approaches for exposure and effects should be similar in measurement units and meaning.

10.4.8   STEP 6: COLLECTION OF DATA NEEDED

A key step in any risk assessment is the collection of data to feed into the extrapolations. The use of an extrapolation method will be possible only with appropriate input data. The data that are needed can be literature data (e.g., laboratory toxicity data) or field data (measured or predicted environmental concentrations), and the extrapolation is then to move these toward common currencies, that is, the estimation of bioavailable concentrations, derived from total concentrations, in both test systems and the field, or species-to-community extrapolation when the concern is the "effect of exposure on biodiversity." Step 6 will often be made together with Step 7, because a method without data is as useless as data without a method.

In extrapolation, specific attention should be paid to data quality; however, this is a subjective concept. There is no such thing as high-quality data. Quality assessment depends on the assessment problem. For example, when the problem is "derivation of an environmental quality criterion (EQC)," one must often apply clearly defined and strict quality criteria to select data: for example, data should be NOECs, and those NOECs should be derived from statistical tests and the like. Quality selection thus results in a limited subset of data considered of sufficient quality for the objective of deriving an EQC. However, when an accident or spill occurs for a substance for which one has no formal EQC and little data (e.g., only LC50s), the assessment must be executed with these data, especially when the only alternative option would be guessing. For extrapolation in this gross context, it can become very clear from such LC50 data whether the situation should be considered "very dangerous," warranting immediate response by the authorities, or "of negligible concern," warranting no response or delayed response. Thus, LC50s are "high-quality data" for this problem. Hence, the circumstances determine the view on data quality. Although "data quality" generally is an important item, this guidance assumes that the data that are used in extrapolation are of sufficient quality for the extrapolation process.

After quality, attention should be paid to data quantity. All extrapolation methods that are in use have a quantitative format, that is, a model. These models have different foundations, the simplest model being a fixed UF and the most complex one being a mechanistic description of the phenomena under investigation. The quantitative format implies that each extrapolation result can be accompanied by the presentation of confidence intervals, although this is not always common. When an extrapolation method is used that is founded on little data, the confidence intervals may be wide. In general, one may expect a trade-off between the amount of input data (and their quality and precision) and the precision in the extrapolation output. For this reason, amongst others, in practice it may be counterproductive to strive for the best possible extrapolation technique per se. The conceptually better technique might have a weak data support. In turn, this may lead to greater uncertainty and higher risk estimates in higher tiers, which is an undesired characteristic of tiered systems. Although the problems of data quality and quantity cannot be solved here, it is important to consider both when choosing amongst alternative methods, and when possible apply the window-of-prediction approach.

10.4.9   STEP 7: WORKING WITH THE SET OF EXTRAPOLATION METHODS

When various extrapolations are needed, assessors have to decide in which order they should use the methods. In the mixture assessment, it was already argued that the approaches that are applied in mixture extrapolation are founded in mechanistic theory on toxicant–receptor interactions at the molecular level. The latter highlights the idea that matrix and media extrapolation should be executed before a mixture toxicity extrapolation is made. Otherwise, one would work with mixture theory developed for molecular interactions at the target site within an organism's tissue while applying this to, for example, total environmental concentrations. It is logical reasoning that should be used to decide on the order of tackling extrapolation problems. The order of addressing extrapolation problems, in general, should be organized along the cause–effect chain, that is,

- matrix and media aspects first,
- substance- and toxicity-related aspects second, and
- biotic aspects third.

A particular feature of working with various extrapolations is that one can obtain results of the subsequently applied methods, all with confidence intervals. In the process, one has to decide how to handle this. The main question is, "How do uncertainties proliferate through the subsequent extrapolations?" Do we eventually end up with an estimate of risk in a confidence interval from 0% to 100%? Little attention has been devoted to this subject. Various mathematical techniques do exist, but (almost) none have been specifically discussed in the context of the subsequent application of extrapolation methods. Maybe the strongest scientific argument for attaining consistency amongst extrapolation methods is associated with this question: when extrapolation methods are inconsistent, the proliferation of uncertainties through the different steps cannot be handled, technically or the results of the assessment become meaningless. An example of pitfalls in the interpretation of extrapolation results obtained by inconsistent approaches is given by Verdonck et al. (2003) on exposure concentration distributions (ECDs), SSDs, and joint probability curves (JPCs). It was shown that the data used to generate ECDs should be consistent in space and time with the data used to construct the SSDs; otherwise, the JPC would be uninterpretable. In particular, this example shows that the exposures and the effects need to be expressed in the same terms, which implies that the extrapolation techniques applied for both sides of the coin should also be consistent.

A final issue in extrapolation practice is the use of various extrapolation methods to address a single problem. This situation occurs when there is no key argument in favor of a particular method. An example is the problem of using either concentration addition or response addition as models to predict mixture risks. In such cases, one could use the concept of "window of prediction," that is, one uses both models and presents the outcomes of both as a window of prediction. In cases where the window is positioned (much) higher or lower than the acceptable limit for risks, the decision on cleanup needs is equivocal and independent of the model chosen. This can be interpreted as double support for a decision. In cases where the window overlaps with the decision criterion, there is a strong justification to go to a next tier. In summary, the window of prediction is a pragmatic approach to cut the Gordian knot in cases where risk assessors have no good reasons to choose between 2 or more (scientifically) competing, possibly appropriate models.

10.4.10 STEP 8: HANDLING AND INTERPRETATION OF ASSESSMENT RESULTS

Risk assessment can be based on various extrapolation methods. When possible, uncertainty should be propagated through the subsequent extrapolation steps, provided that this is technically feasible and scientifically meaningful for the assessment problem. The latter requires scientific consistencies in the types of data and the issues to which they refer, for example, similarity with respect to temporal or spatial dimensions. Eventually, the outcome has to be interpreted. In the interpretation, the assessor can follow the general guidelines for ecological risk assessments again. Various textbooks and legally adopted approaches are available.

The interpretation may be very simple, such as, for example, the observation that the predicted exposure concentration (PEC) is below the no-effect concentration (NEC); in other words, PEC/NEC < 1, which is interpreted as "sufficient protection." Data interpretation may also be quite complex, such as the interpretation of joint probability curves in species-to-community extrapolations where a set of exposure data is used. Also, overinterpretation of data may occur. An example is the interpretation that SSDs would represent an ecological method that generates ecological predictions. Instead, the method is a statistical method used in an ecological context, and it allows ranking the relative hazard of a site on a scale from 0 to 1, or amongst a set of sites on this scale, rather than predicting specific ecological phenomena. Statistically, it only predicts the proportion of test species that would experience a response if they were present in the exposed environment. For all extrapolations, one would be critical about the output, ask whether the output interpretation is supported by the data and methods used as input, and present the issues not handled in the extrapolation process, including the expected effect of that on the outcomes.

10.5   PRACTICAL GUIDANCE ON EXTRAPOLATION

The above sections of the guidance have been conceptual and organizational rather than offering insight regarding the set of practicable extrapolation methods. This book on extrapolation offers theoretical background and practical suggestions for extrapolation techniques. Each of those techniques can be positioned in the generalized tiered system that was proposed above and in the 2 decision matrices, as shown in Table 10.3 and Table 10.4.

A variety of techniques are available. As a consequence, 1 assessment problem could yield a large number of different answers from the combination of different extrapolation techniques. This underscores the importance of risk assessors being aware of the possible inconsistencies that can be created by haphazard and inappropriate selection of extrapolation methodologies.

Assessors should base their selection of methods on clearly defined decision criteria, and they need to communicate the results using clear and transparent language. This includes statements on the extrapolation issues that were considered but not addressed, and the magnitude and direction of the bias that may have been introduced by the extrapolation or lack thereof. In lower tiers and prospective risk assessment, this should lead to setting more appropriate UFs and ensure that lower tier approaches are more conservative than higher tier approaches. All this helps assessors to make informed decisions, on one hand, but it also allows the identification of future research needs, on the other hand, especially when methods are not available.

The overview of the available techniques and their provisional systematic order across tiers and method types does not provide guidance on the decisions that have to be taken. How can an assessor, in practice, get an overview of the extrapolation practices that should be applied, given the assessment problem?

To facilitate working with extrapolation problems, a set of questions is presented. Given the assessment problem, these questions can be asked by the assessor. These questions have been ordered according to the steps identified earlier. Together with Table 10.3 and Table 10.4, these questions show how a risk assessment proceeds. Note that 1 question can occur various times, because of the fact that 1 issue is encountered, for example, from the perspectives of exposure extrapolation (time-varying exposure) and of effects extrapolation (time-varying, age-specific sensitivity).

10.5.1   STEP 1: MOTIVATING THE NEEDS FOR EXTRAPOLATION
AND PROCEDURAL QUESTIONS

10.5.1.1   To Be Answered before Starting Any Assessment

Before starting an assessment, one should clearly define the objectives of the assessment and consider the consequences of that choice regarding the extrapolation practices to be applied. These key issues are defined and handled in this book on extrapolation, and these issues are the basis for any assessment. Weak or nonoperational definitions in this stage make many risk assessments ambiguous. In this stage, a series of questions needs to be answered or addressed to make both the scientific assessment problem and the practical context of addressing this problem clear:

1) What is the assessment endpoint, and what is the protection goal?
   - Is the target of the assessment the derivation of EQCs and/or a decision on the marketing or use of substances (prospective)?
   - Is the target of the assessment a specific assessment of risks in a contaminated ecosystem (retrospective)?
   - By which measures of effect are the protection goals operationally defined?
   The consequences for the design of your assessment are as follows:
   - For the derivation of EQCs and the registration of substances (prospective risk assessment), worst-case assumptions are usually applied in the lower tiers, and UFs may be applied, whereas more specific data are used in higher tiers
   - In retrospective risk assessment, the output should be as accurate as possible at any tier. Tiers differ by the level of sophistication and realism.
   The assessment endpoint can follow from legal requirements or from concerns (e.g., established by scientific reasoning).
2) What is the valued characteristic of the environment that is possibly at risk? This characteristic must be defined, including the definition of the measure of effects on that characteristic that is to be used to judge damage (Suter et al. 1993). An example is "the probability of 10% reduction of abundance" (Suter et al. 1993).

10.5.1.2   To Manage the Extrapolation Processes

Upon defining the assessment endpoint, answers to the following questions either are taken from existing protocols or need to be defined in the assessment context. The relevant questions are as follows:

1) Is there a prescribed procedure to be followed? If so, follow that. If not, proceed with the subsequent questions.
2) If yes, is that procedure tiered?
3) If yes, how are the tiers defined?
4) If no, how could the tiers be defined, given the formulated assessment endpoint?

If the risk assessment protocol that you are following is tiered by prescribed procedures or by your choice:

1) Given the results you obtain in a lower tier, is there a specific trigger defined for progression to a next tier?
2) Can cutoff values be used as an answer to the assessment (i.e., point estimates), especially in prospective assessments? An example is the HC5, the hazardous concentration for 5% of the species, where the cutoff level for the proportion of species that is potentially affected is maximized at 5%.

3) Are there triggers to use probabilistic evaluation in either a prospective or a retrospective context?

4) Are confidence limits helpful in your assessment, especially in the case of retrospective assessments?

5) Can and should a probabilistic approach be applied to exposure and effects, given the assessment endpoint?

And, in the case of a prospective risk assessment:

1) Is there a level of certainty defined that needs to be met?

2) Is there a need to report the confidence bounds and their meanings?

3) Do we want to establish the presence or absence of effects, given a cutoff value for effects?

4) Do we want to establish the presence of an exposure-effect gradient?

5) Do we need to establish a cause–effect relationship — for example, because the polluter needs to be prosecuted, or because emissions need to be reduced?

Answers to these questions provide an outline of the risk assessment process, in terms of requisites of the process itself. The next part is to address the technical extrapolation issues within the process of risk assessment.

10.5.2   STEP 2: PRELIMINARY ASSESSMENT AND RELEVANT EXTRAPOLATION ISSUES

The following questions can be helpful in a preliminary assessment, which is needed to identify the most relevant extrapolation issues. The target of this step is to identify the key items that are likely influencing the outcomes of the risk assessment most, and to discard the numerically less important items, especially in the lower tiers. The important question is whether the extrapolation issues can be ranked according to (numerical) importance. It motivates the need for extrapolation, and the most relevant extrapolation issues (Step 1 and Step 2, as introduced above).

1) Do we have data pertinent to the problematic substances, or is there a need for substance-to-substance extrapolation (e.g., by (Q)SAR)? If so, is there a trigger to consider this quantitatively important?

2) Do we have data pertinent to the matrix or medium in which exposure takes place, or do we need to consider matrix and media extrapolation? If so, is there a trigger to consider this quantitatively important?

3) Do we have data pertinent to the exposed biota, or do we need to consider extrapolation through levels of biological organization? If so, is there a trigger to consider this quantitatively important?

4) In the case of mixtures, do we have data on the overall effect of the pertinent mixture, or do we need to consider mixture extrapolation? If so, is there a trigger to consider this quantitatively important?

5) Do we have data pertinent to the exposure situation to be considered, or are there specific temporal aspects in the exposure regime (intermittent

exposure, exposure gradually changing over time, etc.)? If so, is there a trigger to consider this quantitatively important?

6) Do we have data pertinent to the exposure situation to be considered, or are there specific spatial aspects in the exposure regime (e.g., spatial gradient of substance, or spatial distribution of exposed biota)? If so, is there a trigger to consider this quantitatively important?

Answers to these questions provide preliminary insight into the variables for which extrapolation can be considered most urgent, due to an estimated numerical influence on the assessment results. For example, matrix and media extrapolation might be identified as the most important modifying factor, and the risk assessment might fully focus on this factor in the lowest tier. Information on issues for which no extrapolation will be applied need to be kept in the dossier, for reporting which factors have been considered and what numerical influence (magnitude and direction) is expected.

10.5.3   STEP 3: IDENTIFYING THE (SET OF) POSSIBLE EXTRAPOLATION METHODS

After establishing the likely need for extrapolation on various issues, and not on others, the next task is to list the operational extrapolation methods for each of the important issues. A list of the methods that are presented in the previous chapters of this book is compiled in Table 10.3. The identification of the set of possible extrapolation methods should include a check on the availability of data. If either of these are lacking, the extrapolation for this aspect will not work at all, or will not yield results of sufficient quality. The questions are thus:

1) Which extrapolation methods are available for a problem?
2) Are the data needed to use the extrapolation method available?

Answers to these questions provide insight into the potential chance for successfully addressing the problems for which extrapolation is considered to be needed. A set of specific questions can be helpful for further refining the tasks to be executed.

10.5.3.1   Questions Related to Substances

As compiled from all aspects influencing the issue of exposure and effects of a substance or substance mixture, there is an array of relevant questions:

Associated with the Aspect "Single Substance or Mixture"
1) Is the assessment concerned with a single substance or a mixture?
2) If concerned with a mixture, do we know the components in the mixture?
3) If concerned with a mixture, is it a mixture with a stable composition?

Associated with the Aspect "Toxic Mode of Action (TMoA)"
1) Do we know the primary TMoA of the substance or of all substances?
2) Is the TMoA general (baseline toxicity) or specific (e.g., receptor based)?

Associated with the Toxicity of the Substances
  1) Is the toxicity of the substances known?
  2) To what extent do the available toxicity data match the locally exposed
     biota regarding tested species and field species?

Associated with Modifying Factors
  1) Is there a need for site-specific evaluation?
  2) Matrix and media: do the available toxicity data match the matrix?
  3) Time: is exposure stable in time?
  4) Space: is exposure stable in space?

Associated with Metabolism of the Substance
  1) Does the substance break down?
  2) If so, which metabolites are created along the breakdown pathway?
  3) If so, are the metabolites toxic?
  4) If so, is there a need to consider only the parent substance, or does the
     assessment also concern the metabolites?

Answers to these questions provide insight into the specific actions that are
needed as to the issue of the substances that are present.

10.5.3.2   Questions Related to Matrix and Media Extrapolation

As compiled from all aspects influencing the issue of exposure and due to matrix and
media influences, there is an array of relevant questions:

  1) What is the matrix?
  2) What properties of the matrix could affect availability? If these are sub-
     stance dependent, answer this question for separate substances.
  3) What properties of the matrix could act as (additional) stressors to the
     exposed biota? For example, acid soil both influences the speciation of
     metals (increasing bioavailability) and may cause stress in terrestrial spe-
     cies such as earthworms.
  4) Do properties of the matrix vary in time?
  5) Do properties of the matrix vary in space?
  6) Are there geographical factors to be considered?
  7) Is there a need for site-specific evaluation?
  8) Is there a need for intermedia extrapolation?
  9) Is there a need for intramedium extrapolation?
  10) Are there properties of the media influencing degradation?
  11) Are there other chemicals in the matrix that need to be considered?

Answers to these questions provide insight into the specific actions that are needed
as to the issue of the matrices and media that are involved in the assessment.

10.5.3.3   Questions Related to Biota

As compiled from all aspects influencing the issue of exposure and effects on biota at different levels of biological organization, there is an array of relevant questions:

1) Do we know the routes of exposure of the substances to the biota?
2) Are there specific biota to protect?
3) Is there specific concern for structural or functional endpoints?
4) What types of effects are expected (acute, chronic, biodiversity, and/or function)?
5) Are there multiple data for 1 endpoint in 1 species, and in what way are these data handled (geometric mean, and/or other)?
6) Do we know the spatial distribution of biota in relation to the substance distribution?
7) Are there unexposed subpopulations or habitats nearby from which recolonization and recovery take place?
8) Is recovery relevant?
9) Are indirect effects of concern?
10) Are there behavioral issues that might affect exposure?
11) Is there a sensitive life stage, and if so, which?
12) Is there a need for site-specific evaluation?
13) Are there specific ecological receptors of concern, in view of the presence of substances with a specific TMoA?
14) Are mixture effects to be considered?
15) Is the organization level of the test endpoint equal to the level of the assessment endpoint?
16) Is there a need to consider resistance or selection for tolerance?
17) What is the relationship between temporal change in concentration of the substance and the life history characteristics of the potentially exposed biota?

Answers to these questions provide insight into the specific actions that are needed as to the issue of the exposed biota that are involved in the assessment.

10.5.3.4   Questions Related to Temporal Extrapolation

1) What is the duration of the exposure?
2) What is the duration of the exposure as compared to the life history of the exposed species; that is, can one assign an exposure to be acute or chronic?
3) Is the incipient toxicity level reached during exposure, or is there a need to predict the incipient level from the data?
4) Is the exposure constant or variable over time?
5) Does exposure change over time, for example, due to leaching, breakdown of the substance, or other systematic (pulsing) or stochastic processes?
6) Are there sensitive life stages, and if so, which?

7) What is the timing of exposure as compared to the sensitive life stages of the organism?
8) Can existing data on acute exposure be used to predict chronic effects, or vice versa?
9) Is the response latent?
10) Is there a possibility for reversibility or recovery from responses?
11) Is there seasonal variation in exposure and/or sensitivity of life stages?
12) Is there succession with or without an associated change of sensitivity?
13) Is long-term exposure resulting in adaptation?

10.5.3.5 Questions Related to Spatial Extrapolation

1) What is the spatial distribution of the contaminant concentrations?
2) What is the spatial distribution of the exposed species with regard to both home ranges (as compared to the exposure sites) as well as the presence of refugia (in the case of a metapopulation structure of the exposed species)?
3) Is spatial complexity influencing exposure or effects, and/or does it allow for recovery from unexposed refugia?
4) Is the sensitivity of test species associated with their geographical home ranges? In other words, are sensitivities the same for
   A. tested temperate species and polar and tropical species?
   B. tested northern hemisphere species and southern hemisphere species?
   C. tested Nearctic species and Palearctic species?
   D. lentic and lotic habitats?
   E. saltwater and freshwater habitats?
5) Is the sensitivity in test systems related to the scale of testing (i.e., single-species tests in small test systems, via microcosms to mesocosms and the field)?
6) Can data from small-scale studies be used to predict effects at the landscape scale?
7) Are multiple stressors influencing the sensitivity and responses of the exposed species?

10.5.3.6 Questions Related to the Availability and Type of Data

1) Do the data that can be collected pertain to the problem, that is,
   A. How similar is the exposure period?
   B. How similar are the tested species to those in the exposure situation of concern?
2) Are there no data, limited data, or many data for a problem, and how would method improvement (choice of a higher tier) likely trade off with reduced availability of the relevant data?
3) What is the general quality of the data, either from a scientific viewpoint (methodology of the test is appropriate) or from a regulatory viewpoint (matching with prescribed quality rules)?

10.5.4   Step 4 and Step 5: Assigning Methods to Tiers
and Choosing a Consistent Approach

At this point in the process, the set of extrapolation types has been defined, as well as the set of available methods for each of the extrapolation types. Following the idea of tiering, the available methods can be designated to tiers, according to the most relevant questions:

1) What is the order of handling extrapolation problems? Following the cause-to-effect chain of events, for example, (Q)SARs first, then matrix and media, and then mixture assessment.
2) Can a set of extrapolation methods for all relevant issues be positioned in tiers and levels, so that a methodologically consistent view emerges?

Tables 10.3 and 10.4 can be used for this.

10.5.5   Step 6 and Step 7: Working with Extrapolation
Data and Methods

In these steps, the methods are used, and appropriate data for executing the extrapolations are collected. The methods were identified in the above process, and are described in this book.

Basic questions that may need to be answered are as follows:

1) Is there a need to show the uncertainty or confidence interval of the final result of all extrapolations, and/or of each extrapolation step?
2) What is the list of items for which no extrapolation has been applied, and what is the likely influence (magnitude and direction) of this on the end results?

10.5.6   Step 8: Questions Relating to the Interpretation
and Use of Extrapolation Results

In this step, a short list of questions of major importance can be posed so as to avoid over-interpretation of results, and to clarify the meaning of the output of extrapolation.

Key questions are as follows:

1) What is (known about) the validity of the extrapolation steps and the overall outcome of the whole set of extrapolation steps? In other words, is the outcome of the extrapolation plausible in comparison to facts or other lines of evidence?
2) Will the extrapolation outcome be interpreted as a fixed and final value, like an EQC? If so, what strengths and weaknesses with respect to the outcome can be listed before an outcome is chosen as an EQC?
3) Will the extrapolation outcome be interpreted in comparison to other cases or lines of evidence? For example, in the case of site-specific risk assessment, risk managers might make decisions on the basis of relative

ranking, which relaxes the need to generate absolute outcomes and associated validation criteria — in this case, relative ranking should plausibly represent effect ranking.

10.6   OVERVIEW, DISCUSSION, AND CONCLUSIONS

This chapter provides guidance in using the extrapolation methods listed in this book on extrapolation practices, including considerations to select extrapolation approaches not addressed therein in a way that is founded both in science as well as in daily practices.

A generalized guidance is provided by the following:

- A stepwise protocol to decide on the use of extrapolation methods for an assessment protocol
- A design for assigning extrapolation methods to a scientifically based tiered system, with the tiers being:

  - Tier-0: no extrapolation
  - Tier-1: extrapolation (mostly for protection purposes) based on simple generic approaches (usually uncertainty factor)
  - Tier-2: moderately simple generic approaches, for example, statistics-based extrapolation methods
  - Tier-3: complex specific approaches, for example, methods that merge (semi-)mechanism-based assumptions with a statistics-based approach
  - Tier-4: highly specific approaches, for example, methods using mechanism-based approaches, plus physical models (models that mimic the situation of concern experimentally, such as microcosms or mesocosms)

The proposed tiers are not obligatory but contain extrapolation tools that can be used differently in a number of situations and, in some cases, regulatory protocols, in which certain combinations of extrapolation methods are prescribed as methods that must be used for the assessment, such as in the formal registration of pesticides. The proposed tiered system is based on a scientific classification of the available extrapolation method types. With ideal data and concepts for extrapolation, this scheme may be expected to yield reduced degrees of overestimation of risk when moving up the tiers from Tier-0 to Tier-4 (i.e., risks are more precisely estimated in the higher tiers).

It is often the case that, when using the system, various extrapolation techniques must be applied in sequence, and it is proposed that (in general) the extrapolation should follow the pathway from cause to effect. When applicable, and generally this will be in higher tiers of risk assessment, specific attention can be paid to spatial and temporal modifications of risk.

Various pitfalls can occur when using a tiered system. In the ideal case, the estimated risk appears to be indeed less and less conservative as one increases the tiers. However, the quality and quantity of the available data influence the outcome of a tier, so that it may not always work out this way. Care should be taken in applying tiered systems. An option is to develop and apply methods that show the confidence interval in the outcomes next to the extrapolated values per se. In the case of highly uncertain results, or a tendency for the risk to increase with increasing tiers, this effect would then clearly be visible. A further option is to use a window of prediction when alternative models are available.

There are various dead ends in the generalized tiered system. The tiers are not equally filled out for all extrapolation problems. That is, for some extrapolation issues, the available methods range up to sophisticated mechanistic approaches, whereas for other issues there are "only" empirical- and statistics-based approaches. These dead ends first identify the areas that are ripe for research, especially when the missing items pertain to numerically highly relevant issues. In the case of dead ends, researchers might be forced to develop a tailored extrapolation approach. One of the dead ends in practice seems to be the issue of proliferation of uncertainties through the cause–effect chain of extrapolations when using a set of extrapolation techniques. Showing the confidence intervals that result from the overall set of extrapolation efforts within a risk assessment is necessary to allow the risk assessor to interpret the results of an assessment, to avoid overinterpretation, and to check for a reduction in the overestimation of risks in higher tiers. Application of Monte Carlo methods, and other ways of handling variability and uncertainty, might be the way out.

The critical use of extrapolation methods implies consideration of the issue of validation. It has been remarked that validation of an extrapolation method should be considered in view of the target of an assessment, so that 1 approach can be sufficient for 1 target (e.g., setting quality criteria) but not for others (precise quantification of risk at contaminated sites). Higher tier methods can be used to address the degree of validity of lower tier methods, especially in the case of the higher tier physical models of reality.

A major question to be answered from a practical point of view is whether rejecting an extrapolation method introduces more (numerical) error in the decision-making processes than using an extrapolation method. An example is the discussion on mixture extrapolation, where strict mechanism-based reasoning would call any joint-effect prediction highly disputable. For example, separate assessments for all substances in an ecosystem in which there are 10 substances just below their EQCs would not suggest the presence of risks (all risk quotients < 1, with aggregated risk not considered). However, the facts that most mixture toxicity data suggest the use of mixture extrapolation, and that joint effects of toxicants are often almost equally well predicted (numerically) by either of the 2 competing mixture models, would suggest the same situation to be a case of serious concern.

When there is no guidance to choose the best method, it is always possible to use the window of prediction. Especially for retrospective risk assessments that should result in a decision, 2 or more extrapolation approaches can be used for the same problem (for example, the 2 mixture models), and both results can be compared to the decision criterion. In case both predictions are (by far) lower or higher than the decision criterion, the decision would remain the same, irrespective of the model! Only in cases where the window of prediction overlaps with the decision criterion would further work be necessary.

A final consideration on validation is that many methods have been adopted for formal use, which implies that the method suffices in practice. Wrong predictions would likely have resulted in regulators abandoning the approach. In that sense, various extrapolation methods have reached this practical level of validation. Scientific validation (that is, the model precisely predicts phenomena) is less well developed. It should be pointed out that very precautionary methods of extrapolation may be consistently protective because the uncertainty factors used are large. In this case, the fact that the results are consistent does not mean that they are accurate. In fact, "overregulation" may be harmful because more costly or more environmentally hazardous processes than are needed are selected instead.

It is concluded that

1) An array of extrapolation methods is needed and available to solve a diversity of ecological risk assessment problems.
2) These methods can be used within the context of an existing (tiered) risk assessment protocol (designed for a particular purpose and regulatory adopted as such) as well as in a generalized tiered system, whereby the overestimation of risks can be reduced as one moves up through the tiers.
3) Tiers can be generally described by the use of fixed UFs, statistically based methods, mechanistic methods, and/or experimentation in physical models (e.g., mesocosms).
4) A systematic approach can be followed so as to decide on choosing and using extrapolation methods.
5) Many assessment problems ask for the application of extrapolations for various issues simultaneously and/or subsequently.
6) Validation "by practical use" occurs frequently, especially for the lower tier methods; scientific validation is less well developed; and one extrapolation approach can be considered valid for 1 assessment type, but not for the other.

References

Aamodt A, Plaza E. 1994. Case-based reasoning: foundational issues, methodological variations, and system approaches. AI Comm 7:39–59.

Abdullah AR, Bajet CW, Matin M, Nhan DD, Sulaiman AH. 1997. Ecotoxicology of pesticides in the tropical paddy field ecosystem. Environ Toxicol Chem 16:59–70.

Abercrombie M, Hickman CJ, Johnson ML. 1973. A dictionary of biology. Middlesex (UK): Penguin.

Adams WJ, Rowland CD. 2003. Aquatic toxicology test methods. In: Hoffman DJ, Rattner BA, Burton GA, Cairns J, Jr, editors. Handbook of ecotoxicology. Boca Raton (FL): Lewis Publishers, p 19–43.

Adriaanse PI. 1996. Fate of pesticides in field ditches: the TOXSWA simulation model. No. 90. Wageningen (The Netherlands): Winand Staring Institute.

Ågren GI, Bosatta E. 1998. Theoretical ecosystem ecology: understanding element cycles. Cambridge (UK): Cambridge University Press.

Ahlborg UG, Becking GC, Birnbaum LS, Brouwer A, Derks HJG, Feeley M, Golor G, Hanberg A, Larsen JC, Liem AKD, Safe SH, Schlatter C, Waern F, Younes M, Yrjänheikki E. 1994. Toxic equivalence factors for dioxin-like PCBs. Report on a WHO-ECEH and IPCS consultation, December 1993. Chemosphere 28:1049–1067.

Akcakaya HR, Regan HM. 2002. Population models: metapopulations. In: Pastorok RA, Bartell SM, Ferson S, Ginzburg LR, editors. Ecological modeling in risk assessment: chemical effects on populations, ecosystems, and landscapes. Boca Raton (FL): Lewis Publishers, p 83–96.

Aldenberg T, Jaworska JS. 2000. Uncertainty of the hazardous concentration and fraction affected for normal species distributions. Ecotoxicol Environ Safety 46:1–18.

Aldenberg T, Jaworska JS, Traas TP. 2002. Normal species sensitivity distributions in probabilistic ecological risk assessment. In: Posthuma L, Suter GW, Traas TP, editors. Species sensitivity distributions in ecotoxicology. Boca Raton (FL): Lewis Publishers, p 49–102.

Aldenberg T, Slob W. 1991. Confidence limits for hazardous concentrations based on logistically distributed NOEC toxicity data. Ecotoxicol Environ Safety 18:221–251.

Aldenberg T, Slob W. 1993. Confidence limits for hazardous concentrations based on logistically distributed NOEC toxicity data. Ecotoxicol Environ Safety 25:48–63.

Allen HE. 1999. Integrated approach to assessing the bioavailbility and toxicity of metals in surface waters and sediments. Washington (DC): USEPA.

Allen HE, Fu G, Deng B. 1993. Analysis of acid volatile sulfide (AVS) and simultaneously extracted metals (SEM) for the estimation of potential toxicity in aquatic sediments. Environ Toxicol Chem 12:1441–1453.

Allen HE, Hansen DL. 1996. The importance of trace metal speciation to water quality criteria. Water Environ Res 68:42–54.

Allison J, Brown DS, Novo-Gradac K. US Environmental Protection Agency, Environmental Research Laboratory. 1991. MINTEQA2/PROEFA2: a geochemical assessment model for environmental systems. http://www.epa.gov/ceampubl/mmedia/minteq/index.htm (accessed July 28, 2005).

Altenburger R, Backhaus T, Bödeker W, Faust M, Scholze M, Grimme LH. 2000. Predictability of the toxicity of multiple chemical mixtures to *Vibrio fischeri*: mixtures composed of similarly acting chemicals. Environ Toxicol Chem 19:2341–2347.

Altenburger R, Bödeker W, Faust M, Grimme LH. 1990. Evaluation of the isobologram method for the assessment of mixtures of chemicals: combination effect studies with pesticides in algal biotests. Ecotoxicol Environ Safety 20:98–114.

Altenburger R, Bödeker W, Faust M, Grimme LH. 1996. Regulations for combined effects of pollutants: consequences from risk assessment in aquatic toxicology. Food Chem Toxicol 34:1155–1157.

Altenburger R, Nedza M, Schüürmann G. 2003. Mixture toxicity and its modeling by quantitative structure-activity relationships. Environ Toxicol Chem 22:1900–1915.

Altenburger R, Schmitt H, Schüürmann G. 2005. Algal toxicity of nitrobenzenes: combined effect analysis as a pharmacological probe for similar modes of interaction. Environ Toxicol Chem 24:324–333.

Altenburger R, Walter H, Grote M. 2004. What contributes to the combined effect of a complex mixture? Environ Sci Technol 38:6353–6362.

Andersen ME, Adams H, Hope BK, Powell M. 2004. Risk assessment for invasive species. Risk Anal 24:787–793.

Anderson ME, Barton HA. 1998. The use of biochemical and molecular parameters to estimate dose-response relationships at low levels of exposure. Environ Hlth Perspect 106 Suppl. 1:349–356.

Andersen ME, Dennison JE. 2004. Mechanistic approaches for mixture risk assessments: present capabilities with simple mixtures and future directions. Environ Toxicol Pharmacol 16:1–11.

Ankley GT, DiToro DM, Hansen DJ, Berry WJ. 1996. Technical basis and proposal for deriving sediment quality criteria for metals. Environ Toxicol Chem 15:2056–2066.

Ankley GT, Johnson RD, Toth G, Folmar LC, Detenbeck NE, Bradbury SP. 1997. Development of a research strategy for assessing the ecological risk of endocrine disruptors. Rev Toxicol 1:71–106.

Ankley GT, Mount DR. 1996. Retrospective analysis of the ecological risk of contaminant mixtures in aquatic sediments. Human Ecol Risk Assess 2:434–440.

Ankley GT, Schubauer-Berigan MK. 1995. Background and overview of current standard toxicity identification evaluation procedures. J Aquat Ecosys Hlth 4:133–149.

Australian and New Zealand Environment and Conservation Council and Agriculture and Resource Management Council of Australia and New Zealand [ANZECC ARMCANZ]. 2000. Australian and New Zealand guidelines for fresh and marine water quality. Canberra (Australia): Australian and New Zealand Environment and Conservation Council and Agriculture and Resource Management Council of Australia and New Zealand.

Arizzi Novelli A, Argese E, Tagliapietra D, Bettiol C, Volpi Ghirardini V. 2002. Toxicity of tributyltin and triphenyltin to early life-stages of Paracentrotus lividus (Echinoderma: Echinoidea). Environ Toxicol Chem 21:859–864.

Arnold D, Hill IR, Matthiessen P, Stephenson R, editors. 1991. Guidance document on testing procedures for pesticides in freshwater mesocosms. Brussels (Belgium): SETAC-Europe.

Arrhenius A, Gronvall F, Scholze M, Backhaus T, Blanck H. 2004. Predictability of the mixture toxicity of 12 similarly acting congeneric inhibitors of photosystem II in marine periphyton and epipsammon communities. Aquat Toxicol 86:351–367.

Arts GHP, Buijse-Bogdon LL, Belgers JDM, Van Rhenen-Kersten CH, van Wijngaarden RPA, Roessink I, Maund SJ, van den Brink PJ, Brock TCM. 2006. Ecological impact in ditch mesocosms of simulated spray drift from a crop protection program for potatoes. Int Environ Assess Manag 2:105–125.

Asfaw A, Ellersieck MR, Mayer FL. 2004. Interspecies correlation estimations (ICE) for acute toxicity to aquatic organisms and wildlife. II. User manual and software. No. EPA/600/R-03/106. Gulf Breeze (FL): US Environmental Protection Agency, National Health and Environmental Effects Research Laboratory, Gulf Ecology Division. 14 p.

Ashauer R, Boxall A, Brown C. 2006. Predicting effects on aquatic organisms from fluctuating or pulsed exposure to pesticides. Environ Toxicol Chem 25:1899–1912.

Ashauer R, Boxall A, Brown C. 2007. New ecotoxicological model to simulate survival of aquatic invertebrates after exposure to fluctuating and sequential pulses of pesticides. Environ Sci Technol 41:1480–1486.

Ashford JR. 1981. General models for the joint action of mixtures of drugs. Biometrics 37:457–474.

American Society for Testing and Materials [ASTM]. 1993. SAR/(Q)SAR in the Office of Pollution Prevention and Toxics. Gorsuch JW, Dwyer FJ, Ingersoll CG, La Point TW, editors. Philadelphia (PA): American Society for Testing and Materials.

American Society for Testing and Materials [ASTM]. 2003. Standard test methods for measuring the toxicity of sediment-associated contaminants with freshwater invertebrates. Standard E1706-002, Annual Book of Standards, Volume 11.05. Philadelphia (PA): American Society for Testing and Materials, p 1089–1206.

Attril MJ, Depledge MH. 1997. Community and population indicators of ecosystem health: targeting links between levels of biological organization. Aquat Toxicol 38:183–197.

Auer CM, Zeeman M, Nabholz JV, Clements RG. 1993. SAR: the U.S. regulatory perspective. SAR (Q)SAR Env Res 2:29–38.

Bacci E. 1994. Ecotoxicology of organic contaminants. Boca Raton (FL): Lewis Publishers, 165 p.

Backhaus T, Altenburger R, Bödeker W, Faust M, Scholze M, Grimme LH. 2000. Predictability of the toxicity of a multiple mixture of dissimilarly acting chemicals to Vibrio fischeri. Environ Toxicol Chem 19:2348–2356.

Backhaus T, Faust M, Scholze M, Gramatica P, Vighi M, Grimme LH. 2004. Joint algal toxicity of phenylurea herbicides is equally predicted by concentration addition and independent action. Environ Toxicol Chem 23:258–264.

Bailey HC, Liu DHW, Javitz HA. 1985. Time/toxicity relationships in short-term static, dynamic, and pulse-flow bioassays. In: Bahner RC, Hansen DJ, editors, Aquatic toxicology and hazard assessment: Eighth symposium. Volume ASTM STP 891. Philadelphia (PA): American Society for Testing and Materials, p 193–212.

Bailey RS, Norris RH, Reynoldson TB. 2004. Bioassessment of freshwater ecosystems: using the reference condition approach. Dordrecht (The Netherlands): Kluwer Academic, 167 p.

Baird DJ, Brock TCM, De Ruiter PC, Boxall ABA, Culp JM, Eldridge P, Hommen U, Jak RG, Kidd KA, DeWitt TH. 2001. The food web approach in the environmental management of toxic substances. In: Baird DJ, Burton GA, editors. Ecological variability: separating natural from anthropogenic causes of ecosystem impairment. Pensacola (FL): SETAC Press, p 83–122.

Baldwin WS, Milan DL, LeBlanc GA. 1995. Physiological and biochemical pertubations in Daphnia magna following exposure to the model environmental estrogen diethylstilbestrol. Environ Toxicol Chem 14:945–952.

Barata C, Baird DJ, Soares AMVM. 2002a. Demographic responses of a tropical cladoceran Moinodaphnia macleayi to cadmium: effects of food supply and density. Ecol Appl 12:552–564.

Barata C, Baird DJ, Soares AMVM. 2002b. Determining genetic variability in the distribution of sensitivities to toxic stress among and within field populations of Daphnia magna. Environ Sci Technol 36:3045–3049.

Barbour MT, Yoder CO. 2000. The multimetric approach to bioassessment, as used in the United States of America: assessing the biological quality of freshwaters RIVPACS and other techniques. Ambelside (UK): Freshwater Biological Association, p 281–292.

Barnthouse LW. 2004. Quantifying population recovery rates for ecological risk assessment. Environ Toxicol Chem 23:500–508.

Barron MG. 1990. Bioconcentration: will water-borne organic chemicals accumulate in aquatic animals? Environ Sci Technol 24:1612–1618.

Barron MG, Hansen JA, Lipton J. 2002. Association between contaminant tissue residues and effects in aquatic organisms. Rev Environ Contam Toxicol 173:1–37.

Bartell SM. 2002. Ecosystem models. In: Pastorok RA, Bartell SM, Ferson S, Ginzburg LR, editors. Ecological modeling in risk assessment: chemical effects on populations, ecosystems, and landscapes. Boca Raton (FL): Lewis Publishers, p 107–128.

Bartell SM, Gardner RM, O'Neill RV. 1988. An integrated fates and effects model for estimation of risks in aquatic systems. In: Adams WJ, Chapman GA, Landis WG, editors. Aquatic toxicology and hazard assessment. Volume ASTM STP 971. Philadelphia (PA): American Society for Testing and Materials, p 261–274.

Bartell SM, Pastorok RA, Akçakaya HR, Regan H, Ferson S, Mackay C. 2003. Realism and relevance of ecological models used in chemical risk assessment. Human Ecol Risk Assess 9:907–938.

Bayne BL, Gabbott PA, Widdows J. 1995. Some effects of stress in the adult on the eggs and larvae of Mytilus edulis L. J Mar Biol Ass UK 55:675–689.

Beaumont AR, Tserpes G, Budd MD. 1987. Some effects of copper on the veliger larvae of the mussel *Mytilus edulis* and the scallop *Pecten maximus* (Mollusca: Bivalvia). Mar Environ Res 21:299–309.

Bedaux JJM, Kooijman SALM. 1994. Statistical analysis of bioassays, based on hazard modeling. Environ Ecol Stat 1:303–314.

Belanger SE. 1997. Literature review and analysis of biological complexity in model stream ecosystems: influence of size and experimental design. Ecotoxicol Environ Safety 36:1–16.

Belanger SE. 2003. Lessons learned during 15 years of research at the Procter & Gamble Experimental Stream Facility (ESF). SETAC Globe July–August 2003, 39–40.

Belanger SE, Bowling JW, Lee DM, LeBlanc EM, Kerr KM, McAvoy DC, Christman SC, Davidson DH. 2002. Integration of aquatic fate and ecological responses to linear alkyl benzene sulfonate (LAS) in model stream ecosystems. Ecotoxicol Environ Safety 52:150–171.

Belanger SE, Meiers EM, Bausch RG. 1995. Direct and indirect ecotoxicological effects of alkyl sulfate and alkyl ethoxysulfate on macroinvertebrates in stream mesocosms. Aquat Toxicol 33:65–87.

Belden JB, Gilliom RS, Martin JD, Lydy MJ. 2007. Relative toxicity and occurrence patterns of pesticide mixtures in streams draining agricultural watersheds dominated by corn and soybean production. Int Environ Assess Manag 3:90–100.

Bérard A, Pelte T, Druart J-C. 1999. Seasonal variations in the sensitivity of Lake Geneva phytoplankton community structure to atrazine. Archiv fuer Hydrobiologie 145:277–295.

Bergman HL, Dorward King EJ. 1997. Reassessment of metals criteria for aquatic life protection. Pensacola (FL): SETAC Press. 114 p.

Biek R, Funk CW, Maxell BA, Mills LS. 2002. What is missing in amphibian decline research: insights for ecological sensitivity analysis. Conserv Biol 16:728–734.

Biever RC, Giddings JM, Kiamos M, Annunziato MF, Meyerhoff R, Racke K. 1994. Effects of chlorpyrifos on aquatic microcosms over a range of off-target drift exposure levels. Brighton (UK): Brighton Crop Protection Conference — Pests and Diseases, p 1367–1372.

Biggs BJF, Hickey CW. 1994. Periphyton responses to a hydraulic gradient in a regulated river in New Zealand. Freshwater Biol 32:49–59.

Birnbaum LS. 1999. TEFs: A practical approach to a real-world problem. Human Ecol Risk Assess 5:13–24.

Blanck H. 2002. A critical review of procedures and approaches used for assessing pollution-induced community tolerance (PICT) in biotic communities. Human Ecol Risk Assess 8:1003–1034.

Blanck H, Admiraal W, Cleven RFMJ, Guasch H, van den Hoop MAGT, Ivorra N, Nyström B, Paulsson M, Petterson RP, Sabater S, Tubbing GMJ. 2003. Variability in zinc tolerance, measured as incorporation of radio-labeled carbon dioxide and thymidine, in periphyton communities sampled from 15 European river stretches. Arch Environ Contam Toxicol 44:17–29.

Blanck H, Dahl B. 1996. Pollution-induced community tolerance (PICT) in marine periphyton in a gradient of tri-n-butyltin (TBT) contamination. Aquat Toxicol 35:59–77.

Bliss CI. 1939. The toxicity of poisons applied jointly. Ann Appl Biol 26:585–615.

Bodar CW, Pronk ME, Sijm DT. 2005. The European Union risk assessment on zinc and zinc compounds: the process and the facts. Int Environ Assess Manag 1:301–319.

Bödeker W, Altenburger R, Faust M, Grimme LH. 1990. Methods for the assessment of mixtures of plant protection substances (pesticides): mathematical analysis of combination effects in phytopharmacology and ecotoxicology. Nachrichtenblatt Deutsche Pflanzenschutzdienst (Braunschweig) 70–78.

Boesten JJTI. 2000. Modeller subjectivity in estimating pesticide parameters for leaching models using the same laboratory data set. Agric Water Manage 44:389–409.

Boeston JJTI, Köpp H, Adriaanse PI, Brock TCM, Forbes VE. 2007. Conceptual model for improving the link between exposure and effects in the aquatic risk assessment of pesticides. *Ecotoxicology and Environmental Safety* 66: 291–308.

Bol J, Verhaar HJM, van Leeuwen CJ, Hermens JLM. 1993. Predictions of the aquatic toxicity of high-production-volume-chemicals. Part A: introduction and methodology. The Hague (The Netherlands): Netherlands Ministry of Housing Spatial Planning and the Environment (VROM).

Bolt HM, Mumtaz MM. 1996. Risk assessment of mixtures and standard setting: working towards practical compromises. Food Chem Toxicol 34:1179–1181.

Boltzman L. 1981. Vorlesung uber gastheorie I und II Teil. Reprint of 1896–1898 edition. Graz (Austria): Akademische Druck.

Bond W. 2001. Keystone species: hunting the snark? Science 292:63–64.

Boone MD, Bridges CM. 1999. The effect of temperature on the potency of carbaryl for survival of tadpoles of the green frog (*Rana clamitans*). Environ Toxicol Chem 18:1482–1484.

Borgert CJ, Quill TF, McCarty LS. 2004. Can mode of action predict mixture toxicity for risk assessment? Toxicol Appl Pharmacol 201:85–96.

Bossuyt BTA, De Schamphelaere KAC, Janssen CR. 2004. Using the biotic ligand model for predicting the acute sensitivity of cladoceran dominated communities to copper in natural surface waters. Environ Sci Technol 38:5030–5037.

Bossuyt BTA, Janssen CR. 2004. Copper toxicity to different field-collected cladoceran species: intra- and inter-species sensitivity. Environ Pollut 136:145–154.

Box GEP, Hunter WG, Hunter JS. 1978. Statistics for experimenters. New York (NY): John Wiley. 653 p.

Boxall ABA, Brown CD, Barrett KL. 2002. Higher-tier laboratory methods for assessing the aquatic toxicity of pesticides. Pest Manag Sci 58:637–648.

Boyce M. 1992. Population variability analysis. Annu Rev Ecol Syst 23:481–506.

Boyle TP, Fairchild JF, Robinson-Wilson EF, Haverland PS, Lebo JA. 1996. Ecological restructuring in experimental aquatic mesocosms due to the application of diflubenzuron. Environ Toxicol Chem 15:1806–1814.

Brack W. 2003. Effect-directed analysis: a promising tool for the identification of organic toxicants in complex mixtures? Ann Bioanal Chem 377:397–407.

Bradbury SP, Russom CL, Ankley GT, Schultz TW, Walker JD. 2003. Overview of data and conceptual approaches for deivation of quantitative structure-activity relationships for ecotoxicological effects of organic chemicals. Environ Toxicol Chem 22:1789–1798.

Brandes LH, den Hollander H, van de Meent D. 1996. SimpleBox 2.0: a nested multimedia
    fate model for evaluating the environmental fate of chemicals. RIVM No. 719101 029.
    Bilthoven (The Netherlands): National Institute of Public Health and the Environment
    (RIVM), 156 p. http://www.rivm.nl (accessed December 28, 2007) and HA.den.Hol-
    lander@rivm.nl.
Breck JE. 1988. Relationships among models for acute toxicity effects: applications to fluctu-
    ating concentrations. Environ Toxicol Chem 7:775–778.
Brix KV, DeForest D, Adams WJ. 2001. Assessing acute and chronic copper risks to freshwa-
    ter aquatic life using species sensitivity distributions for different taxonomic groups.
    Environ Toxicol Chem 20:1846–1856.
Brock TCM. 1998. Assessing chemical stress in aquatic ecosystems: remarks on the need of
    an integrative approach. Aquat Ecol 32:107–111.
Brock TCM. 2001. Risk assessment and mitigation measures for pesticides: are all patches
    of freshwater habitat equal? In: Streloke M, editor. Workshop on risk assessment and
    risk mitigation measures in the context of the authorization of plant protection prod-
    ucts (WORMM). Volume 383. Berlin (Germany): Mitteilungen aus der Biologischen
    Bundesanstalt für Land- und Forstwirtschaft, p 68–72.
Brock TCM, Arts GHP, Maltby L, van den Brink PJ. 2006. Aquatic risks of pesticides, eco-
    logical protection goals and common aims in EU legislation. Int Environ Assess
    Manag 2:e20–e46.
Brock TCM, Budde BJ. 1994. On the choice of structural parameters and endpoints to indi-
    cate responses of freshwater ecosystems to pesticide stress. In: Hill IR, Heimbach F,
    Leeuwangh P, Matthiessen P, editors. Freshwater field tests for hazard assessment of
    chemicals. Boca Raton (FL): CRC Press, p 19–56.
Brock TCM, Crum SJH, Deneer JW, Heimbach F, Roijackers RMM, Sinkeldam JA. 2004.
    Comparing aquatic risk assessment methods for the photosynthesis-inhibiting herbi-
    cides metribuzin and metamitron. Environ Pollut 130:403–426.
Brock TCM, Lahr J, van den Brink PJ. 2000a. Ecological risks of pesticides in freshwa-
    ter ecosystems part 1: herbicides. No. 088. Wageningen (The Netherlands): Alterra,
    124 p.
Brock TCM, Ratte HT. 2002. Ecological risk assessment for pesticides: discussion paper for
    the CLASSIC workshop. In: Streloke M, editor. Community-level aquatic systems
    studies: interpretation studies: CLASSIC. Pensacola (FL): SETAC, p 44.
Brock TCM, Roijackers RMM, Rollon R, Bransen F, van der Heyden L. 1995. Effects of
    nutrient loading and insecticide application on the ecology of Elodea-dominated
    freshwater microcosms. II. Responses of macrophytes, periphyton and macroinver-
    tebrate grazers. Archiv fuer Hydrobiologie 134:53–74.
Brock TCM, van den Bogaert M, Bos AR, van Breukelen SWF, Reiche R, Terwoert J, Suyker-
    buyk REM, Roijackers RMM. 1992. Fate and effects of the insecticide Dursban 4E in
    indoor Elodea-dominated and macrophyte-free freshwater ecosystems: II. Second-
    ary effects on community structure. Arch Environ Contam Toxicol 23:391–409.
Brock TCM, van Wijngaarden RPA, van Geest GJ. 2000b. Ecological risks of pesticides in
    freshwater ecosystems part 2: insecticides. No. 089. Wageningen (The Netherlands):
    Alterra, 141 p.
Brock TCM, Vet JJRM, Kerkhofs MJJ, Lijzen J, van Zuilekom WJ, Gijlstra R. 1993. Fate
    and effects of the insecticide Dursban 4E in indoor Elodea-dominated and macro-
    phyte-free freshwater model ecosystems: III. aspects of ecosystem functioning. Arch
    Environ Contam Toxicol 25:160–169.
Brockway DL, Smith PD, Stancil FE. 1984. Fate and effects of atrazine in small aquatic
    microcosms. Bull Environ Contam Toxicol 32:345–353.
Broderius S, Kahl M. 1985. Acute toxicity of organic chemical mixtures to the fathead min-
    now. Aquat Toxicol 6:307–322.

Broomhall S. 2002. The effects of endosulfan and variable water temperature on survivorship and subsequent vulnerability to predation in *Litoria citropa* tadpoles. Aquat Toxicol 61:243–250.

Bro-Rasmussen F, Lokke H. 1984. Ecoepidemiology: a casuistic discipline describing ecological disturbances and damages in relation to their specific causes; exemplified by chlorinated phenols and chlorophenoxy acids. Reg Toxicol Pharmacol 4:391–399.

Bryan ND, Robinson V, Livens F, Hesketh N, Jones M, Lead J. 1997. Metal-humic interactions: a random structural modelling approach. Geochim Cosmochim Acta 61:805–820.

Burgess RM. 2000. Characterizing and identifying toxicants in marine waters: a review of marine toxicity identification evaluations (ties). Int J Environ Pollut 13:2–33.

Burnison BK, Hodson PV, Nuttley DJ, Efler S. 1996. A bleached-kraft mill effluent fraction causing induction of a fish mixed-function oxygenase enzyme. Environ Toxicol Chem 15:1524–1531.

Burns LA. 1997. Exposure analysis modeling system: user's guide for EXAMS II, version 2.97.5. No. EPA/600/R-97/047. Washington (DC): US Environmental Protection Agency, Office of Research and Development, p 1–106.

Burton GA, Chapman P, Smith E. 2002. Weight of evidence approaches for assessing ecosystem impairment. Human Ecol Risk Assess 8:1657–1674.

Cairns J, Niederlehner BR, Smith EP. 1992. The emergence of functional attributes as endpoints in ecotoxicology. In: Burton GA, Jr, editor. Sediment toxicity assessment. Boca Raton (FL): Lewis Publishers, p 111–128.

Cairns J, Jr. 1990. The genesis of biomonitoring in aquatic ecosystems. Environ Profess 14:186–196.

Cairns JJ, van der Schalie WH. 1980. Biological monitoring. Part I: early warning systems. Water Res 14:1179–1196.

Calabrese EJ, Baldwin LA. 1993. Performing ecological risk assessments. Boca Raton (FL): Lewis Publishers, 257 p.

Calabrese EJ, Baldwin LA. 2001. Hormesis: a generalizable and unifying hypothesis. Crit Rev Toxicol 31:353–424.

Calamari D, Marchetti R. 1973. The toxicity of mixtures of metals and surfactants to rainbow trout (*Salmo gairdneri* Rich). Water Res 7:1453–1464.

Calow P. 1998a. Ecological risk assessment: risk for what? How do we decide? Ecotoxicol Environ Safety 40:15–18.

Calow P. 1998b. Handbook of environmental risk assessment and management. Oxford (UK): Blackwell Science, 590 p.

Calow P, Townsend CR. 1981. Resource utilization in growth. In: Townsend CR, Calow P, editors. Physiological ecology: an evolutionary approach to resource use. Sunderland (MA): Sinauer Associates, p 220–244.

Campbell PGC. 1995. Interactions between trace metals and aquatic organisms: a critique of the free-ion activity model. In: Tessier A, Turner DP, editors. Metal speciation and bioavailability in aquatic systems. Chichester (UK): John Wiley, p 45–102.

Campbell PJ, Arnold D, Brock T, Grandy N, Heger W, Heimbach F, Maund SJ, Streloke M, editors. 1999. Guidance document: higher-tier aquatic risk assessment for pesticides. Brussels (Belgium): SETAC-Europe, 179 p.

Caprio MA, Tabashnik BE. 1992. Gene flow accelerates local adaptation among finite populations: simulating the evolution of insecticide resistance. J Econ Entomol 85:611–620.

Carlisle DM, Clements WH. 1999. Sensitivity and variability of metrics used in biological assessments of running waters. Environ Toxicol Chem 18:285–291.

Carlsen L. 2003. An introduction to (Q)SAR for regulatory applications. Unpublished manuscript.

Carpenter SR, Kitchell JF, Hodgson JR. 1985. Cascading trophic interactions and lake productivity. Bioscience 35:634–639.

Carroll S. 2002a. Population models: life history. In: Pastorok RA, Bartell SM, Ferson S, Ginzburg LR, editors. Ecological modeling in risk assessment: chemical effects on populations, ecosystems, and landscapes. Boca Raton (FL): Lewis Publishers, p 55–64.

Carroll S. 2002b. Population models: food webs. In: Pastorok RA, Bartell SM, Ferson S, Ginzburg LR, editors. Ecological modeling in risk assessment: chemical effects on populations, ecosystems, and landscapes. Boca Raton (FL): Lewis Publishers, p 97–106.

Castillo LE, de la Cruz E, Ruepert C. 1997. Ecotoxicology and pesticides in tropical aquatic ecosystems of Central America. Environ Toxicol Chem 16:41–51.

Caswell H. 1996. Demography meets ecotoxicology: untangling the population level effects of toxic substances. In: Newman MC, Jagoe CH, editors. Ecotoxicology: a hierarchical treatment. Boca Raton (FL): Lewis Publishers, p 255–292.

Caswell H. 2001. Matrix population models: construction, analysis, and interpretation. Sunderland (MA): Sinauer Associates, 722 p.

Caughley G. 1994. Directions in conservation biology. J Anim Ecol 63:215–244.

Chapin RE. 2004. Nature is complex: our orchestra seats at the most wonderful show on earth. Toxicol Sci 80:1–2.

Chapman J. 2001. The revised Australian and New Zealand water quality guidelines for toxicants: approach to their derivation and application. Aust J Ecotoxicol 7:95–108.

Chapman JC, Napier GM, Sunderam RIM, Wilson SP. 1993. The contribution of ecotoxicological research to environmental research. Aust Biol 6:72–81.

Chapman PM. 1986. Sediment quality criteria from the sediment quality triad: An example. Environ Toxicol Chem 3:957–964.

Chapman PM. 1996. Presentation and interpretation of sediment quality triad data. Ecotoxicol 5:327–339.

Chapman PM. 2000. Whole-effluent toxicity testing: usefulness, level of protection, and risk assessment. Environ Toxicol Chem 19:3–13.

Chapman PM. 2007. Determining when contamination is pollution: weight of evidence determinations for sediments and effluents. Environ Int 33:492–501, 2007.

Chapman PM, McDonald BG, Lawrence GS. 2002. Weight of evidence issues and frameworks for sediment quality (and other) assessments. Human Ecol Risk Assess 8:1489–1516.

Chapman PM, Riddle MJ. 2003. Missing and needed: polar marine ecotoxicology. Mar Poll Bull 46: 927–928.

ChemSilico. 2005. CSpKaTM. http://www.chemsilico.com/index.html (accessed July 23, 2005).

Chen CY, Chen SL, Christensen ER. 2005. Individual and combined toxicity of nitriles and aldehydes to *Raphidocelis subcapitata*. Environ Toxicol Chem 24:1067–1073.

Christensen V, Pauly D. 1992. A guide to ECOPATH II software system (version 2.1). Manilla (the Phillipines): ICLARM.

Clements RG, Nabholz JV, Zeeman MG, Auer CM. 1995. The application of structure-activity relationships (SARs) in the aquatic toxicity evaluation of discrete organic chemicals. SAR (Q)SAR Env Res 3:203–215.

Clements WH. 1997. Ecological significance of endpoints used to assess sediment quality. In: Ingersoll CG, Dillon T, Biddinger GR, editors. Ecological risk assessments of contaminated sediments. Pensacola (FL): SETAC Press, p 123–134.

Clements WH, Carlisle DM, Lazorchak JM, Johnson PC. 2000. Heavy metals structure benthic communities in Colorado mountain streams. Ecol Appl 10:626–638.

Cleuvers M. 2003. Aquatic ecotoxicity of pharmaceuticals including the assessment of combination effects. Toxicol Lett 142:185–194.

Cohen SM, Garland EM, Cano M, St. John MK, Khachab M, Wehner JM, Arnold LL. 1995. Effects of sodium ascorbate, sodium saccharin and ammonium chloride on the male rat urinary bladder. Carcinogenesis 16:2743–2750.

Collins JF, Brown JP, Alexeef GV, Salmon AG. 1998. Potency equivalency factors for some polycyclic aromatic hydrocarbons and polycyclic aromatic hydrocarbon derivatives. Reg Toxicol Pharmacol 28:45–54.

Commission of the European Communities. 2000. Communication from the Commission on the Precautionary Principle. No. COM(2000)1. Brussels (Belgium): European Commission, 29 p.

Compton R, Sigal EA. 1999. The use of toxic equivalency factors (TEFs) in ecological risk assessment: strengths and limitations. Human Ecol Risk Assess 5:33–42.

Connell JH, Sousa WP. 1983. On the evidence needed to judge ecological stability or persistence. Am Nat 121:789–824.

Connor PM. 1972. Acute toxicity of heavy metals to some marine larvae. Mar Poll Bull 3:190–192.

Cook RB, Suter GW, Sain ER. 1999. Ecological risk assessment in a large river-reservoir: 1 introduction and background. Environ Toxicol Chem 18:581–588.

Cossins AR, Bowler K. 1987. Temperature biology of animals. New York (NY): Chapman & Hall, 339 p.

Courtney LA, Clements WH. 2000. Sensitivity to acidic pH in benthic invertebrate assemblages with different histories of exposure to metals. J NA Benthol Soc 19:112–127.

Cowan CE, Mackay D, Feijtel TCJ, van de Meent D, Di Guardo A, Davies J, Mackay DN, editors. 1995. The multi-media fate model: a vital tool for predicting the fate of chemicals. Pensacola (FL): SETAC Press, 78 p.

Cox CB, Moore PD. 1993. Biogeography: an ecological and evolutionary approach. Oxford (UK): Blackwell Scientific, 298 p.

Crane M, Newman MC, Chapman PF, Fenlon J, editors. 2002. Risk assessment with time to event models. Boca Raton (FL): Lewis Publishers, 175 p.

Crane M, Whitehouse P, Comber S, Watts C, Giddings JM, Moore DRJ, Grist E. 2003. Evaluation of probabilistic risk assessment of pesticides in the UK: chlorpyrifos use on top fruit. Pest Manag Sci 59:512–526.

Cronin MTD, Walker JD, Jaworska JS, Comber MHI, Watts CD, Worth AP. 2003. Use of (Q)SARs in international decision-making frameworks to predict ecological effects and environmental fate of chemical substances. Environ Hlth Perspect 111:1376–1390.

Crosby DG. 1998. Environmental toxicology and chemistry. New York (NY): Oxford University Press, 336 p.

Crum SJH, Brock TCM. 1994. Fate of chlorpyrifos in indoor microcosms and outdoor experimental ditches. In: Hill IR, Heimbach F, Leeuwang P, Matthiessen P, editors. Freshwater field tests for hazard assessment of chemicals. Boca Raton (FL): Lewis Publishers, p 315–322.

Culp JM, Lowell RB, Cash KJ. 2000a. Integrating mesocosm experiments with field and laboratory studies to generate weight-of-evidence risk assessments for large rivers. Environ Toxicol Chem 19:1167–1173.

Culp JM, Podemski CL, Cash KJ, Lowell RB. 2000b. A research strategy for using stream microcosms in ecotoxicology: integrating experiments at different levels of biological organization with field data. J Aquat Ecosys Stress Recovery 7:167–176.

Cuppen JGM, Crum SJH, van den Heuvel HH, Smidt RA, van den Brink PJ. 2002. The effects of a mixture of two insecticides on freshwater microcosms. I. Fate of insecticides and responses of macroinvertebrates. Ecotoxicol 11:19–34.

Cuppen JGM, van den Brink PJ, van der Woude H, Zwaardemaker N, Brock TCM. 1997. Sensitivity of macrophyte-dominated freshwater microcosms to chronic levels of the herbicide linuron. Ecotoxicol Environ Safety 38:25–35.

Daam MA, van den Brink PJ. 2003. Effects of three pesticides that differ in mode of action on the ecology of small indoor aquatic microcosms: an evaluation of the effects of the insecticide chlorpyrifos, the herbicide atrazine and the fungicide carbendazim. Report No. 788. Wageningen (The Netherlands): Alterra.

Danish Environmental Protection Agency. 2001. Report on the advisory list for self-classification of dangerous substances. Environmental Project No. 636 2001. http://www.mst.dk/udgiv/publications/2001/87-7944-694-9/pdf/87-7944-695-7.pdf (accessed July 28, 2005).

David CPC. 2003. Establishing the impact of acid mine drainage through metal bioaccumulation and taxa richness of benthic insects in a tropical Asian stream (the Philippines). Environ Toxicol Chem 22:2952–2959.

Davies PI, Cook LSJ. 1993. Catastrophic macroinvertebrate and sublethal effects on brown trout, *Salmo trutta,* caused by cypermethrin spraying on a Tasmanian stream. Aquat Toxicol 27:201–224.

DeAngelis DL. 1996. Indirect effects: concepts and approaches from ecological theory. In: Baird DJ, Maltby L, Greig-Smith PW, Douben PET, editors. ECOtoxicology: ecological dimensions. London (UK): Chapman & Hall, p 25–41.

De Coen WM, Janssen CR. 2003. The missing biomarker link: relationships between effects on the cellular energy allocation biomarker of toxicant-stressed *Daphnia magna* and corresponding population characteristics. Environ Toxicol Chem 22:1632–1641.

De Groot A, Peijnenburg W, van den Hoop M, Ritsema R, van Veen RPM. 1998. Heavy metals in Dutch field soils: an experimental and theoretical study on equilibrium partitioning. RIVM No. 607220001. Bilthoven (The Netherlands): National Institute of Public Health and the Environment (RIVM). 46 p. http://www.rivm.nl (accessed December 28, 2007).

den Hollander H, van Eijkeren JCH, van de Meent D. 2004. SimpleBox 3.0: multimedia mass balance model for evaluating the fate of chemicals in the environment. Bilthoven (The Netherlands): National Institute of Public Health and the Environment (RIVM). http://www.rivm.nl (accessed December 28, 2007) and HA.den.Hollander@rivm.nl.

Deneer JW. 2000. Toxicity of mixtures of pesticides in aquatic systems. Pest Manage Sci 56:516–520.

Deneer JW, Budde BJ, Weijers A. 1999. Variations in the lethal body burdens of organophosphorus compounds in the guppy. Chemosphere 38:1671–1683.

Deneer JW, Sinnige TL, Seinen W, Hermens JLM. 1988. The joint acute toxicity to *Daphnia magna* of industrial organic chemicals at low concentrations. Aquat Toxicol 12:33–38.

De Ruiter PC, Neutel AM, Moore JC. 1994. Modeling food webs and nutrient cycling in agroecosystems. TREE 9:378–383.

De Ruiter PC, Neutel A-M, Moore JC. 1995. Energetics, patterns of interaction strength, and stability in real ecosystems. Science 269:1257–1260.

De Schamphelaere KAC, Janssen CR. 2002. A biotic ligand model predicting acute copper toxicity to *Daphnia magna*: the effects of calcium, magnesium, sodium, potassium and pH. Environ Sci Technol 36:48–54.

De Vlaming V, Connor V, DiGiorgio C, Bailey HC, Deanovic LA, Hinton DE. 2000. Application of whole-effluent toxicity test procedures to ambient water quality assessment. Environ Toxicol Chem 19:42–62.

De Wolf W, de Bruijn JHM, Seinen W, Hermens JLM. 1992. Influence of biotransformation on the relationship between bioconcentration factors and octanol-water partition coefficients. Environ Sci Technol 26:1197–1201.

De Zwart D. 2002. Observed regularities in SSDs for aquatic species. In: Posthuma L, Suter GW, Traas TP, editors. Species sensitivity distributions in ecotoxicology. Boca Raton (FL): CRC Press, p 133–154.

De Zwart D. 2005. Ecological effects of pesticide use in the Netherlands: modeled and observed effects in the field ditch. Int Environ Assess Manag 1:123–134.

De Zwart D, Dyer SD, Posthuma L, Hawkins CP. 2006. Use of predictive models to attribute potential effects of mixture toxicity and habitat alteration on the biological condition of fish assemblages. Ecol Appl 16:1295–1310.

De Zwart D, Posthuma L. 2005. Complex mixture toxicity for single and multiple species: proposed methodologies. Environ Toxicol Chem 24:2665–2676.

De Zwart D, Sterkenburg A. 2002. Toxicity-based assessment of water quality. In: Posthuma L, Suter GW, Traas T, editors. Species sensitivity distributions in ecotoxicology. Boca Raton (FL): CRC Press, p 383–402.

Díaz-Ravina M, Bååth E. 2001. Response of soil bacterial communities pre-exposed to different metals and reinoculated in an unpolluted soil. Soil Biol Biochem 33:241–248.

Di Giulio RT, Benson WH, Sanders BM, van Veld PA. 1995. Biochemical mechanisms: metabolism, adaptation, and toxicity. In: Rand GM, editor. Fundamentals of aquatic toxicology, 2nd ed. Washington (DC): Taylor & Francis, p 523–561.

Di Toro DM, Allen HE, Bergman HL, Meyer JS, Paquin RR, Santore RC. 2001a. Biotic ligand model of the acute toxicity of metals. I. Technical basis. Environ Toxicol Chem 20:2383–2396.

Di Toro DM, Hansen D, McGrath J, Berry WJ. 2001b. Predicting the toxicity of metals in sediments using organic carbon normalized SEM and AVS. Draft manuscript.

Di Toro DM, Paquin PR, Santore R, Wu KB. 1997. Chemistry of silver bioavailability: a model of acute silver toxicity to fish. 5th International Argentum Conference on Transport, Fate and Effects of Silver in the Environment, Hamilton, Ontario, Canada, p 191–204.

Di Toro DD, Zarba CS, Hansen DJ, Berry DW, Swartz RC, Cowan CE, Pavlou SP, Allen HE, Nelson TA, Paquin PR. 1991. Technical basis for establishing sediment quality criteria for nonionic organic chemicals using equilibrium partitioning. Environ Toxicol Chem 10:1541–1583.

Dodds WK, Lopez AJ, Bowden WB, Gregory S, Grimm NB, Hamilton SK, Hershey AE, Marti E, McDowell WH, Meyer JL, Morrall DD, Mulholland PJ, Peterson BJ, Tank JL, Valett HM, Webster JR, Wolheim WM. 2002. N uptake as a function of concentration in streams. J NA Benthol Soc 21:206–220.

Doe KG, Burton GA, Ho KT. 2003. Porewater toxicity testing: an overview. In: Carr RS, Nipper M, editors. Porewater toxicity testing: biological, chemical, and ecological considerations. Pensacola (FL): SETAC Press, p 125–142.

Domsch KH, Jagnow G, Anderson TH. 1983. An ecological concept for the assessment of side-effects of agrochemicals on soil microorganisms. Res Rev 86:65–105.

Donker MH, Abdel-Lafeif HM, Khalil MA, Bayoumi BM, van Straalen NM. 1998. Temperature, physiological time, and zinc toxicity in the isopod Porcellio scaber. Environ Toxicol Chem 17:1558–1563.

Drescher K, Bödeker W. 1995. Assessment of the combined effects of substances: the relationship between concentration addition and independent action. Biometrics 51:716–730.

Dudgeon D. 2000. The ecology of tropical Asian rivers and streams in relation to biodiversity conservation. Annu Rev Ecol Syst 31:239–263.

Dyer SD, Belanger SE. 1999. Determination of the sensitivity of stream mesocosms. Environ Toxicol Chem 18:2903–2907.

Dyer SD, Belanger SE, Carr GJ. 1997. An initial evaluation of the use of Euro/North American fish species for tropical effects assessments. Chemosphere 35:2767–2781.

Dyer SD, Peng C, McAvoy DC, Fendinger NJ, Masscheleyn P, Castillo L, Lim JMU. 2003. The influence of untreated wastewater to aquatic communities in the Balatuin River, the Philippines. Chemosphere 52:43–53.

Eckert J. 2000. Wie verwandt sind die Lebensgemeinschaften in aquatischen Mesokosmen und Gewässern in ihrer Umgebung? (How are communities from freshwater mesocosms related to natural ponds in their surrounding?) [Diploma thesis]. Aachen (Germany): Aachen University, 56 p.

Ecological Committee on FIFRA Risk Assessment Methods (ECOFRAM) USEPA. 1999. ECOFRAM aquatic and terrestrial final draft reports. http://www.epa.gov/oppefed1/ecorisk/index.htm (accessed June 1, 1999).

European Food Safety Authority. 2006. Opinion of the scientific panel on plant protection products and their residues on a request from EFSA on the final report of the FOCUS Working Group on Landscape and Mitigation Factors in Ecological Risk Assessment. EFSA J 437:1–30. http://www.efsa.europa.eu/EFSA/efsa_locale-1178620753812_home.htm (accessed December 28, 2007).

Ehrlich PR, Ehrlich AH. 1981. Extinction: the causes and consequences of the disappearance of species. New York (NY): Random House, 305 p.

Ellersieck MR, Asfaw A, Mayer FL, Krause GF, Sun K, Lee G. 2003. Acute-to-chronic estimation (ACE) v 2.0 with time-concentration effect models. No. EPA600/R-03/107. Washington (DC): US Environmental Protection Agency, 20 p.

Ellis D. 1989. Environments at risk: case histories of impact assessment. New York (NY): Springer-Verlag, 329 p.

Elmegaard N, Jagers op Akkerhuis GAJM. 2000. Safety factors in pesticide risk assessment: differences in species sensitivity and acute-chronic relations. NERI Technical. No. 325. Silkeborg (Denmark): National Environmental Research Institute, Department of Terrestrial Ecology, 60 p.

Elzerman AW, Coates JT. 1987. Hydrophobic organic compounds on sediments: equilibria and kinetics of sorption. In: Hites RA, Eisenreich SJ, editors. Sources and fates of aquatic pollutants. Washington (DC): American Chemical Society, p 263–285.

Emans HJB, van de Plassche EJ, Canton JH, Okkerman PC, Sparenburg PM. 1993. Validation of some extrapollation methods used for effect assessment. Environ Toxicol Chem 12:2139–2154.

Emlen JM, Duda JJ, Kirchhoff M, Freeman FC. 2006. Fitting population models from field data II: new developments and a reapplication to old data. Ecol Model 192:557–570.

Englund G, Cooper SD. 2003. Scale effects and extrapolation in ecological experiments. Adv Ecol Res 33:161–202.

Environment Canada. 1997. Environmental assessments of priority substances under the Canadian Environmental Protection Act guidance manual version 1.0. March 1996, No. EPS/2/CC/3E. Ottawa: Environment Canada.

Environment Canada Existing Substances Branch. 2003. Guidance manual for the categorization of organic and inorganic substances on Canada's Domestic Substances List. http://www.ec.gc.ca/substances/ese/eng/psap/guidman2.cfm (accessed July 28, 2005).

Environment Canada Existing Substances Branch. 2005. Existing substances evaluation. http://www.ec.gc.ca/substances/ese/eng/esehome.cfm (accessed December 15, 2005).

Environment Canada. N.d. Domestic Substances List: CEPA Section 71 notices. http://www.ec.gc.ca/substances/ese/eng/dsl/notices_avis.cfm (accessed January 8, 2008).

Erickson RJ, Benoit DA, Mattson VR, Nelson HP, Jr, Leonard EN. 1996. The effects of water chemistry on the toxicity of copper to fathead minnows. Environ Toxicol Chem 15:181–193.

Eriksson L, Jaworska JS, Worth AP, Cronin MTD, McDowell RM, Gramatica P. 2003. Methods for reliability and uncertainty assessment and applicability evaluations of classification- and regression-based (Q)SARs. Environ Hlth Perspect 111:1361–1375.

Escher BI, Hermans JLM. 2002. Modes of action in ecotoxicology: their role in body burdens, species sensitivity, (Q)SARs and mixture effects. A critical review. Environ Sci Technol 36:4201–4216.

EUFRAM. 2005. EUFRAM report, volume 1. Introducing probabilistic methods into the ecological risk assessment of pesticides. No. Version 6. York (UK): EUFRAM, 50 p. http://www.eupra.com/report.pdf (accessed December 28, 2007).

European Centre for Ecotoxicology and Toxicology of Chemicals [ECETOC]. 1993. Aquatic toxicity data evaluation. Technical. No. 56. Brussels (Belgium): ECETOC, 66 p.

European Centre for Ecotoxicology and Toxicology of Chemicals [ECETOC]. 2001. Aquatic toxicity of mixtures. Technical Report No. 80. Brussels (Belgium): ECETOC, 64 p.

European Chemicals Bureau. 2005. Existing chemicals. http://ecb.jrc.it/existing-chemicals/ (accessed July 28, 2005).

European Commission. 1996. Technical guidance documents in support of Directive 93/67/EEC on risk assessment of new notified substances and regulation (EC) No. 1488/94 on risk assessment of existing substances (Parts I, II, III, and IV). No. CR-48-96-001, 002, 003, 004-EN-C. Luxembourg: Office for Official Publications of the European Community, 739 p. http://ecb.jrc.it/DOCUMENTS/TECHNICAL_GUIDANCE_DOCUMENT/EDITION_1/tgdpart4.pdf (accessed December 28, 2007).

European Commission. 2001. White paper. Strategy for a future chemicals policy. http://europa.eu.int/comm/environment/chemicals/0188_en.pdf (accessed December 2, 2003).

European Inland Fisheries Advisory Commission [EIFAC]. 1980. Combined effects on freshwater fish and other aquatic life of mixtures of toxicants in water. Rome (Italy): Food and Agriculture Organisation.

European Inland Fisheries Advisory Commission [EIFAC]. 1987. Revised report on combined effects on freshwater fish and other aquatic life of mixtures of toxicants in water. No. EIFAC Tech Pap 37, Rev 1. Rome (Italy): European Inland Fisheries Advisory Commission, Working Party on Water Quality Criteria for European freshwater fish, 75 p.

European Union. 1991. Council Directive of 15 July 1991 concerning the placing of plant protection products on the market (91/414/EEC). Brussels (Belgium): European Union.

European Union. 1997. Council Directive 97/57/EC of September 21, 1997; establishing Annex VI to Directive 91/414/EEC concerning the placing of plant protection products on the market. Off J Europ Commun L265:87–109.

European Union. N.d. REACH & GHS. http://ec.europa.eu/enterprise/reach/index_en.htm (accessed January 8, 2008).

Fahrig L, Freemark K. 1995. Landscape-scale effects of toxic events for ecological risk assessment. In: Cairns J, Jr, Niederlehner BR, editors. Ecological toxicity testing: scale, complexity, and relevance. Boca Raton (FL): Lewis Publishers, p 193–208.

Faust M, Altenburger R, Backhaus T, Blanck H, Bödeker W, Gramatica P, Hamer V, Scholze M, Vighi M, Grimme LH. 2001. Predicting the joint algal toxicity of multicomponent s-triazine mixtures at low-effect concentrations of individual toxicants. Aquat Toxicol 56:13–32.

Faust M, Altenburger R, Backhaus T, Blanck H, Bödeker W, Gramatica P, Hamer V, Scholze M, Vighi M, Grimme LH. 2003. Joint algal toxicity of 16 dissimilarly acting chemicals is predictable by the concept of independent action. Aquat Toxicol 63:43–63.

Faust M, Altenburger R, Backhaus T, Bödeker W, Scholze M, Grimme LH. 2000. Predictive assessment of the aquatic toxicity of multiple chemical mixtures. J Environ Qual 29:1063–1068.

Faust M, Altenburger R, Bödeker W, Grimme LH. 1994. Algal toxicity of binary combinations of pesticides. Bull Environ Contam Toxicol 53:134–141.

Ferretti JA, Calesso DF, Lazorchak JM, Dolce TJ, Arnold J, Goodfellow WL, Smith ME, Serbst J. 2004. Interlaboratory comparison of a reduced volume marine sediment toxicity test method using the amphipod *Ampelisca abdita*. Environ Toxicol Chem 23:632–637.

Ferson S. 2002. Population models: scalar abundance. In: Pastorok RA, Bartell SM, Ferson S, Ginzburg LR, editors. Ecological modeling in risk assessment: chemical effects on populations, ecosystems, and landscapes. Boca Raton (FL): Lewis Publishers, p 37–54.

Finlay BJ, Maberley SC, Cooper JI. 1997. Microbial diversity and ecosystem function. Oikos 80:209–213.

Finney DJ. 1971. Probit analysis. London (UK): Cambridge University Press, 333 p.

FOCUS. 2001. FOCUS surface water scenarios in the EU evaluation process under 91/414/EEC. No. SANCO/4802/2001-rev. 2. Brussels (Belgium): FOCUS Working Group on Surface Water Scenarios, 245 p. http://viso.ei.jrc.it/focus/ (accessed December 28, 2007).

FOCUS. 2005. Landscape and mitigation factors in aquatic risk assessment. Vol. 1. Extended summary and recommendations. No. SANCO/10422/2005. Brussels (Belgium): FOCUS Working Group on Landscape and Mitigation Factors in Ecological Risk Assessment, 133 p.

Forbes VE, editor. 1998. Genetics and ecotoxicology. Ann Arbor (MI): Taylor & Francis, 231 p.

Forbes VE, Calow P. 1999. Is the per capita rate of increase a good measure of population-level effects in ecotoxicology? Environ Toxicol Chem 18:1544–1556.

Forbes VE, Calow P. 2002. Species sensitivity distributions revisited: a critical appraisal. Human Ecol Risk Assess 8:473–492.

Forbes VE, Calow P, Sibly RM. 2001a. Are current species extrapolation models a good basis for ecological risk assessment? Environ Toxicol Chem 20:442–447.

Forbes VE, Sibly RM, Calow P. 2001b. Determining toxicant impacts on density-limited populations: a critical review of theory, practice and results. Ecol Appl 11:1249–1257.

Forbes VE, Sibly RM, Linke-Gamenick I. 2003. Joint effects of a toxicant and population density on population dynamics: an experimental study using Capitella sp I (Polychaeta). Ecol Appl 13:1094–1103.

Fore LS. 2003. Biological assessment of mining disturbance on stream invertebrates in mineralized areas of Colorado. In: Simon TP, editor. Biological response signatures: indicator patterns using aquatic communities. Boca Raton (FL): CRC Press, p 347–370.

Fox GA. 1991. Practical causal inference for ecoepidemiologists. J Toxicol Environ Hlth 33:359–373.

Frampton GK, van den Brink PJ, Gould PJL. 2000. Effects of spring precipitation on a temperate arable collembolan community analysed using principal response curves. Appl Soil Ecol 14:231–248.

FreshwaterLife. N.d. The FreshwaterLife information portal. http://www.freshwaterlife.org (accessed January 7, 2008).

Friberg-Jensen U, Wendt-Rasch L, Woin P, Christoffersen K. 2003. Effects of the pyrethroid insecticide cypermethrin on a freshwater community studied under field conditions. I. Direct and indirect effects on abundance measures of organisms at different trophic levels. Aquat Toxicol 63:357–371.

Gale SA, Smith SV, Lim RP, Feffree RA, Petocz P. 2003. Insights into the mechanisms of copper tolerance of a population of black-banded rainbowfish (*Melanotaenia nigrans*) (Richardson) exposed to mine leachate, using 64/67Cu. Aquat Toxicol 62:135–153.

Ganzelmeier H, Rautmann D, Spangenberg R, Streloke M, Herrmann M, Wenzelburger H-J, Walter H-F. 1995. Studies on the spray drift of plant protection products. Berlin (Germany): Blackwell Wissenschafts-Verlag, 111 p.

Gaston KJ, Williams PH, Eggleton P, Humphries CJ. 1995. Large-scale patterns of biodiversity: spatial variation in family richness. Proc R Soc Lond B Biol Sci 260:149–154.

Gentile JH, Solomon KR, Butcher JB, Harrass M, Landis WG, Power M, Rattner BA, Warren-Hicks WJ, Wenger R. 1999. Linking stressors and ecological responses. In: Foran JA, Ferenc SA, editors. Multiple stressors in ecological risk and impact assessment. Pensacola, FL: SETAC Press, p 27–50.

Ghilarov AM. 2000. Ecosystem functioning and intrinsic value of biodiversity. Oikos 90:408–412.

Giddings JM, Anderson TA, Hall LW, Jr, Kendall RJ, Richards RP, Solomon KR, Williams WM. 2005. A probabilistic aquatic ecological risk assessment of atrazine in North American surface waters. Pensacola (FL): SETAC Press, 432 p.

Giddings JM, Brock TCM, Heger W, Heimbach F, Maund SJ, Norman SM, Ratte HT, Schafers C, Streloke M, editors. 2002. Community-level aquatic systems studies: interpretation studies: CLASSIC. Pensacola (FL): SETAC Press, 44 p.

Giesy JP. 2001. Hormesis: does it have relevance at the population-, community- or ecosystem-levels of organization? Belle 10:14–17.

Giesy JP, Graney RL. 1989. Recent developments in and intercomparisons of acute and chronic bioassays and bioindicators. Hydrobiologia 188–189:21–60.

Giesy JP, Solomon KR, Coates JR, Dixon KR, Giddings JM, Kenaga EE. 1999. Chlorpyrifos: ecological risk assessment in North American aquatic environments. Rev Environ Contam Toxicol 160:1–129.

Golding LA, Timperley MH, Evans CW. 1997. Non-lethal responses of the freshwater snail Potamopyrgus antipodarum to dissolved arsenic. Environ Monit Assess 47:239–254.

Gorsuch JW, Janssen CR, Lee CM, Reiley MC, editors. 2002. The biotic ligand model for metals: current research, future directions, regulatory implications. Comp Biochem Physiol C 133:1–343.

Gotelli NJ. 2001. A primer of ecology. Sunderland (MA): Sinauer Associates, 265 p.

Government of Canada. 2001. A Canadian perspective on the precautionary approach/ principle discussion document. Ottawa: Government of Canada, 20 p. http://www. pco-bcp.gc.ca/raoics-srdc/docs/precaution/Discussion/discussion_e.htm (accessed December 28, 2007).

Graham JD. 1999. Making sense of the precautionary principle. Risk Persp, September 1999, 1–6.

Grant A. 1998. Population consequences of chronic toxicity: incorporating density dependence into the analysis of life table response experiments. Ecol Model 105:325–335.

Grant A. 2002. Pollution-tolerant species and communities: intriguing toys or invaluable monitoring tools? Human Ecol Risk Assess 8:955–970.

Grapentine L, Anderson J, Boyd D, Burton GA, DeBarros C, Johnson G, Marvin C, Milani D, Painter S, Pascoe T, Reynoldson T, Richman L, Solomon K, Chapman PM. 2002. A decision-making framework for sediment assessment developed for the Great Lakes. Human Ecol Risk Assess 8:1641–1655.

GREAT-ER. 2005. A GIS assisted model for environmental risk assessment and management of chemicals in river basins. http://www.great-er.org/pages/home.cfm (accessed December 15, 2005).

Greco WR, Bravo G, Parsons JC. 1995. The search for synergy: a critical review from a response surface perspective. Pharm Rev 47:331–385.

Green DWJ, Williams KA, Pascoe D. 1986. The acute and chronic toxicity of cadmium to different life history stages of the freshwater crustacean Asellus aquaticus (L.). Arch Environ Contam Toxicol 15:465–471.

Greulich K, Pflugmacher S. 2003. Differences in susceptibility of various life stages of amphibians to pesticide exposure. Aquat Toxicol 65:329–336.

Grimme LH, Faust M, Boedeker W, Altenburger R. 1996. Aquatic toxicity of chemical substances in combination: still a matter of controversy. Human Ecol Risk Assess 2:426–433.

Grist EPM, Leung KMY, Wheeler JR, Crane M. 2002. Better bootstrap estimation of hazardous concentration thresholds for aquatic assemblages. Environ Toxicol Chem 21:1515–1524.

Groenendijk D. 1999. Dynamics of metal adaptation in riverine chironomids [PhD thesis]. Amsterdam (The Netherlands): University of Amsterdam, 159 p.

Groom MJ, Pascual MA. 1997. The analysis of population persistence: an outlook on the practice of viability analysis. In: Fiedler PA, Karevia PM, editors. Conservation biology for the coming decade. New York (NY): Chapman & Hall, p 4–27.

Gross TS, Arnold BS, Sepulveda MS, McDonald K. 2003. Endocrine disrupting chemicals and endocrine active agents. In: Hoffman DJ, Rattner BA, Burton GA, Cairns JJ, editors. Handbook of ecotoxicology. Boca Raton (FL): Lewis Publishers, p 1033–1098.

Grote M, Brack W, Walter H, Altenburger R. 2005. Confirmation of cause-effect relationships in effect-directed analysis of complex contaminated samples. Environ Toxicol Chem 24:1420–1427.

Groten JP, Schoen ED, Feron VJ. 1996. Use of factorial designs in combination toxicity studies. Food Chem Toxicol 34:1083–1089.

Groten JP, Schoen ED, van Bladeren PJ, Kuper FCF, van Zorge JA, Feron VJ. 1997. Subacute toxicity of a combination of nine chemicals in rats: detecting interactive effects with a two-level factorial design. Fund Appl Toxicol 36:15–29.

Groten JP, Feron VJ, Suhnel J. 2001. Toxicology of simple and complex mixtures. Trends in Pharmacological Sciences 22:316–322.

Gruessner B, Watzin MC. 1996. Response of aquatic communities from a Vermont stream to environmentally realistic atrazine exposure in laboratory microcosms. Environ Toxicol Chem 15:410–419.

Guasch H, Sabater S. 1998. Light history influences the sensitivity to atrazine in periphytic algae. J Phycol 34:233–241.

Gunderson LH. 2000. Ecological resilience in theory and application. Ann Rev Ecol Syst 31:425–439.

Gurney WSC, McCauley E, Nisbet RM, Murdoch WW. 1990. The physiological ecology of Daphnia: Formulation and tests of a dynamic model of growth and reproduction. Ecol 71:716–732.

Gustavson K, Møhlenberg F, Schlüter L. 2003. Effects of exposure duration of herbicides on natural stream periphyton communities and recovery. Arch Environ Contam Toxicol 45:48–58.

Gustavson K, Wängberg S-Å. 1995. Tolerance induction and succession in microalgal communities exposed to copper and atrazine. Aquat Toxicol 32:283–302.

Haddad S, Charest G, Tardif G, Krishnan K. 2000. Physiologically based modeling of the maximal effect of metabolic interactions on the kinetics of components of complex chemical mixtures. J Toxicol Environ Hlth A 61:209–223.

Haddad S, Krishnan K. 1998. Physiological modeling of toxicokinetic interactions: implications for mixture risk assessment. Environ Hlth Perspect 106:1377–1384.

Haddad S, Tardif R, Charest-Tardif G, Krishnan K. 1999. Physiological modeling of the toxicokinetic interactions in a quaternary mixture of aromatic hydrocarbons. Toxicol Appl Pharmacol 161:249–257.

Hall LWJ, Anderson RD. 1995. The influence of salinity on the toxicity of various classes of chemicals to aquatic biota. Crit Rev Toxicol 25:281–346.

Hall LWJ, Giddings JM. 2000. The need for multiple lines of evidence for predicting site-specific ecological effects. Human Ecol Risk Assess 6:679–710.

Hall RJ, Swineford D. 1980. Toxic effects of endrin and toxaphene on the southern leopard frog (*Rana sphenocephala*). Environmental Pollution A 23:53–65.

Hambright KD. 1994. Can zooplanktivorous fish really affect lake thermal dynamics? Angewand Limnol 22:2560–2565.

Hamelink JL, Bergman HL, Kimerle RA, Landrum PF. 1994. Mechanistic understanding of bioavailability. Boca Raton (FL): Lewis Publishers.

Hamers T, Aldenberg T, van de Meent D. 1996. Definition report: indicator effects toxic substances (Itox). No. 607128 001. Bilthoven (The Netherlands): National Institute of Public Health and the Environment (RIVM), 95 p.

Hammers-Wirtz M, Ratte HT. 2000. Offspring fitness in *Daphnia*: is the *Daphnia* reproduction test appropriate for extrapolating effects on the population level? Environ Toxicol Chem 19:1856–1866.

Hanazato T, Yasuno M. 1990. Influence of time of application of an insecticide on recovery patterns of a zooplankton community in experimental ponds. Arch Environ Contam Toxicol 19:77–83.

Handersen S, Wratten SD. 2000. Sensitivity of aquatic life stages of *Xanthocnemis zealandica* (Odonata: Zygoptera) to azinphosmethyl and carbaryl. NZ J Mar Freshwat Res 34:117–123.

Hansen SR, Garton RR. 1982. The effects of diflubenzuron on a complex laboraory stream community. Arch Environ Contam Toxicol 11:1–10.

Hanski I, Gyllenberg M. 1993. Two general metapopulation models and the core-satellite species hypothesis. Am Nat 132:360–382.

Hanson ML, Sanderson H, Solomon KR. 2003a. Variation, replication, and power analysis of *Myriophyllum spp.* microcosm toxicity data. Environ Toxicol Chem 22:1318–1329.

Hanson ML, Sibley PK, Mabury SA, Muir DCG, Solomon KR. 2003b. Field level evaluation and risk assessment of the toxicity of dichloracetic acid to the aquatic macrophytes *Lemna gibba, Myriophyllum spicatum,* and *Myriophyllum sibericum*. Ecotoxicol Environ Safety 55:46–63.

Hanson ML, Solomon KR. 2002. New technique for estimating thresholds of toxicity in ecological risk assessment. Environ Sci Technol 36:3257–3264.

Harkey GA, Lydy MJ, Kukkonen J, Landrum PF. 1994. Feeding selectivity and assimilation of PAH and PCB in *Diporeia spp.* Environ Toxicol Chem s13:1445–1455.

Harremoes P. 2002. Lessons from the history of innovations and failures. Koninkljke Hollandsche Maatschappij der Wetenschappen. Symposium ter Gelegenheid van het 250-Jarig Jubileum.

Hart A, editor. 2001. Probabilistic risk assessment for pesticides in Europe: implementation and research needs. A report from the European Workshop on Probabilistic Risk Assessment for the Environmental Impacts of Plant Protection Products (EUPRA), No. QLK5-2000-31035. York (UK): Central Science Laboratory, 109 p.

Harwell MA, Cooper W, Flaak R. 1992. Prioritizing ecological and human welfare risks from environmental stresses. Environ Managment 16:451–464.

Harwell MA, Myers V, Young T, Bartuska A, Gassman N, Gentile J, Harwell C, Appelbaum S, Barko J, Causey B, Johnson C, McLean A, Smola RT, Templet P, Tosini S. 1999. A framework for an ecosystem integrity report card. Bioscience 49:543–556.

Hawker DW, Connell DW. 1986. Bioconcentration of lipophilic compounds by some aquatic organisms. Ecotoxicol Environ Safety 11:184–197.

Hawkins BA. 2001. Ecology's oldest pattern. TREE 16:470.

Health Council of The Netherlands. 1997. The food web approach in ecotoxicological risk assessment. No. 1997/14E. Rijswijk (The Netherlands): Standing Committee on Ecotoxicology.

Hector A, Schmid B, Beierkuhnlein C, Caldeira MC, Diemer M, Dimitrakopoulos PG, Finn JA, Freitas H, Giller PS, Good J, Harris R, Högberg P, Huss-Danell K, Joshi J, Jumpponen A, Körner C, Leadley PW, Loreau M, Minns A, Mulder CPH, O'Donovan G,

Otway SJ, Pereira JS, Prinz A, Read DJ, Scherer-Lorenzen M, Schulze E-D, Siamantziouras A-SD, Spehn EM, Terry AC, Troumbis AY, Woodward FI, Yachi S, Lawton JH. 1999. Plant diversity and productivity experiments in European grasslands. Science 286:1123–1126.

Heming TA, McGuinness EJ, George LM, Blumhagen KA. 1988. Effects of pulsed- and spiked-exposure to methoxychlor on early life stages of rainbow trout. Bull Environ Contam Toxicol 40:764–770.

Hendley P, Holmes C, Kay S, Maund SJ, Travis KZ, Zhang M. 2001. Probabilistic risk assessment of cotton pyrethroids: III. A spatial analysis of the Mississippi cotton landscape. Environ Toxicol Chem 20:669–678.

Hendriks AJ, Heikens A. 2001. The power of size. 2. Rate constants and equilibrium ratios for accumulation of inorganic substances related to species weight. Environ Toxicol Chem 20:1421–1437.

Hendriks JA, van Der Linde A, Cornelissen G, Sijm DTHM. 2001. The power of size. 1. Rate constants and equilibrium ratios for accumulation of organic substances related to octanol-water partition ratio and species weight. Environ Toxicol Chem 20:1399–1420.

Heneghan PA, Biggs J, Jepson PC, Kedwards T, Maund SJ, Sherratt TN, Shillabeer N, Stickland TR, Williams P] Oregon State University, Department of Entomology. 1999. Pond-FX: ecotoxicology from pH to population recovery, 1st ed. http://www.ent.orst. edu/PondFX (accessed May 23, 2007).

Henriques W, Jeffers RD, Lacher TE, Jr, Kendall RJ. 1997. Agrochemical use of banana plantations in Latin America: Perspectives on ecological risk. Environ Toxicol Chem 16:91–99.

Hermens J, Canton H, Steyger N, Wegman R. 1984. Joint effects of a mixture of 14 chemicals on mortality and inhibition of reproduction of *Daphnia magna*. Aquat Toxicol 5:315–322.

Hermens JLM. 1991. (Q)SAR in environmental sciences and drug design. Sci Tot Environ 109–110:1–7.

Hermens JLM, Broekhuyzen E, Canton JH, Wegman R. 1985. (Q)SARs and mixture toxicity studies of alcohols and chlorohydrocarbons: Effects on growth of *Daphnia magna*. Aquat Toxicol 6:209–217.

Hertzberg RC, Teuschler LK. 2002. Evaluating quantitative formulas for dose–response assessment of chemical mixtures. Environ Hlth Perspect 110:965–970.

Heugens EHW, Jager T, Creyghton R, Kraak MHS, Hendriks AJ, van Straalen NM, Admiraal W. 2003. Temperature-dependent effects of cadmium on *Daphnia magna*: accumulation versus sensitivity. Environ Sci Technol 37:2145–2151.

Hickey CW, Clements WH. 1998. Effects of heavy metals on benthic macroinvertebrate communities in a New Zealand stream. Environ Toxicol Chem 17:2338–2346.

Hickey CW, Roper DS, Buckland S. 1995a. Metal concentrations of resident and transplanted freshwater mussels *Hyridella menziesi* (Unionacea: Hyriidae) and sediments in the Waikato River, New Zealand. Sci Tot Environ 175:163–177.

Hickey CW, Roper DS, Holland PT, Trower TM. 1995b. Accumulation of organic contaminants in two sediment dwelling shellfish with contrasting feeding modes: deposit- (*Macomona liliana*) and filter-feeding (*Austrovenus stutchburyi*). Arch Environ Contam Toxicol 29:221–231.

Hickie BE, McCarty LS, Dixon DG. 1995. A residue-based toxicokinetic model for pulse-exposure toxicity in aquatic systems. Environ Toxicol Chem 14:2187–2197.

Hill AB. 1965. The environment and disease: association or causation? Proc R Soc Med 58:295–300.

Hill IR, Heimbach F, Leeuwangh P, Matthiessen P, editors. 1994. Freshwater field tests for hazard assessment of chemicals. Boca Raton (FL): CRC Press. 561 p.

Ho KT, McKinney RA, Kuhn A, Pelletier MC, Burgess RM. 1997. Identification of acute toxicants in New Bedfordshire harbor sediments. Environ Toxicol Chem 16:551–558.

Hoare K, Beaumont AR, Davenport J. 1995. Variation among populations in the resistance of *Mytilus edulis* embryos to copper: adaptation to pollution? Mar Ecol Prog Ser 120:155–161.

Hodson PV, Maj MK, Efler S, Burnison BK, van Heiningen ARP, Girard R, Carey JH. 1997. MFO induction in fish by spent cooking liquors from kraft pulp mills. Environ Toxicol Chem 16:908–916.

Hoffmann AA, Parsons PA. 1991. Evolutionary genetics and environmental stress. Oxford (UK): Oxford Science Publishers, 296 p.

Hoffmann AA, Sørensen JG, Loeschcke V. 2003. Adaptation of *Drosophila* to temperature extremes: bringing together quantitative and molecular approaches. J Therm Biol 28:175–216.

Hommen U. 1998. Ökosystemmodelle von Modellökosystemen. Aachen (Germany): Shaker Verlag, 199 p.

Hooper MJ. 2003. Avian toxicity testing. In: Newman MC, Unger MA, editors. Fundamentals of ecotoxicology, 2nd ed. Boca Raton (FL): Lewis Publishers.

Hose GC, Lim RP, Hyne RV, Pablo F. 2002. A pulse of endosulfan contaminated sediment affects macroinvertebrates in artificial streams. Ecotoxicol Environ Safety 51:44–52.

Hose GC, van den Brink PJ. 2004. Confirming the species sensitivity distribution concept for endosulfan using laboratory, mesocosm and field data. Arch Environ Contam Toxicol 47:511–520.

Howard PH]. 1998. User's guide for the SMILES CAS program. http://esc.syrres.com/interkow/docsmile.htm (accessed July 7, 2003).

Hubbell SP. 2001. The unified neutral theory of biodiversity and biogeography. Princeton (NJ): Princeton University Press, 448 p.

Huggett RJ, Kimerle RA, Mehrle PM, Bergman H, editors. 1992. Biomarkers: biochemical, physiological and histological markers of anthropogenic stress. Boca Raton (FL): Lewis Publishers, 347 p.

Huijbregts MAJ, van de Meent D, Goedkoop M, Spriensma R. 2002. Ecotoxicological impacts in life cycle assessment. In: Posthuma L, Suter GW, Traas TP, editors. Species sensitivity distributions in ecotoxicology. Boca Raton (FL): Lewis Publishers, p 421–436.

Hurd MK, Perry S, Perry WB. 1996. Nontarget effects of a test application of diflubenzuron to the forest canopy on stream macroinvertebrates. Environ Toxicol Chem 15:1344–1351.

Hurlbert SH. 1975. Secondary effects of pesticides on aquatic ecosystems. Res Rev 57:81–148.

Huryn AD. 1996. Temperature-dependent growth and life cycle of *Deleatidium* (Ephemeroptera: Leptophlebiidae) in two high-country streams in New Zealand. Freshwater Biol 36:351–361.

Huston MA. 1997. Hidden treatments in ecological experiments: re-evaluating the ecosystem function of biodiversity. Oecologia 110:449–460.

Hutchinson T, Scholz N, Guhl W. 1998a. Analysis of the ECETOC Aquatic Toxicity (EAT) database, IV: comparative toxicity of chemical substances to freshwater versus saltwater organisms. Chemosphere 36:143–153.

Hutchinson T, Solbe J, Kloepper-Sams P. 1998b. Analysis of the ECETOC Aquatic Toxicity (EAT) database, III: comparative toxicity of chemical substances to different life stages of aquatic organisms. Chemosphere 36:129–142.

Hutchinson TH, Brown R, Brugger KE, Campbell PM, Holt M, Lange R, McCahon P, Tattersfield LJ, van Egmond R. 2000. Ecological risk assessment of endocrine disruptors. Environ Hlth Perspect 108:1107–1014.

HydroQual Inc. 2005. The Biotic Ligand Model and BLM-Monte. http://www.hydroqual. com/wr_blm.html (accessed February 15, 2006).

Hyne RV, Maher WA. 2003. Invertebrate biomarkers: links to toxicosis that predict population decline. Ecotoxicol Environ Safety 54:366–374.

International Programme on Chemical Safety [IPCS]. 2002. Global assessment of the state-of-the-science of endocrine disruptors. No. WHO/PCS/EDC/02.2. Geneva, Switzerland: International Programme on Chemical Safety of the World Health Organization, 180 p. http://www.who.int/pcs (accessed December 28, 2007).

Ivorra N. 2000. Metal induced succession in benthic diatom consortia [PhD thesis]. Amsterdam (The Netherlands): University of Amsterdam, 157 p.

Jager DT. 2003. Worming your way into bioavailability: Modelling the uptake of organic chemicals in earthworms. PhD thesis. Utrecht (The Netherlands): University of Utrecht, 218 p.

Jager T, Heugens EHW, Kooijmans SALM. 2006. Making sense of ecotoxicological test results: towards application of process-based models. Ecotoxicol 15:305–314.

Jagoe RH, Newman MC. 1997. Bootstrap estimation of community NOEC values. Ecotoxicol 6:293–306.

Jamil K. 2001. Bioindicators and biomarkers of environmental pollution and risk assessment. Plymouth (UK): Science Publishers, 204 p.

Janes N, Playle RC. 1995. Modeling silver binding to gills of rainbow trout (*Oncorhynchus mykiss*). Environ Toxicol Chem 14:1847–1858.

Janssen RPT, Peijnenburg WJGM, Posthuma L, van den Hoop MAGT. 1997. Equilibrium partitioning of heavy metals in Dutch field soils. I. Relationships between metal partition coefficients and soil characteristics. Environ Toxicol Chem 16:2470–2478.

Jaworska JS, Comber M, Auer C, van Leeuwen CJ. 2003. Summary of a workshop on regulatory acceptance of (Q)SARs for human health and environmental endpoints. Environ Hlth Perspect 111:1358–1361.

Jenkins DG, Buikema AL. 1990. Response of a winter plankton food web to simazine. Environ Toxicol Chem 9:693–705.

Jensen A, Forbes VE, Parker ED. 2001. Variation in cadmium uptake, feeding rate, and life-history effects in the gastropod *Potamopyrgus antipodarum*: linking toxicant effects on individuals to the population level. Environ Toxicol Chem 20:2503–2513.

Jensen KIN, Stephenson GR, Hunt LA. 1977. Detoxification of atrazine in three gramineae subfamilies. Weed Sci 25:212–220.

Jessen RJ. 1978. Statistical survey techniques. New York (NY): John Wiley.

Johnson AK. 2002. Landscape ecotoxicology and assessment of risks at multiple scales. Human Ecol Risk Assess 8:127–146.

Johnson AR, Rodgers JH. 2005. Scaling in ecotoxicology: theory, evidence and research needs. Aquat Ecosys Hlth Manage 8:353–362.

Johnson BT. 1986. Potential impact of selected agricultural chemical contaminants on a northern prairie wetland: a microcosm evaluation. Environ Toxicol Chem 5:473–485.

Johnson JC, Borgatti SP, Luczkovich LL, Everett MG. 2003. Network role analysis in the study of food webs: an application of regular role correlation. J Soc Struct 2. http://www.cmu.edu/joss/content/articles/volume2/JohnsonBorgatti.html (accessed December 28, 2007).

Jones CG, Lawton JH, Shachak M. 1994. Organisms as ecosystem engineers. Oikos 69:373–386.

Jones DS, Barnthouse LW, Suter GW, Efroymson RA, Field JM, Beauchamp JJ. 1999. Ecological risk assessment in a large river-reservoir: 3. Benthic invertebrates. Environ Toxicol Chem 18:599–609.

Jonker MJ. 2003. Joint toxic effects on *Caenorhabditis elegans* on the analysis and interpretation of mixture toxicity data [PhD thesis]. Wageningen (The Netherlands): Wageningen University and Research Centre.

Jorgensen SE. 2002. Explanation of ecological rules and observation by application of ecosystem theory and ecological models. Ecol Model 158:241–248.

Jorgensen SE, Verdonschot P, Lek S. 2002. Explanation of the observed structure of functional feeding groups of aquatic macro-invertebrates by an ecological model and the maximum exergy principle. Ecol Model 158:223–231.

Jurgensen TA, Hoagland KD. 1990. Effects of short-term pulses of atrazine on attached algal communities in a small stream. Arch Environ Contam Toxicol 19:617–623.

Jüttner I, Peither A, Lay JP, Kettrup A, Ormerod SJ. 1995. An outdoor mesocosm study to assess ecotoxicological effects of atrazine on a natural plankton community. Arch Environ Contam Toxicol 29:435–441.

Kaiser KLE, Dearden C, Klein W, Schultz TW. 1999. A note of caution to users of ECOSAR. Water Qual Res J Canada 34:179–182.

Kallander DB, Fisher SW, Lydy MJ. 1997. Recovery following pulsed exposure to organophosphorus and carbamate insecticides in the midge, *Chironomus riparius*. Arch Environ Contam Toxicol 33:29–33.

Karickhoff SW, Brown DS. 1979. Determination of octanol/water distribution coefficients, water solubilities, and sediment/water partition coefficients for hydrophobic organic pollutants. No. EPA-660/4-79/032. Athens (GA): US Environmental Protection Agency.

Kasai F, Hanazato T. 1995a. Effects of the triazine herbicide, simetryn, on freshwater plankton communities in experimental ponds. Environ Pollut 19:197–202.

Kasai F, Hanazato T. 1995b. Genetic changes in phytoplankton communities exposed to the herbicide simetryn in outdoor experimental ponds. Arch Environ Contam Toxicol 28:154–160.

Kaushik NK, Stephenson GL, Solomon KR, Day KE. 1985. Impact of permethrin on zooplankton communities using limnocorrals. Canadian Journal of Fisheries and Aquatic Science 42:77–85.

Kedwards TJ, Maund SJ, Chapman PF. 1999. Community level analysis of ecotoxicological field studies: II. Replicated-design studies. Environ Toxicol Chem 18:158–171.

Kelley RA, Spofford WO, Jr. 1977. Application of an ecosystem model to water quality management: The Delaware estuary. In: Hall CAS, Day JW, editors. Ecosystem modeling in theory and practice. New York (NY): John Wiley, 707 p.

Kenaga EE, Goring CAI. 1980. Relationship between water solubility, soil sorbtion, octanol-water partitioning, and concentration of chemicals in biota. In: Eaton JC, Parrish PR, Hendricks A, editors. Aquat toxicol. Volume ASTM STP 707. Philadelphia (PA): American Society for Testing and Materials, p 78–115.

Kennedy JH, Ammann LP, Aller WT, Warren JE, Hosmer AJ, Cairns SH, Johnson PC, Graney RL. 1999. Using statistical power to optimize sensitivity of analysis of variance for microcosms and mesocosms. Environ Toxicol Chem 18:113–117.

Kersting K. 1994. Functional endpoints in field testing. In: Matthiessen P, editor. Freshwater field tests for hazard assessment of chemicals. Boca Raton (FL): CRC Press, p 57–81.

Kersting K, van den Brink PJ. 1997. Effects of the insecticide Dursban®4E (active ingredient chlorpyrifos) in outdoor experimental ditches: III Responses of ecosystem metabolism. Environ Toxicol Chem 16:251–259.

Kersting K, van Wijngaarden RPA. 1999. Effects of pulsed treatment with the herbicide afalon (active ingredient linuron) on macrophyte dominated mesocosms I. Resonses of ecosysem metabolism. Environ Toxicol Chem 18:2859–2865.

Kiffney PM, Clements WH. 1996. Size-dependent response of macroinvertebrates to metals in experimental streams. Environ Toxicol Chem 15:1352–1356.

Kindig AC, Conquest LL, Taub FB. 1983. Differential sensitivity of new versus mature synthetic microcosms to streptomycin sulfate treatment. In: Bishop WE, Cardwell RD, Heidolph BB, editors. Aquatic toxicology and hazard assessment: sixth symposium. Volume ASTM STP 802. Philadelphia, (PA): American Society for Testing and Materials, p 192–203.

Lavelle P, Bignell D, Lepage M, Wolters V, Roger P, Ineson P, Heal OW, Dhillion S. 1997. Soil function in a changing world: the role of invertebrate ecosystem engineers. Europ J Soil Biol 33:159–193.

Lawton JH. 1994. What do species do in ecosystems? Oikos 71:367–374.

LeBlanc GA. 1984. Interspecies relationships in acute toxicity of chemical to aquatic organisms. Environ Toxicol Chem 3:47–60.

Lee G, Ellersieck MR, Mayer FL, Krause GF. 1995. Predicting chronic lethality of chemicals to fishes from acute toxicity test data: multifactor probit analysis. Environ Toxicol Chem 14:345–349.

Lee JH, Landrum CH, Koh CH. 2002. Prediction of time-dependent PAH toxicity in *Hyalella azteca* using a damage assessment model. Environ Sci Technol 36:3131–3138.

Leistra M, Zweers AJ, Warinton JS, Crum SJH, Hand LH, Beltman WHJ, Maund SJ. 2003. Fate of the insecticide lambda-cyhalothrin in ditch enclosures differing in vegetation density. Pest Manag Sci 60:75–84.

Lemly AD. 2002. Selenium assessment in aquatic ecosystems: a guide for hazard evaluation and water quality criteria. New York (NY): Springer-Verlag, 161 p.

Leonard AW, Hyne RV, Lim RP, Chapman JC. 1999. Effects of endosulfan runoff from cotton fields on macroinvertebrates in the Namoi River. Ecotoxicol Environ Safety 42:125–134.

Leonard AW, Hyne RV, Lim RP, Leigh KA, Le J, Beckett R. 2001. Fate and toxicity of endosulfan in Namoi River water and bottom sediment. J Environ Qual 30:750–759.

Leonard AW, Hyne RV, Lim RP, Pablo F, van den Brink PJ. 2000. Riverine endosulfan concentrations in the Namoi River, Australia: link to cotton field runoff and macroinvertebrate population densities. Environ Toxicol Chem 19:1540–1551.

Leung KMY, Morritt D, Wheeler JR, Whitehouse P, Sorokin N, Toy R, Holt R, Crane M. 2002. Can saltwater toxicity be predicted from freshwater data? Mar Poll Bull 42:1007–1013.

Leuven RSEW, Poudevigne I. 2002. Riverine landscape dynamics and ecological risk assessment. Freshwater Biol 47:845–865.

Liess M. 1998. Significance of agricultural pesticides on stream macroinvertebrate communities. Verh Internat Verein Limnol 26:1245–1249.

Liess M, Schulz R. 1996. Chronic effects of short-term contamination with the pyrethroid insecticide on the caddisfly *Limnephilus lunatus*. Hydrobiologia 324:99–106.

Liess M, von der Ohe PC. 2005. Analyzing effects of pesticides on invertebrate communities in streams. Environ Toxicol Chem 24:954–965.

Lincoln RJ, Boxshall GA, Clark PF. 1983. A dictionary of ecology, evolution and systematics. Cambridge (UK): Cambridge University Press.

Linders J, Adriaanse P, Allen R, Capri E, Gouy V, Hollis J, Jarvis N, Klein M, Lolos P, Maier W-M, Maund S, Pais C, Russell M, Smeets L, Teixeira J-L, Vizantinopoulos S, Yon D. 2002. FOCUS surface water scenarios in the EU evaluation process under 91/414/EEC. No. SANCO/4802/2001-rev.1. Brussels (Belgium): FOCUS Working Group on Surface Water Scenarios, 228 p.

Ling N, Hickey CW, Burton GA. 1998. Are Antarctic marine organisms sensitive to anthropogenic pollutants? NZ Nat Sci 23:106.

Linke-Gamenick I, Forbes VE, Sibly RM. 1999. Density-dependent effects of a toxicant on life-history traits and population dynamics of a capitellid polychaete. Mar Ecol Prog Ser 184:139–148.

Linke-Gamenick I, Vismann B, Forbes VE. 2000. Effects of fluoranthene and ambient oxygen levels on survival and metabolism in three sibling species of *Capitella* (Polychaeta). Mar Ecol Prog Ser 194:169–177.

Linkov I, Burmistrov D, Cura J, Bridges TS. 2002. Risk-based management of contaminated sediments: considerations of spatial and temporal patterns in exposure modeling. Environ Sci Technol 36:238–246.

Lipnick RL. 1993. Baseline toxicity (Q)SAR models: a means to assess mechanism of toxicity for aquatic organisms and mammals. In: La Point TW, editor. Environmental toxicology and risk assessment STP 1216. Volume 2. Philadelphia (PA): American Society for Testing and Materials, p 610–619.

Little EE, Dwyer FJ, Fairchild JF, De Lonay AJ, Zajicek JL. 1993. Survival of bluegill and their behavioral responses during continuous and pulsed exposures to esfenvalerate, a pyrethroid insecticide. Environ Toxicol Chem 12:871–879.

Lloyd R. 1961. The toxicity of mixtures of zinc and copper sulfates to rainbow trout (*Salmo gairdneri* Richardson). Ann Appl Biol 49:535–538.

Loewe S, Muischnek H. 1926a. Über Kombinationswirkungen. Arch Exp Pathol Pharmak 114:313–326.

Loewe S, Muischnek H. 1926b. Über Kombinationswirkungen. 1. Mitteilung: Hilfsmittel der Fragestellung. Archiv für experimentelle Pathologie und Pharmakologie, p 313–326.

Lowe-McConnell R. 1994. The changing ecosystem of Lake Victoria, East Africa. Freshwater Forum 4:76–89.

Ludovisi A, Poletti A. 2003. Use of thermodynamic indices as ecological indicators of the development state of lake ecosystems 2. Exergy and specific exergy indices. Ecol Model 159:223–238.

Luoma SN. 1977. Detecting of trace contaminant effects in aquatic ecosystems. J Fish Res Board Can 34:436–439.

Luoma SN. 1983. Bioavailability of trace metals to aquatic organisms: a review. Sci Tot Environ 28:1–22.

Luoma SN, Fisher NS. 1997. Uncertainties in assessing contaminant exposure from sediments: bioavailabililty. In: Biddinger G, Dillon T, editors. Ecological risk assessment of contaminated sediments. Pensacola (FL): SETAC Press.

Mackay CE, Pastorok RA. 2002. Landscape models: aquatic and terrestrial. In: Pastorok RA, Bartell SM, Ferson S, Ginzburg LR, editors. Ecological modeling in risk assessment. Boca Raton (FL): Lewis Publishers, p 149–180.

Mackay D, Wan-Ying S, Kuo-Ching M. 1997. Illustrated handbook of physical-chemical properties and environmental fate of organic chemicals. Boca Raton (FL): Lewis Publishers, 704 p.

MacRae RK. 1994. The copper binding affinity of rainbow trout (*Oncorhynchus mykiss*) and brook trout (*Salvelinus fontinalis*) gills [MSc thesis]. Laramie University of Wyoming, 51 p.

Maise S. 2001. Natural variability of zooplankton and phytoplankton in outdoor aquatic microcosms [PhD thesis]. Weihenstephan (Germany): Technical University of München, 199 p.

Maltby L, Blake N, Brock TCM, van den Brink PJ. 2002. Addressing interspecific variation in sensitivity and the potential to reduce this source of uncertainty in ecotoxicological assessments. No. DEFRA project code PN0932. London (UK): UK Department for Environment, Food and Rural Affairs, 22 p.

Maltby L, Blake NN, Brock TCM, van den Brink PJ. 2005. Insecticide species sensitivity distributions: the importance of test species selection and relevance to aquatic ecosystems. Environ Toxicol Chem 24:379–388.

Maltby L, Kedwards TJ, Forbes VE, Grasman K, Kammenga JE, Munns WR, Ringwood AH, Weis JS, Wood SN. 2001. Linking individual-level responses and population-level responses. In: Baird DJ, Burton GA, editors. Ecological variability: separating natural from anthropogenic causes of ecosystem impairment. Pensacola (FL): SETAC Press, p 27–82.

Mancini JL. 1983. A method for calculating effects, on aquatic organisms, of time-varying concentrations. Water Res 10:1355–1362.

[Mangels G] Waterborne Environmental, Inc. 2001. The development of MUSCRAT (multiple scenario risk assessment tool): a software tool for conducting surface water exposure assessments. http://www.waterborne-env.com/modeling/model_download_muscrat.html (accessed January 2, 2007).

Markert BA, Breure AM, Zechmeister HG. 2003. Bioindicators and biomonitors: principles, concepts and applications. Oxford (UK): Elsevier, 997 p.

Maruya KA, Lee RF. 1998. Biota-sediment accumulation and trophic transfer factors for extremely hydrophobic polychlorinated biphenyls. Environ Toxicol Chem 17:2463–2469.

Matthews RA, Landis WG, Matthews GB. 1996. The community conditioning hypothesis and its application to environmental toxicology. Environ Toxicol Chem 15:597–603.

Mauer BA, Holt RD. 1996. Effects of chronic pesticide stress on wildlife populations in complex landscapes: processes at multiple scales. Environ Toxicol Chem 15:420–426.

Maund SJ, Hamer MJ, Warinton JS, Kedwards TJ. 1998. Aquatic ecotoxicology of the pyrethroid insecticide lambda-cyhalothrin: considerations for higher-tier aquatic risk assessment. Pestic Sci 54:408–417.

May RM. 1974a. Biological populations with nonoverlapping generations: stable points, stable cycles and chaos. Science 186:645–647.

May RM. 1974b. Stability and complexity in model ecosystems. Princeton (NJ): Princeton Universiy Press.

Mayer FL, Bengtsson BE, Hamilton SJ. 1988. Effects of pulp mill and ore smelter effluents on vertebrae of fourhorn sculpin: laboratory and field comparisons. In: Adams WJ, Chapman GA, Landis WG, editors. Aquatic toxicology and hazard assessment. Volume 971. Philadelphia (PA): American Society for Testing and Materials, p 406–419.

Mayer FL, Ellersieck MR, Krause GP, Sun K, Lee G, Buckler DR. 2001. Time-concentration-effect models in predicting chronic toxicity from acute toxicity data. In: Crane M, Newman MC, Chapman PF, Fenlon J, editors. Risk assessment with time to event models. Boca Raton (FL): Lewis Publishers, p 39–67.

Mayer FL, Sun K, Lee G, Ellersieck MR, Krause GF. 1999. User guide: acute to chronic estimation (with software). No. EPA/600/R-98/152. Washington (DC): US Environmental Protection Agency, 18 p.

Mayer FL, Jr, Ellersieck ME. 1986. Manual of acute toxicity: interpretation and data base for 410 chemicals and 66 species of freshwater animals. No. 160. Washington (DC): U.S. Fish and Wildlife Service. 560 p.

Mayer FL, Jr, Krause GF, Buckler RD, Ellersieck MR, Lee G. 1994. Predicting chronic lethality of chemicals to fishes from acute toxicity test data: concepts and linear regression analysis. Environ Toxicol Chem 13:671–678.

Mazzo N, Dardano B, Marticorena A. 1998. Interclonal variation in response to simazine stress in *Lemna gibba* (Lemnaceae). Ecotoxicol 7:151–160.

McCahon CP, Pascoe D. 1988. Use of *Gammarus pulex* (L.) in safety evaluation tests: culture and selection of a sensitive life stage. Ecotoxicol Environ Safety 15:245–252.

McCahon CP, Whiles AJ, Pascoe D. 1989. The toxicity of cadmium to differenrt larval instars of the trichopteran larvae *Agapetus fuscipes* Curtis and the importance of lifecycle information to the design of toxicity tests. Hydrobiologia 185:153–162.

McCarthy JF, Shugart LR. 1992. Biomarkers of environmental contamination. Boca Raton (FL): Lewis Publishers, 457 p.

McCarty LS, Mackay D. 1993. Enhancing ecotoxicological modeling and assessment. Environ Sci Technol 27:1719–1728.

McCarty LS, Mackay D, Smith AD, Ozburn GW, Dixon GD. 1992. Residue-based interpretation of toxicity and bioconcentration (Q)SARs from aquatic bioassays: neutral narcotic organic. Environ Toxicol Chem 11:917–930.

McCarty LS, Munkittrick KR. 1996. Environmental biomarkers in aquatic toxicology: fiction, fantasy, or functional? Human Ecol Risk Assess 2:268–274.

McCarty LS, Power M, Munkittrick KR. 2002. Bioindicators versus biomarkers in ecological risk assessment. Human Ecol Risk Assess 8:159–164.

McCauley E, Murdoch WW, Nisbet RM, Gurney WSC. 1990. The physiological ecology of Daphnia: development of a model of growth and reproduction. Ecol 71:1425–1459.

Meier PG, Choi K, Sweet L. 2000. Acute and chronic life cycle toxicity of acenaphtene and 2,4,6-trichlorophenol to the midge *Paranytarsus parthenogeneticus* (Diptera: Chironomidae). Aquat Toxicol 51:31–44.

Melancon MJ. 2003. Bioindicators of contaminant exposure and effect in aquatic and terrestrial monitoring. In: Hoffman DJ, Rattner BA, Burton GA, Cairns J, Jr, editors. Handbook of ecotoxicology. Boca Raton (FL): Lewis Publishers, p 257–278.

Menzies C, Henning MH, Cura J, Finkelstein K, Gentille J, Maughan J, Mitchell D, Petron S, Potocki B, Svirsky S, Tyler P. 1996. A weight-of-evidence approach for evaluating ecological risks: report of the Massachusetts Weight-of-Evidence Workgroup. Human Ecol Risk Assess 2:277–304.

Mesman M, Posthuma L. 2003. Ecotoxicity of toxicant mixtures in soils: recommendations for application in the Dutch regulatory context, as derived from a scientific review on approaches, models and data. No. 711701035. Bilthoven (The Netherlands): National Institute of Public Health and the Environment (RIVM), 70 p.

Meyer JS, Gulley DD, Goodrich MS, Szmania DC, Brooks AS. 1995. Modelling toxicity due to intermittent exposure of rainbow trout and common shiners to monochloroamine. Environ Toxicol Chem 14:165–175.

Meyland WM, Howard PH. 1998. User's guide for the ECOSAR class program. North Syracuse (NY): USEPA.

Millward RN, Klerks PL. 2002. Contaminant-adaptation and community tolerance in ecological risk assessment: introduction. Human Ecol Risk Assess 8:921–932.

Mineau P. 1991. Cholinesterase-inhibiting insecticides: their impact on wildlife and the environment. In: Welling W, editor. Chemicals in agriculture. Volume 2. Amsterdam (The Netherlands): Elsevier Science, p 348.

Moffet MF, Anderson LE, Corry T, Hanratty MP, Heinis LJ, Knuth ML, Liber K, O'Halloran SL, Schmude KL, Stay FS, Tanner DK. 1995. Effects, persistence and distribution of diflubenzuron in littoral enclosures. Duluth, (MN): USEPA, Mid-Continent Ecology Division.

Mol JH, Ramlal JS, Lietar C, Verloo M. 2001. Mercury contamination in freshwater, estuarine and marine fishes in relation to small-scale gold mining in Suriname, South America. Environ Res 86:183–197.

Moore DRJ, Breton RL, MacDonald DB. 2003. A comparison of model performance for six quantitative structure-activity relationship packages that predict acute toxicity to fish. Environ Toxicol Chem 22:1799–1809.

Moore DW, Schluchter MD, Scott GI. 1990. Use of hazard models in evaluating the effect of exposure duration on the acute toxicity of three pesticides. In: Landis WG, Van der Schalie H, editors. Aquatic toxicology and risk assessment. Volume ASTM STP 1096. Philadelphia (PA): American Society for Testing and Materials, p 247–263.

Moorhead DL, Kosinski RJ. 1986. Effects of atrazine on the productivity of artificial stream algal communities. Bull Environ Contam Toxicol 37:330–336.

Morel FM. 1983. Principles of aquatic chemistry. New York (NY): Wiley Interscience, 588 p.

Mount DI, Anderson-Carnahan L. 1989. Methods for aquatic toxicity identification evaluations: phase II toxicity identification procedures. No. EPA/600/3-88-035. Duluth (MN): US Environmental Protection Agency, 32 p.

Mount DI, Norberg-King TJ, Ankley GT, Burkhard LP, Durhan EJ, Schubauer-Berigan MK, Lukasewycz M. 1993. Methods for aquatic toxicity identification evaluations: phase III toxicity confirmation procedures for samples exhibiting acute and chronic toxicity. No. EPA/600/R-92/081. Duluth (MN): US Environmental Protection Agency, 32 p.

Mulder C, Aldenberg T, De Zwart D, van Wijnen HJ, Breure AM. 2004. Evaluating the impact of pollution on plant-lepidoptera relationships. Environmetrics 16:357–373.

Mullins JA, Carsel RF, Scarbrough JE, Ivery AM. 1993. PRZM-2: a model for predicting pesticide fate in the crop root zone and unsaturated soil zones: program and user's manual for release 2.0. No. EPA/600/R-93/046. Athens (GA): US Environmental Protection Agency.

Mulvey M, Diamond SA. 1991. Genetic factors and tolerance acquisition in populations exposed to metals and metalloids. In: Newman MC, McIntosh AW, editors. Metal ecotoxicology: concepts and applications. Chelsea (MI): Lewis Publishers, p 301–321.

Munkittrick KR, McCarthy LH, Servos MR, Van Der Kraak GJ. 1998. An overview of recent studies on the potential of pulp-mill effluents to alter reproductive parameters in fish. J Toxicol Environ Hlth B 1:347–371.

Munkittrick KR, Van der Kraak GJ, McMaster ME, Portt CB. 1992. Response of hepatic MFO activity and plasma sex steroids to secondary treatment of bleached kraft mill effluent and mill shutdown. Environ Toxicol Chem 11:1427–1439.

Munkittrick KR, Van der Kraak GJ, McMaster ME, Portt CB, van den Heuvel MR, Servos MR. 1994. Survey of receiving water environmental impacts associated with discharges from pulp mills. II gonad size, liver size, hepatic EROD activity and plasma sex steroid levels in white sucker. Environ Toxicol Chem 13:1089–1102.

Munro IC, Kennepohl E, Kroes R. 1999. A procedure for the safety evaluation of flavouring substances. Food Chem Toxicol 37:207–232.

Murray K, Linder P. 1983. Fulvic acids: structure and metal binding. I. A random molecular model. J Soil Sci 34:511–523.

Murty A. 1986. Toxicity of pesticides to fish: joint action of pesticide mixtures. Boca Raton (FL): CRC Press.

Muschal M, Warne MS. 2003. Risk posed by pesticides to aquatic organisms in rivers or northern inland New South Wales, Australia. Human Ecol Risk Assess 9:1765–1787.

Myers N, Mittermeier RA, Mittermeier CG, de Fonseca GAB, Kent J. 2000. Biodiversity hotspots for conservation priorities. Nature 403:853–858.

Nabholz JV, Clements RG, Zeeman MG. 1997. Information needs for risk assessment in EPA's Office of Pollution Prevention and Toxics. Ecol Appl 7:1094–1098.

Nabholz JV, Clements RG, Zeeman MG, Osborn KC, Wedge R. 1993. Validation of structure activity relationships used by the USEPA's Office of Pollution Prevention and Toxics for the environmental hazard assessment of industrial chemicals. In: Gorsuch JW, Dwyer FJ, Ingersoll CG, LaPoint TW, editors. Environmental toxicology and risk assessment. Volume 2. Philadelphia (PA): American Society for Testing and Materials, p 571–591.

Nabholz, JV. Personal communication, February, 2002.

Naddy RB, Klaine SJ. 2001. Effect of pulse frequency and interval on the toxicity of chlorpyrifos to *Daphnia magna*. Chemosphere 45:497–506.

Naeem S. 1998. Species redundancy and ecosystem reliability. Conserv Biol 12:39–45.

Naiman RJ, Turner MG. 2000. A future perspective on North America's freshwater ecosystems. Ecol Appl 10:958–970.

Naito W, Miyamoto K, Nakanishi J, Masunaga S, Bartell SM. 2003. Evaluation of an ecosystem model in ecological risk assessment of chemicals. Chemosphere 53:363–375.

Napier GM. 1992. Application of laboratory-drived data to natural aquatic ecosystems [PhD thesis]. Sydney, Australia: Macquarie University.

National Research Council of Canada [NRCC]. 1987. Pyrethroids: their effects on aquatic and terrestrial ecosystems. No. 24376. Ottawa (Canada): National Research Council of Canada, Associate Committee on Scientific Criteria for Environmental Quality, 248 p.

Nestler JM, Goodwin RA. 2000. Simulating population dynamics in an ecosystem context using Eulerian-Langrarian hybrid models (CEL HYBRID models). No. ERDC/EL TR-00-4. Washington (DC): U.S. Army Corps of Engineers, 45 p. http://el.erdc. usace.army.mil/elpubs/pdf/trel00-4.pdf (accessed December 28, 2007).

Newman MC. 1995. Quantitative methods in aquatic ecotoxicology. Boca Raton (FL): Lewis Publishers, 426 p.

Newman MC. 1996. Ecotoxicology as a science. In: Newman MC, Jagoe CH, editors. Ecotoxicology: a hierarchical treatment. Boca Raton (FL): Lewis Publishers, p 1–9.

Newman MC. 2001. Population ecotoxicology. New York (NY): Wiley, 228 p.

Newman MC, Owenby DR, Mézin LCA, Powell DC, Christensen TRL, Lerberg SB, Anderson B-A. 2000. Applying species-sensitivity distributions in ecological risk assessment: assumptions of distribution type and sufficient numbers of species. Environ Toxicol Chem 19:508–515.

Newman MC, Unger MA, editors. 2003. Fundamentals of ecotoxicology. 2nd ed. Boca Raton (FL): Lewis Publishers, 402 p.

Niederlehner RB, Cairns J, Jr, Smith EP. 1998. Modeling acute and chronic toxicity of nonpolar narcotic chemicals and mixtures to *Ceriodaphnia dubia*. Ecotoxicol Environ Safety 39:136–146.

Nielsen SN. 1992. Application of maximum exergy in structural dynamic models [PhD thesis]. Roskilde (Denmark): Roskilde University.

Niemi GJ, DeVore P, Detenbeck N, Taylor D, Lima A, Pastor J, Yount JD, Naiman RJ. 1990. Overview of case studies on recovery of aquatic systems from disturbance. Environ Managment 14:571–587.

Nisbet RM, Muller EB, Lika K, Kooijman SALM. 2000. From molecules to ecosystems through dynamics energy budget models. J Anim Ecol 69:913–926.

Norberg-King TJ, Mount DI, Armato JR, Jersen DA, Thompson JA. 1991. Methods for aquatic toxicity identification evaluation: phase I toxicity characterization procedures. 2nd ed. Duluth (MN): US Environmental Protection Agency. No. EPA/600/6-91/003. http://www.epa.gov/cgi-bin/claritgw?op-Display&document=clserv:ORD:0235;&rank=4&template=epa (accessed December 28, 2007).

Norris RH, Georges A. 1993. Analysis and interpretation of benthic macroinvertebrate surveys. In: Rosenberg DM, Resh VH, editors. Freshwater biomonitoring and benthic macroinvertebrates. New York (NY): Chapman & Hall, p 234–286.

Norwood WP, Borgmann U, Dixon DG, Wallace A. 2003. Effects of metals mixtures on aquatic biota: A review of observations and methods. Human Ecol Risk Assess 9:795–811.

Nyström B, Paulsson M, Almgren K, Blank H. 2000. Evaluation of the capacity for development of atrazine tolerance in periphyton from a Swedish freshwater site as determined by inhibition of photosynthesis and sulfolipid synthesis. Environ Toxicol Chem 19:1324–1331.

Oberdorff T, Hugueny B, Guégan J-F. 1997. Is there an influence of historical events on contemporary fish species richness in rivers? Comparisons between Western Europe and North America. J Biogeogr 24:461–467.

Odum EP. 1971. Fundamentals of ecology. Philadelphia (PA): W. B. Saunders, 574 p.

Odum EP. 1981. The effects of stress on the trajectory of ecological succession. In: Barrett GW, Rosenberg R, editors. Stress effects on natural ecosystems. New York (NY): John Wiley.

Odum EP. 1992. Great ideas in ecology for the 1990s. Bioscience 42:532–545.

Ohio EPA. 1996. Dissolved metals criteria. Great Lakes Initiative Issue Paper. Colombus (OH): Ohio Environmental Pollution Agency.

Okkerman PC, van de Plassche EJ, Emans HJB, Canton JH. 1993. Validation of some extrapolation methods with toxicity data derived from multiple species experiments. Ecotoxicol Environ Safety 25:341–359.

Olson DM, Dinerstein E. 1998. The Global 200: A representation approach to conserving the earth's most biologically valuable ecoregions. Conserv Biol 12:502–515.

O'Neill RV, DeAngelis DL, Waide JB, Allen TFH. 1986. A hierarchical concept of the ecosystem. Princeton (NJ): Princeton University Press. 253 p.

Organisation for Economic Co-Operation and Development [OECD]. 1992. Report of the OECD workshop on the extrapolation of laboratory aquatic toxicity data to the real environment. Environmental Monograph. No. 59. OCDE/GD (92)169. Paris (France): Organisation for Economic Co-Operation and Development

Organisation for Economic Co-Operation and Development [OECD]. 1994. USEPA/EC joint project on the evaluation of (quantitative) structure activity relationships. No. OECD/GD(94)28. Paris (France): Organisation for Economic Co-Operation and Development, 368 p.

Organisation for Economic Co-Operation and Development [OECD]. 2001. Guidance document on the use of the harmonized system for the classification of chemicals which are hazardous for the aquatic environment. OECD Series on Testing and Assessment. No. 27; ENV/JM/MONO. Paris (France): Organization of Economic Cooperation and Development, 8 p.

Organisation for Economic Co-Operation and Development [OECD]. 2007. Manual for investigation of HPV chemicals, chapter 4: initial assessment of data. Paris (France): Organisation for Economic Co-Operation and Development. http://www.oecd.org/document/7/0,2340,en_2649_34379_1947463_1_1_1_1,00.html.

Organisation for Economic Co-Operation and Development [OECD]. 2003. Ad-hoc expert group on (Q)SARs: summary conclusions of the 1st meeting and draft work plan. No. ENV/JM. Paris (France): Organization of Economic Cooperation and Development, 18 p.

Organisation for Economic Co-operation and Development [OECD]. 2005a. OECD principles for the validation, for regulatory purposes, of (quantitative) structure-activity relationship models. http://www.oecd.org/document/23/0,2340,en_2649_34365_33957015_1_1_1_1,00.html (accessed December 15, 2005).

Organisation for Economic Co-operation and Development [OECD]. 2005b. OECD's database on chemical risk assessment models. http://webdomino1.oecd.org/comnet/env/models.nsf (accessed December 15, 2005).

Organisation for Economic Co-operation and Development [OECD]. N.d.-a. Description of OECD work on investigation of high production volume chemicals. http://www.oecd.org/document/21/0,2340,en_2649_34379_1939669_1_1_1_1,00.html (accessed January 8, 2008).

Organisation for Economic Co-operation and Development [OECD]. N.d.-b. OECD quantitative structure-activity relationships [(Q)SARs] project. http://www.oecd.org/document/23/0,2340,en_2649_34365_33957015_1_1_1_1,00.html (accessed January 8, 2008).

Paasivirta J. 1991. Chemical ecotoxicology. Boca Raton (FL): Lewis Publishers, 210 p.

Pagenkopf GK. 1983. Gill surface interaction model for trace-metal toxicity to fishes: Role of complexation, pH and water hardness. Environ Sci Technol 17:342–347.

Pandovani L, Capri E, Trevisan M. 2004. Landscape-level approach to assess aquatic exposure via spray drift for pesticides: a case study in a Mediterranean area. Environ Sci Technol 38:3239–3246.

Paquin PR, Santore RC, Farley K, Di Toro DM, Wu KB, Mooney KG, Winfield RP. 2003. Metals in aquatic systems: a review of exposure, bioaccumulation, and toxicity models. Pensacola (FL): SETAC Press, 168 p.

[Park RA] US Environmental Protection Agency. 1999. AQUATOX for Windows: a modular toxic effects model for aquatic ecosystems [computer program]. Washington (DC): US Environmental Protection Agency. http://www.epa.gov/ost/models/aquatox/ (accessed December 28, 2007).

Parker R Environmental Fate and Effects Division, Office of Pesticide Programs, USEPA. 1999. GENEEC [computer program]. Version 1.3. Washington, DC: US Environmental Protection Agency.

Parkhurst BR, Warren-Hicks WJ, Cardwell RD, Volosin JS, Etchison T, Butcher JB, Covington SM. 1996. Aquatic ecological risk assessment: a multi-tiered approach to risk assessment. No. 91-AER-1. Alexandria (VA): Water Environment Research Foundation.

Parkhurst DL, Thorstenson D, Plummer L. 1990. PHREEQE: a computer program for geochemical calculations. No. USGS/WRI-80-96. Reston (VA): US Geological Survey, 195 p. http://water.usgs.gov/software/phreeqe.html (accessed December 28, 2007).

Parrott JL, Chong-Kit R, Rokosh DA. 1999. EROD induction in fish: a tool to measure environmental exposure in impact assessment of hazardous aquatic contaminants. In: Rao SS, editor. Impact assessment of hazardous aquatic contaminants: concepts and approaches. Boca Raton (FL): Lewis Publishers, p 99–117.

Parrott JL, Hodson PV, Servos MR, Huestis SL, Dixon DG. 1995. Relative potencies of polychlorinated dibenzo-p-dioxins and dibenzofurans for inducing mixed function oxygenase activity in rainbow trout. Environ Toxicol Chem 14:1041–1050.

Pastorok RA, Akçakaya HR, Regan H, Ferson S, Bartell SM. 2003. Role of ecological modelling in risk assessment. Human Ecol Risk Assess 9:939–972.

Pastorok RA, Bartell SM, Ferson S, Ginzburg LR, editors. 2002. Ecological modeling in risk assessment. Boca Raton (FL): Lewis Publishers, 302 p.

Patra PWR. 2000. Effects of temperature on the toxicity of chemicals to Australian fish and invertebrates [PhD thesis]. Sydney, Australia: University of Technology.

Peakall D, editor. 1992. Animal biomarkers as pollution indicators. London (UK): Chapman & Hall, 302 p.

Perez KT, Morrison GE, Davey EW, Lackie NF, Soper AE, Blasco RJ, Winslow DL, Johnson RL, Murphy PG, Heltshe JF. 1991. Influence of size on fate and ecological effects of Kepone in physical models. Ecol Appl 1:237–248.

PERPEST. N.d. PERPEST. http://www.perpest.alterra.nl (accessed January 7, 2008).

Perry JN, Smith RH, Woiwod IP, Morse DR, editors. 2000. Chaos in real data: the analysis of non-linear dynamics from short ecological time series. Dordrecht (The Netherlands): Kluwer, 236 p.

Persoone G, Janssen CR. 1994. Field validation of predictions based on laboratory toxicity tests. In: Hill IR, Heimbach F, Leeuwangh P, Matthiessen P, editors. Freshwater field tests for hazard assessment of chemicals. Boca Raton (FL): CRC Press, p 379–397.

Peters EC, Gassman NJ, Firman JC, Richmond RH, Power EA. 1997. Ecotoxicology of tropical marine ecosystems. Environ Toxicol Chem 16:12–40.

Petersen JE, Hastings A. 2001. Dimensional approaches to scaling experimental ecosystems: designing mousetraps to catch elephants. Am Nat 157:324–333.

Peterson BJ, Wolheim WM, Mulholland PJ, Webster JR, Meyer JL, Tank JL, Marti E, Bowden WB, Vallett HM, Hershey AE, McDowell WH, Dodds WK, Hamilton SK, Gregory S, Morrall DD. 2001. Control of nitrogen export from watersheds by headwater streams. Science 292:86–90.

Plackett RL, Hewlett PS. 1952. Quantal responses to mixtures of poisons. Roy Stat Soc B 14:141–163.

Playle RC, Dixon DG, Burnison K. 1993a. Copper and cadmium binding to fish gills: estimates of metal-gill stability constants and modelling of metal accumulation. Can J Fish Aquat Sci 50:2678–2687.

Playle RC, Dixon DG, Burnison K. 1993b. Copper and Cadmium binding to fish gills: modification by dissolved organic carbon and synthetic ligands. Can J Fish Aquat Sci 50:2667–2677.

Playle RC, Gensemer RW, Dixon DG. 1992. Copper accumulation on gills of fathead minnows: influence of water hardness, complexation and pH on the gill microenvironment. Environ Toxicol Chem 11:381–391.

Poff NL, Olden JD, Vieira NKM, Finn DS, Simmons MP, Kondratieff BC. 2006. Functional trait niches of North American lotic insects: trait-based ecological applications in light of phylogenetic relationships. J NA Benthol Soc 25:730–755.

PondFX. N.d. PondFX. http://www.ent.orst.edu/PondFX (accessed January 7, 2008).

Popper KR. 1959. The logic of scientific discovery. New York (NY): Hutchinson.

Popper KR. 1979. Truth, rationality and the growth of scientific knowledge. Frankfurt am Main: V. Klostermann, 61 p.

Posthuma L, Baerselman R, van Veen RPM, Dirven-Van Breemen EM. 1997. Single and joint toxic effects of copper and zinc on reproduction of *Enchytraeus crypticus* in relation to sorption of metals in soils. Ecotoxicol Environ Safety 38:108–121.

Posthuma L, De Zwart D, Wintersen A, Lijzen J, Swartjes F, Cuypers C, van Noort P, Harmsen J, Groenenberg BJ. 2006. Beslissen over Bagger op Bodem. Deel 1. Systeembenadering, Model en Praktijkvoorbeelden. No. 711701044. Bilthoven (The Netherlands): National Institute for Public Health and the Environment, 110 p. http://www.rivm.nl/bibliotheek/rapporten/711701044.html (accessed December 28, 2007).

Posthuma L, Traas TP, De Zwart D, Suter GW, II. 2002a. Conceptual and technical outlook on species sensitivity distributions. In: Posthuma L, Suter, SW, Traas TP, editors. Species sensitivity distributions in ecotoxicology. Boca Raton (FL): Lewis Publishers, p 475–510.

Posthuma L, Traas TP, Suter GW, editors. 2002b. Species sensitivity distributions in risk assessment. Boca Raton (FL): CRC Press, 564 p.

Posthuma L, van Gestel CAM, Smit CE, Bakker DJ, Vonk JW. 1998. Validation of toxicity data and risk limits for soils: final report. No. 607505004. Bilthoven (the Netherlands): National Institute of Public Health and the Environment (RIVM).

Posthuma L, Weltje L, Antón-Sánchez FA. 1996. Joint toxic effects of cadmium and pyrene on reproduction and growth of the earthworm *Eisenia andrei*. No. 607506001. Bilthoven (the Netherlands): National Institute of Public Health and the Environment (RIVM).

Postma JF, Davids C. 1995. Tolerance induction and life-cycle changes in cadmium exposed *Chronimus riparius* (Diptera) during consecutive generations. Ecotoxicol Environ Safety 30:195–202.

Postma JF, Mol S, Larsen H, Admiraal W. 1995a. Life-cycle changes and zinc shortage in cadmium tolerant midges, *Chrironomus riparius* (Diptera), reared in the absence of cadmium. Environ Toxicol Chem 14:117–121.

Postma JF, van Kleunen A, Admiraal W. 1995b. Alterations in life-history traits of *Chironomus riparius* (Diptera) obtained from metal contaminated rivers. Arch Environ Contam Toxicol 29:469–475.

Power ME, Brozovic N, Bode C, Zilberman D. 2005. Spatially explicit tools for understanding and sustaining inland water ecosystems. Front Ecol Environ 3:47–55.

Pratt JR, Bowers NJ. 1992. Variability of community metrics: detecting changes in structure and function. Environ Toxicol Chem 11:451–457.

Pratt JR, Bowers NJ, Niederlehner BR, Cairns J, Jr. 1988. Effects of atrazine on freshwater microbial communities. Arch Environ Contam Toxicol 17:449–457.

Pratt JR, Cairns J, Jr. 1996. Ecotoxicology and the redundancy problem: understanding effects on community structure and function. In: Newman MC, Jagoe CH, editors. Ecotoxicology: a hierarchical treatment. Boca Raton (FL): Lewis Publishers, p 347–370.

Preston BL. 2001. Indirect effects in aquatic ecotoxicology: implications for ecological risk assessment. Environ Managment 29:311–323.

Preston BL. 2002. Indirect effects in aquatic ecotoxicology: implications for ecological risk assessment. Environ Managment 29:311–323.

Preston BL, Shackelford J. 2002. Risk-based analysis of environmental monitoring data: application to heavy metals in North Carolina surface waters. Environ Managment 30:279–293.

Probst M, Berenzen N, Lentzen-Godding A, Schulz R. 2005. Scenario-based simulation of runoff-related pesticide entries into small streams on a landscape level. Ecotoxicol Environ Safety 62:145–159.

Prusha BA, Clements WH. 2004. Landscape attributes, dissolved organic C, and metal bio-accumulation in aquatic macroinvertebrates (Arkansas River Basin, Colorado). J NA Benthol Soc 23:327–339.

Pusey BJ, Arthington AH, MacClean J. 1994. The effects of a pulsed application of chlor-pyrifos on macroinvertebrate communities in an outdoor artificial stream system. Ecotoxicol Environ Safety 27:221–250.

Quinn JM, Davies-Colley RJ, Hickey CW, Vickers ML, Ryan PA. 1992. Effects of clay dis-charges on streams 2. Benthic invertebrates. Hydrobiologia 248:235–247.

Quinn JM, Hickey CW. 1993. Effects of sewage stabilization lagoon effluent on stream inver-tebrates. J Aquat Ecosys Hlth 2:205–219.

Quinn JM, Hickey CW. 1994. Hydraulic parameters and benthic invertebrate distributions in two gravel-bed New Zealand rivers. Freshwater Biol 32:489–500.

Raiffa H. 1982. The art and science of negotiation. Cambridge (MA): Belknap Press of Harvard University Press, 373 p.

Rajagopal S, van der velde G, van der Gaag M, Jenner HA. 2003. How effective is intermit-tent chlorination to control adult mussel fouling in cooling water systems? Water Res 37:329–338.

Rand GM, Petrocelli SR, editors. 1985. Fundamentals of aquatic toxicology. New York (NY): Hemisphere Publishing, 666 p.

Ray S, Berec L, Straskraba M, Jorgensen SE. 2001. Optimization of exergy and implications of body sizes of phytoplankton and zooplankton in an aquatic ecosystem model. Ecol Model 140:219–234.

Regan HM. 2002. Population models: individual based. In: Pastorok RA, Bartell SM, Ferson S, Ginzburg LR, editors. Ecological modeling in risk assessment: chemical effects on populations, ecosystems, and landscapes. Boca Raton (FL): Lewis Publishers, p 65–82.

Reinert KH, Giddings JM, Judd L. 2002. Effects analysis of time-varying or repeated expo-sures in aquatic ecological risk assessment of agrochemicals. Environ Toxicol Chem 21:1977–1992.

Ren S. 2003. Ecotoxicity prediction using mechanism- and non-mechanism-based (Q)SARs: a preliminary study. Chemosphere 53:1053–1065.

Renner R. 1997. Rethinking water quality standards for metals toxicity. Environ Sci Technol 31:466–468.

Reynoldson TB, Rodriguez P. 1998. Field methods and interpretation for sediment bioas-sessment. In: Murdoch A, Azcue J, Murdoch P, editors. Manual of bioassessment of aquatic sediment quality. Boca Raton (FL): CRC Press.

Richards SM, Kendall RJ. 2002. Biochemical effects of chlorpyrifos on two developmental stages of *Xenopus laevis*. Environ Toxicol Chem 21:1826–1835.

Richter BD, Mathews R, Harrison DL, Eigington R. 2003. Ecological sustainable water man-agement: managing river flows for ecological integrity. Ecol Appl 13:206–224.

Riley D (Ed). 1993. Principles of risk assessment. Wageningen (The Netherlands): Winand Staring Centre for Integrated Land, Soil and Water Research.

Ringwood AH. 1992. Comparative sensitivity of gametes and early developmental stages of a sea urchin species (*Echinometra mathaei*) and a bivalve species (Isognomon califor-nicum) during metal exposures. Arch Environ Contam Toxicol 22:288–295.

Robinson P. 2003. Use of (Q)SARs in the categorization of discrete organic substances on Canada's domestic substances list (DSL). 30th annual Aquatic Toxicity Workshop, September 2003, Ottawa, ON, Canada.

Roessink I, Arts GHP, Belgers JDM, Bransen F, Maund SJ, Brock TCM. 2005. Effects of lambda-cyhalothrin in two ditch microcosm systems of different trophic status. Environ Toxicol Chem 24:1684–1696.

Ross AH, Nisbet RM. 1990. Dynamic models of growth and reproduction of the mussel *Mytilus edulis* L. Funct Ecol 4:777–787.

Ross HLB. 1996. The interaction of chemical mixtures and their implications on water quality guidelines. Sydney (Australia): University of Technology.

Ross HLB, Warne MSJ. 1997. Most chemical mixtures have additive aquatic toxicity. Third Annual Conference of the Australasian Society for Ecotoxicology, July 17–19, 1997, Brisbane, Australia, 30 p.

Rulis AM. 1996. Making regulatory decisions across the food ingredient spectrum. Drug Metabol Rev 28:197–208.

Rulis AM, Tarantino LM. 1996. Food ingredient review at FDA: recent data and initiatives to improve the process. Reg Toxicol Pharmacol 24:224–231.

Rusin PA, Rose JB, Haas CN, Gerba CP. 1997. Risk assessment of opportunistic bacterial pathogens in drinking water. Rev Environ Contam Toxicol 152:57–83.

Russom CL, Breton RL, Walker JD, Bradbury SP. 2003. An overview of the use of quantitative structure-activity relationships for ranking and prioritizing large chemical inventories for environmental risk assessments. Environ Toxicol Chem 22:

Safe S. 1990. Polychlorinated biphenyls (PCBs), dibenzo-p-dioxins (PCDDs), dibenzofurans (PCDFs), and related compounds: environmental and mechanistic considerations which support the development of toxic equivalency factors (TEFs). Crit Rev Toxicol 21:51–62.

Safe S. 1998. Hazard and risk assessment of chemical mixtures using the toxic equivalency factor approach. Environ Hlth Perspect 106:1051–1062.

Salvito DT, Senna RJ, Federle TW. 2002. A framework for prioritizing fragrance materials for aquatic risk assessment. Environ Toxicol Chem 21:1301–1308.

Sanders HO. 1970. Pesticide toxicities to tadpoles of the western chorus frog *Pseudacris triseriata* and Fowler's toad Bufo woodhousii fowleri. Copeia 1970:246–251.

Sanders HO. 1972. Toxicity of some insecticides to four species of malacostracon crustaceans. No. Technical Paper 66. Washington (DC): US Fish and Wildlife Service, 19 p.

Sanderson H. 2002. Pesticide studies: replicability of micro/mesocosms. Environ Sci Poll Res 9:429–435.

Sanderson H, Johnson DJ, Wilson CJ, Brain RA, Solomon KR. 2003. Probabilistic hazard assessment of environmentally occurring pharmaceuticals toxicity to fish, daphnids, and algae by ECOSAR screening. Toxicol Lett 144:385–395.

Sanderson H, Thomsen M. 2007. Ecotoxicological quantitative structure-activity relationships for pharmaceuticals. Bull Environ Contam Toxicol 79:331–335.

Sandheinrich M. 2003. The role of behavior in ecotoxicology. In: Newman MC, Unger MA, editors. Fundamentals of ecotoxicology, 2nd ed. Boca Raton (FL): Lewis Publishers, p 165–167.

Santore RC, Di Toro DM, Paquin PR, Allen HE, Meyer JS. 2001. Biotic ligand model of the acute toxicity of metals. 2. Application to acute copper toxicity in freshwater fish and Daphnia. Environ Toxicol Chem 20:2397–2402.

Santore RC, Driscoll CT. 1995. The chess model for calculating chemical equilibria in soils and solutions. Soil Sci Soc Am Spec Pubs 42:357–375.

SAS Institute. 2002. Statistical analysis software [computer program]. Version 8. Cary (NC): SAS.

Sauer JR, Pendleton GW. 1995. Population modelling and its role in toxicological studies. In: Hoffman DJ, Rattner BA, Burton RA, Jr, Cairns J, Jr, editors. Handbook of ecotoxicology, 1st ed. Boca Raton (FL): Lewis Publishers/CRC Press, p 681–702.

Sauer JR, Pendleton GW. 2003. Population modelling. In: Hoffman DJ, Rattner BA, Burton GA, Jr, Cairns J, Jr, editors. Handbook of ecotoxicology, 2nd ed. Boca Raton (FL): CRC Press, p 324–371.

Schaëfers C. 2002. Community level study with copper in aquatic microcosms. Study report prepared for the International Copper Association. Smallenberg (Germany): Fraunhofer Institute for Molecular Biology and Applied Ecology, 119 p.

Schaffer ML. 1981. Minimum population sizes for species conservation. Bioscience 31:131–134.

Scheffer M. 1998. Ecology of shallow lakes. London (UK): Chapman & Hall, 357 p.

Scheffer M, Szabo S, Gragnani A, van Nes EH, Rinaldi S, Kautsky N, Norberg J, Roijackers RMM, Franken RJM. 2003. Floating plant dominance as a stable state. Proc Nat Acad Sci US 100:4040–4045.

Scheringer M, Steinbach D, Escher B, Hungerbuhler K. 2002. Probabilistic approaches in the effect assessment of toxic chemicals: what are the benefits and limitations? Environ Sci Poll Res 9:307–314.

Schindler DW. 1998. Replication versus realism: the need for ecosystem-scale experiments. Ecosystems 1:323–334.

Schmieder PK, Ankley G, Mekenyan O, Walker JD, Bradbury S. 2003. Quantitative structure-activity relation models for prediction of estrogen receptor binding affinity of structurally diverse chemicals. Environ Toxicol Chem 22:1844–1854.

Scholze M, Boedeker W, Faust M, Backhaus T, Altenburger R, Grimme LH. 2001. A general best-fit method for concentration-response curves and the estimation of low-effect concentrations. Environ Toxicol Chem 20:448–457.

Schouten AJ, Bogte JJ, Driven van Breemen EM, Rutgers M. 2003. Site specific ecological risk assessment: application of the TRIAD approach. Part 2. No. 722701032. Bilthoven (The Netherlands): National Institute of Public Health and the Environment (RIVM), 90 p.

Schroer AFW, Belgers JDM, Brock TCM, Matser AM, Maund SJ, van den Brink PJ. 2004a. Comparison of laboratory single species and field population-level effects of the pyrethroid lambda-cyhalothrin on freshwater invertebrates. Arch Environ Contam Toxicol 46:324–335.

Schroer AFW, Belgers D, Brock TCM, Matser A, Maund SJ, van den Brink PJ. 2004b. Comparison of laboratory single species and field population-level effects of the pyrethroid insecticide lambda-cyhalothrin on freshwater invertebrates. Arch Environ Contam Toxicol 46:324–335.

Schubauer-Berigan MK, Ankley GT. 1991. The contribution of ammonia, metals and nonpolar organic compounds to the toxicity of sediment interstitial water from an Illinois River tributary. Environ Toxicol Chem 10:925–939.

Schulz R. 2004. Field studies on exposure, effects, and risk mitigation of aquatic nonpoint-source insecticide pollution: a review. J Environ Qual 33:419–448.

Schulz R, Liess M. 2000. Toxicity of fenvalerate to caddisfly larvae: chronic effects of 1- vs 10-h pulse exposure with constant doses. Chemosphere 41:1511–1517.

Schulz R, Liess M. 2001. Toxicity of aqueous-phase and suspended particle-associated fenvalerate: chronic effects after pulse-dosed exposure of Limnephilus lunatus (Trichoptera). Environ Toxicol Chem 20:185–190.

Schwarz RC, Schults DW, Ozretich RW, Lamberson JO, Cole FA, DeWitt TH, Redmond MS, Ferraro SP. 1995. Sigma PAH: a model to predict the toxicity of polynuclear aromatic hydrocarbon mixtures in field-collected sediments. Environ Toxicol Chem 14:1977–1978.

Schwarzenbach RP, Gschwend PM, Imboden DM. 1993. Environmental organic chemistry. New York (NY): John Wiley, 681 p.

Scribner EA, Goolsby DA, Thurman EM, Meyer MT, Pomes ML. 1994. Concentrations of selected herbicides, two triazine metabolites, and nutrients in storm runoff from nine stream basins in the Midwest United States, 1990–92. Open File Report. No. 94-396. Lawrence (KS): US Geological Survey.

Seguin F, Leboulanger C, Rimet F, Druart JC, Berard A. 2001. Effects of atrazine and nicosulfuron on phytoplankton in systems of increasing complexity. Arch Environ Contam Toxicol 40:198–208.

Selck H, Forbes VE. 2003. Uptake, depuration and toxicity of dissolved and sediment-bound fluoranthene in the polychaete, *Capitella sp.* I. Environ Toxicol Chem 22:2354–2363.

Selck H, Forbes VE, Forbes TL. 1998. The toxicity and toxicokinetics of cadmium in Capitella sp. I: relative importance of water and sediment as routes of cadmium uptake. Mar Ecol Prog Ser 164:167–178.

Selck H, Riemann B, Christoffersen K, Forbes VE, Gustavson K, Hansen BW, Jacobsen JA, Kusk OK, Petersen S. 2002. Comparing sensitivity of ecotoxicological effect endpoints between laboratory and field. Ecotoxicol Environ Safety 52:97–112.

Servos MR. 1999. Review of the aquatic toxicity, estrogenic responses and bioaccumulation of alkylphenols and alkylphenol polyethoxylates. Water Qual Res J Canada 34:123–177.

Sharma Shanti S, Schat H, Vooijs R, van Heerwaarden LM. 1999. Combination toxicology of copper, zinc, and cadmium in binary mixtures: concentration-dependent antagonistic, nonadditive, and synergistic effects on root growth in *Silene vulgaris*. Environ Toxicol Chem 18:348–355.

Shea D. 1988. Deriving sediment quality criteria. Environ Sci Technol 22:1256–1261.

Shelly A, Ford D, Bruce B. 2000. Quality assurance of environmental models. NRCSE-TRS. No. 042. Seattle (WA): National Research Center for Statistics and the Environment, 69 p.

Sherratt TN, Jepson PC. 1993. A metapopulation approach to modeling the long-term impact of pesticides on invertebrates. J Appl Ecol 30:696–705.

Sherratt TN, Roberts G, Williams P, Whitfield M, Biggs J, Shillabeer N, Maund SJ. 1999. A life-history approach to predicting the recovery of aquatic invertebrate populations after exposure to xenobiotic chemicals. Environ Toxicol Chem 18:2512–2518.

Sherry J, Gamble A, Hodson PV, Solomon KR, Hock B, Marx A, Hansen P. 1999. Vitellogenin induction in fish as an indicator of exposure to environmental estrogen. In: Rao SS, editor. Impact assessment of hazardous aquatic contaminants: concepts and approaches. Boca Raton, FL: Lewis Publishers, p 123–160.

Sibley PK, Chappel MJ, George TK, Solomon KR, Liber K. 2000. Integrating effects of stressors across levels of biological organization: example using organophosphorus insecticide mixtures in field-level exposures. J Ecosys Stress Revovery 7:117–130.

Sibley PK, Harris ML, Bestari KT, Steele TA, Robinson RD, Gensemer RW, Day KE, Solomon KR. 2001a. Response of zooplankton communities to liquid creosote in freshwater microcosms. Environ Toxicol Chem 20:394–405.

Sibley PK, Harris ML, Bestari KT, Steele TA, Robinson RD, Gensemer RW, Day KE, Solomon KR. 2001b. Response of phytoplankton communities to liquid creosote in freshwater microcosms. Environ Toxicol Chem 20:2785–2791.

Sibley PK, Kaushik NK, Kreutzweiser R. 1991. Impact of a pulse application of permethrin on the macroinvertebrate community of a headwater stream. Environ Pollut 70:35–55.

Sibly RM, Williams TD, Jones MB. 2000. How environmental stress affects density dependence and carrying capacity in a marine copepod. J Appl Ecol 37:388–397.

Sijm DTHM, Van Wezel AP, Crommentuijn T. 2002. Environmental risk limits in the Netherlands. In: Posthuma L, Suter GW, Trass TP, editors. Species sensitivity distributions in ecotoxicology. Boca Raton (FL): CRC Press, p 221–253.

Silva E, Rajapakse N, Kortenkamp A. 2002. Something from "nothing": eight weak estrogenic chemicals combined at concentrations below NOECs produce significant mixture effects. Environ Sci Technol 36:1751–1756.

Simon TP. 2003. Biological response signatures: toward the detection of cause-and-effect and diagnosis in environmental disturbance. In: Simon TP, editor. Biological response signatures: indicator patterns using aquatic communities. Boca Raton (FL): CRC Press, p 3–12.

Sipes IG, Gandolfi AJ. 1986. Biotransformation of toxicants. In: Klaasen CD, Amdur MO, Doull J, editors, Casarett and Doull's toxicology: the basic science of poisons, 3rd ed. New York (NY): Macmillan, p 64–98.

Smit CE, Schouten AJ, van den Brink PJ, van Esbroek MLP, Posthuma L. 2002. Effects of zinc contamination on the natural nematode community in outdoor soil mesocosms. Arch Environ Contam Toxicol 42:205–216.

Snel JFH, Vos JH, Gylstra R, Brock TCM. 1998. Inhibition of photosystem II (PSII) electron transport as a convenient endpoint to assess stress of the herbicide linuron on freshwater plants. Aquat Ecol 32:113–123.

Snell TW, Serra M. 2000. Using probability of extinction to evaluate the ecological significance of toxicant effects. Environ Toxicol Chem 19:2357–2363.

Society for Environmental Toxicology and Chemistry [SETAC]. 1994. Pesticide risk and mitigation. Final Report of the Aquatic Risk Assessment and Mitigation Dialog Group. Pensacola (FL): SETAC Foundation for Environmental Education, 220 p.

Solomon KR, Baker DB, Richards P, Dixon KR, Klaine SJ, La Point TW, Kendall RJ, Giddings JM, Giesy JP, Hall LWJ, Weisskopf C, Williams M. 1996. Ecological risk assessment of atrazine in North American surface waters. Environ Toxicol Chem 15:31–76.

Solomon KR, Stephenson GL, Kaushik NK. 1989. Effects of methoxychlor on zooplankton in freshwater enclosures: influence of enclosure size and number of applications. Environ Toxicol Chem 8:659–670.

Solomon KR, Takacs P. 2002. Probabilistic risk assessment using species sensitivity distributions. In: Posthuma L, Suter GW, Traas TP, editors. Species sensitivity distributions in ecotoxicology. Boca Raton (FL): CRC Press, p 285–313.

Sousa WP. 1984. The role of disturbance in natural communities. Annu Rev Ecol Syst 15:353–391.

Spacie A. 1994. Interactions of organic pollutants with inorganic solid phases: are they important to bioavailability? In: Hamelink JL, Landrum PF, Bergman HL, Benson WH, editors. Bioavailability, physical, chemical and biological interactions. Boca Raton (FL): Lewis Publishers, p 12–31.

Sparks TH, Scott WA, Clarke RT. 1999. Traditional multivariate techniques: potential for use in ecotoxicology. Environ Toxicol Chem 18:128–137.

Spencer M, Ferson S. 1998. Ecological risk assessment for structured populations. Setauket (NY): Applied Biomathematics.

Sprague JB. 1970. Measurement of pollutant toxicity to fish. II. Utilizing and applying bioassay results. Water Res 4:3–32.

Sprague JB, Logan WJ. 1979. Separate and joint toxicity to rainbow trout of substances used in drilling fluids for oil exploration. Environ Pollut 19:269–281.

Sprague JB, Ramsay BA. 1965. Lethal levels of mixed copper-zinc solutions for juvenile salmon. J Fish Res Bd Can 22:425–432.

Spromberg JA, John BM, Landis WG. 1998. Metapopulation dynamics: indirect effects and multiple distinct outcomes in ecological risk assessment. Environ Toxicol Chem 17:1640–1649.

Stay FS, Flum TE, Shannon LJ, Yount JD. 1989. An assessment of the precision and accuracy of SAM and MFC microcosms exposed to toxicants. In: Cowgill UM, Williams LR, editors. Aquatic toxicology and hazard assessment. ASTM STP 1027, Volume 12. Philadelphia (PA): American Society of Testing and Materials, p 189–203.

Stay FS, Larsen DP, Katko A, Rohm CM. 1985. Effects of atrazine on community level responses in Taub microcosms. In: Boyle TP, editor. Validation and predictability of laboratory methods for assessing the fate and effects of contaminants in aquatic ecosystems. ASTM Special Technical Publication No. 865. Philadelphia (PA): American Society of Testing and Materials, p 75–90.

Stearns SS. 1992. The evolution of life histories. Oxford (UK): Oxford University Press, 262 p.

Steen R. 2002. Overzicht van het voorkomen van bestrijdingsmiddelen in de Maas, Afdedamde Maas en Bommelerwaard. In: Kruine R, editor. Belasting van de Afgedamde Maas door bestrijdingsmiddelen. Volume 395. Wageningen (The Netherlands): Alterra-rapport, p 78–90.

Steenbergen NTTM, Iaccino F, de Winkel M, Reijnders L, Peijnenburg W. 2005. Development of a biotic ligand model and a regression model predicting acute copper toxicity to the earthworm Aporrectodea caliginosa. Environ Sci Technol 39:5694–5702.

Stephan CE. 2002. Use of species sensitivity distributions in the derivation of water quality criteria for aquatic life by the U.S. Environmental Protection Agency. In: Posthuma L, Suter GW, Traas TP, editors. Species sensitivity distributions in ecoloxicology. Boca Raton (FL): CRC Press, p 211–254.

Stephan CE, Rogers JW. 1985. Advantages of using regression analysis to calculate results of chronic toxicity tests. In: Bahner RC, Hansen DJ, editors. Aquatic toxicology and hazard assessment: eighth symposium. Vol. ASTM STP 891. Phillanelphia (PA): American Society for Testing and Materials, p 328–338.

Stephenson GL, Kaushik NK, Solomon KR, Day KE, Hamilton P. 1986. Impact of methoxychlor on freshwater plankton communities in limnocorrals. Environ Toxicol Chem 5:587–603.

Stephenson GR, Ferris IG, Holland PT, Nordberg M. 2006. Glosary of terms relating to pesticides (IUPAC Recommendations 2006, © IUPAC). Pure Appl Chem 78:2075–2154.

Straub JO. 2002. Environmental risk assessment for new human pharmaceuticals in the European Union according to the draft guideline/discussion paper of January 2001. Toxicol Lett 135:231–237.

Stuijfzand SC, Poort L, Greve GD, van der Geest HG, Kraak MHS. 2000. Variables determining the impact of diazinon on aquatic insects: taxon, developmental stage and exposure time. Environ Toxicol Chem 19:582–587.

Sun K, Krause GF, Mayer FL, Ellersieck MR, Basu AP. 1995. Predicting chronic lethality of chemicals to fishes from acute toxicity test data: theory of accelerated life testing. Environ Toxicol Chem 14:1745–1752.

Suter G, Antcliffe BL, Davis W, Dyer S, Gerritson J, Linder G, Munkittrick KR, Rankin E. 1999a. Conceptual approaches to identify and assess multiple stressors. In: Ferenc SA, Foran JA, editors. Multiple stressors in ecological risk and impact assessment. Pensacola (FL): SETAC Press, p 1–26.

Suter GW, Barnthouse LW, Bartell SM, Mill T, Mackay D, Patterson S. 1993. Ecological risk assessment. Boca Raton (FL): Lewis Publishers, 538 p.

Suter GW, Barnthouse LW, Efroymson RA, Jager H. 1999b. Ecological risk assessment in a large river-reservoir: 2. Fish community. Environ Toxicol Chem 18:589–598.

Suter GW, Efroymson RA, Sample BE, Jones DS. 2000. Ecological risk assessment for contaminated sites. Boca Raton (FL): CRC Press, Lewis Publishers, 438 p.

Suter GW, Traas TP, Posthuma L. 2002. Issues and practices in the derivation and use of species sensitivity distributions. In: Posthuma L, Suter GW, Traas TP, editors. Species sensitivity distributions in ecoloxicology. Boca Raton (FL): CRC Press, p 437–474.

Suter GW, II. 1996. Abuse of hypothesis testing statistics in ecological risk assessment. Human Ecol Risk Assess 2:331–347.

Swartjes FA. 1999. Risk-based assessment of soil and groundwater quality in the Netherlands: standards and remediation urgency. Risk Anal 19:1235–1249.

Swartz RC, Di Toro DM. 1997. Sediments as complex mixtures: an overview of methods to assess ecotoxicological significance. In: Ingersoll CG, Dillon T, Biddinger GR, editors. Ecological risk assessment of contaminated sediments. Pensacola (FL): SETAC Press.

Swartzman GL, Taub FB, Meador J, Huang C, Kindig A. 1990. Modeling the effect of algal biomass on multispecies aquatic microcosms' response to copper toxicity. Aquat Toxicol 17:93–118.

Takhtajan A. 1986. Floristic regions of the world. Cronquist A, Crovello TJ, translators. Berkeley: University of California Press. 422 p.

Tanaka Y, Nakanishi J. 2001. Model selection and parameterizatiopn of the concentration-response functions for population-level effects. Environ Toxicol Chem 20:1857–1865.

Taub FB, Kindig AC, Meador JP, Swartzman GL. 1991. Effects of "seasonal succession" and grazing on copper toxicity in aquatic microcosms. Verh Internat Verein Limnol 24:2205–2214.

Teske ME, Scott RL. 2000. AgDRIFTTM: An update of the aerial spray model AGDISP. http://www.agdrift.com (accessed October 23, 2000).

Teuschler LK, Rice GE, Wilkes CR, Lipscomb JC, Power FW. 2004. A feasibility study of cumulative risk assessment methods for drinking water disinfection by-product mixtures. Journal of Toxicology and Environmental Health Part A 67:755–777.

Thakali S, Allen HE, Di Toro DM, Ponizovsky AA, Rooney CP, Zhao F-J, McGrath SP. 2006b. A terrestrial biotic ligand model. 1. Development and application to Cu and Ni toxicities to barley root elongation in soils. Environ Sci Technol 40:7085–7093.

Thakali S, Allen HE, Di Toro DM, Ponizovsky AA, Rooney CP, Zhao F-J, McGrath SP, Criel P, van Eeckhout H, Janssen CR, Oorts K, Smolders E. 2006a. A terrestrial biotic ligand model. 2. Application to Ni and Cu toxicities to plants, invertebrates, and microbes in soil. Environ Sci Technol 40:7094–7100.

Thompson DG, Holmes SB, Wainio-Keizer K, MacDonald L, Solomon KR. 1993a. Impact of hexazinone and metsulfuron methyl on the zooplankton community of a boreal forest lake. Environ Toxicol Chem 12:1709–1717.

Thompson DG, Holmes SB, Thomas D, MacDonald L, Solomon KR. 1993b. Impact of hexazinone and metsulfuron methyl on the phytoplankton community of a mixed-wood/boreal forest lake. Environ Toxicol Chem 12:1695–1707.

Tillitt DE. 1999. The toxic equivalents approach for fish and wildlife. Human Ecol Risk Assess 5:25–32.

Tillitt DE, Solomon KR, Mihaich EM, Cobb GP, Touart L, Kubiak TJ. 1998. Role of exposure assessment in characterizing risk of endocrine-disrupting substances in wildlife. In: Suk W, editor. Principles and processes for evaluating endocrine disruption in wildlife. Pensacola (FL): SETAC Press, p 39–68.

Tilman D. 1996. Biodiversity: population versus ecosystem stability. Ecol 77:350–363.

Tilman D, Knops J, Wedlin D, Reich P, Ritchie M, Siemann E. 1997. The influence of functional diversity and composition on ecosystem processes. Science 277:1300–1302.

Tilman D, Wedlin D, Knops J. 1996. Productivity and sustainability influenced by biodiversity in grassland ecosystems. Nature 379:718–720.

Timmerman K, Andersen O. 2003. Bioavailability of pyrene to the deposit-feeding polychaete Arenicola marina: importance of sediment versus water uptake routes. Mar Ecol Prog Ser 246:163–172.

Tipping E, Hurley M. 1992. WHAM-A computer equilibrium model and code for waters, sediments, and soils in corporating a discrete site/electrostatic model of ion-binding by humic substances. *Comput Geo Sci* 20:973–1023.

Tipping E. 1994. WHAM: a chemical equilibrium model and computer code for waters, sediments, and soils incorporating a discrete site/electrostatic model of ion-binding by humic substances. Comput Geosci 20:973–1022.

Tipping E. 1998. Humic ion-binding model VI: an improved description of the interactions of protons and metal ions with humic substances. Aquat Geochem 4:3–48.

Topping CJ. 1997. Predicting the effect of landscape heterogeneity on the distribution of spiders in agroecosystems using a population dynamics driven landscape-scale simulation model. Biol Agric Hort 15:325–336.

Topping CJ, Hansen TS, Jensen TS, Jepsen JU, Nikolajsen F, Odderskær P. 2003. ALMaSS, an agent-based model for animals in temperate European landscapes. Ecol Model 167:65–82.

Traas TP. 2003. Evaluatie van groep: en somnormen in het kader van Integrale Normstelling Stoffen. No. 601501 014. Bilthoven (The Netherlands): National Institute of Public Health and the Environment (RIVM), 43 p.

Traas TP. 2004. Food web models in ecotoxicological risk assessment [PhD thesis]. Utrecht (The Netherlands): University of Utrecht, 231 p.

Traas TP, Janse JH, Aldenberg T, Brock TCM. 1998. A food web model for fate and direct and indirect effects of Dursban 4E (active ingredient chlorpyrifos) in freshwater microcosms. Aquat Ecol 179–190.

Traas TP, Janse JH, van den Brink PJ, Brock TCM, Aldenberg T. 2004. A freshwater food web model for the combined effects of nutrients and insecticide stress and subsequent recovery. Environ Toxicol Chem 23:521–529.

Traas TP, van de Meent D, Posthuma L, Hamers T, Kater BJ, De Zwart D, Aldenberg T. 2002. The potentially affected fraction as a measure of ecological risk. In: Posthuma L, Suter GW, Traas TP, editors. The use of species sensitivity distributions in ecotoxicology. Boca Raton (FL): CRC Press.

Ulanowicz RE University of Maryland, Chesapeake Biological Laboratory. 1987. NETWRK4: a package of computer algorithms to analyse ecological flow networks [computer program]. Solomons (MD): http://www.cbl.cees.edu/~ulan/ntwk/network.html (accessed December 28, 2007).

Underwood AJ. 1992. Beyond BACI: the detection of environmental impacts on populations in the real, but variable, world. J Exp Mar Biol Ecol 161:145–178.

United Nations Economic Commission for Europe. 2004. Globally Harmonized System of Classification and Labelling of Chemicals (GHS). http://www.unece.org/trans/danger/publi/ghs/ghs_welcome_e.html (accessed January 8, 2008).

Urban DJ, Cook NJ. 1986. Standard evaluation procedure for ecological risk assessment. No. EPA/540/09-86/167. Washington (DC): Hazard Evaluation Division, Office of Pesticide Programs, US Environmental Protection Agency.

US Environmental Protection Agency [USEPA]. 1992. Framework for ecological risk assessment. Washington (DC): US Environmental Protection Agency, 48 p.

US Environmental Protection Agency [USEPA]. 1994. USEPA/EC joint project on the evaluation of (quantitative) structure activity relationships. Washington (DC): US Environmental Protection Agency. http://intranet.epa.gov/oppthome/testsite/MPDSAR/index.html (accessed December 28, 2007).

US Environmental Protection Agency [USEPA]. 1995a. The use of the benchmark dose approach in health risk assessment. Risk Assessment Forum. Washington (DC): US Environmental Protection Agency, 77 p.

US Environmental Protection Agency [USEPA]. 1995b. Final water quality guidance for Great Lakes system. Federal Register, March 23, 1995, p 15366–15425.

US Environmental Protection Agency [USEPA]. 1998. Guidelines for ecological risk assessment. Risk Assessment Forum. Washington (DC): US Environmental Protection Agency.

US Environmental Protection Agency [USEPA]. 2000a. Guidance for the data quality objectives process. Washington (DC): US Environmental Protection Agency, Office of Environmental Information, 100 p. http://www.epa.gov/quality/qs-docs/g4-final.pdf (accessed December 28, 2007).

US Environmental Protection Agency [USEPA]. 2000b. Methods for measuring the toxicity and bioaccumulation of sediment-associated contaminants with freshwater invertebrates. Washington (DC): US Environmental Protection Agency, p 175–237.

US Environmental Protection Agency [USEPA]. 2000c. Stressor identification guidance document. No. EPA 822-B-00-025. Washington (DC): US Environmental Protection Agency, 228 p. http://www.epa.gov/ost/biocriteria/stressors/stressorid.pdf (accessed December 28, 2007).

US Environmental Protection Agency [USEPA]. 2000d. Supplementary guidance for conducting health risk assessment of chemical mixtures. EPA/630/R-00/002. Washington (DC): US Environmental Protection Agency.

US Environmental Protection Agency [USEPA], Office of Pesticide Programs, Environmental Fate and Effects Division. 2001. Environmental effects database (EEDB). ECOTOX Database System. http://www.epa.gov/ecotox/ (accessed July 23, 2005).

US Environmental Protection Agency [USEPA]. 2002a. Terrestrial (soil-core) microcosm test guideline. Washington (DC): US Environmental Protection Agency.

US Environmental Protection Agency [USEPA]. 2002b. Generic freshwater (laboratory) microcosm test guideline. Washington (DC): US Environmental Protection Agency.

US Environmental Protection Agency [USEPA]. 2002c. Site-specific aquatic (laboratory) microcosm test guideline. Washington (DC): US Environmental Protection Agency.

US Environmental Protection Agency [USEPA]. 2003. A summary of general assessment factors for evaluating the quality of scientific and technical information. Washington (DC): US Environmental Protection Agency, 18 p.

US Environmental Protection Agency [USEPA], Office of Prevention, Pesticides and Toxic Substances. 2005. Chemical screening tool for exposures and environmental releases (ChemSTEER). http://www.epa.gov/oppt/p2framework/docs/exprisk.htm (accessed December 2005).

US Environmental Protection Agency [USEPA], Office of Prevention, Pesticides and Toxic Substances. 2007. EPI suite 3.12. http://www.epa.gov/oppt/exposure/docs/episuitedl. htm (accessed May 23, 2007).

US Environmental Protection Agency [USEPA]. N.d. High production volume (HPV) challenge. http://www.epa.gov/chemrtk/ (accessed January 8, 2008).

Usseglio-Polatera P, Bournaud M, Richoux P, Tachet H. 2000. Biological and ecological traits of benthic freshwater macroinvertebrates: relationships and definition of groups with similar traits. Freshwater Biol 43:175–205.

van de Meent D. 1993. SimpleBox: a generic multimedia fate evaluation model. No. 672720 001. Bilthoven (the Netherlands): National Institute of Public Health and the Environment (RIVM), 107 p.

van den Berg M, Birnbaum L, Bosveld ATC, Brunström B, Cook P, Feeley M, Giesy JP, Hanberg A, Hasegawa R, Kennedy SW, Kubiak T, Larsen JC, Leeuwen FXR, Liem AKD, Nolt C, Peterson RE, Poellinger L, Safe S, Schrenk D, Tillitt D, Tysklind M, Younes M, Wærn F, Zacharewski T. 1998. Toxic equivalency factors (TEFs) for PCBs, PCDDs, PCDFs for humans and wildlife. Environ Hlth Perspect 106:775–792.

van den Brink PJ, Blake N, Brock TCM, Maltby L. 2006a. Predictive value of species sensitivity distributions for effects of herbicides in freshwater ecosystems. Human Ecol Risk Assess 12:645–674.

van den Brink PJ, Brock TCM, Posthuma L. 2002a. The value of the species sensitivity distribution concept for predicting field effects: (non-)confirmation of the concept using semi-field experiments. In: Posthuma L, Suter GW, Traas TP, editors. Species sensitivity distributions in ecoloxicology. Boca Raton (FL): CRC Press, p 155–198.

van den Brink PJ, Brown CD, Dubus IG. 2006b. Using the expert model PERPEST to translate measured and predicted pesticide exposure data into ecological risks. Ecol Modell 191:106–117.

van den Brink PJ, Hartgers EM, Fettweis U, Crum SJH, van Donk E, Brock TCM. 1997. Sensitivity of macrophyte-dominated freshwater microcosms to chronic levels of the herbicide linuron I. Primary producers. Ecotoxicol Environ Safety 38:13–24.

van den Brink PJ, Hartgers EM, Gylstra R, Bransen F, Brock TCM. 2002b. Effects of a mixture of two insecticides in freshwater microcosms. II. Responses of plankton and ecological risk assessment. Ecotoxicol 11:181–197.

van den Brink PJ, Kuyper MC. 2001. HERBEST, a model describing the effects and recovery of populations in stressed aquatic ecosystems. 11th annual meeting of SETAC-Europe, Madrid, Spain.

van den Brink PJ, Roelsma J, van Ness EH, Scheffer M, Brock TCM. 2002c. PERPEST model, a case-based reasoning approach to predict ecological risks of pesticides. Environ Toxicol Chem 21:2500–2506. http://www.perpest.alterra.nl (accessed December 28, 2007).

van den Brink PJ, Ter Braak CJF. 1998. Multivariate analysis of stress in experimental ecosystems by principle response curves and similarity analysis. Aquat Ecol 32:163–178.

van den Brink PJ, Ter Braak CJF. 1999. Principal response curves: analysis of time-dependent multivariate responses of biological community to stress. Environ Toxicol Chem 18:138–148.

van den Brink PJ, van den Brink NW, Ter Braak CJF. 2003. Multivariate analysis of ecotoxicological data using ordination: demonstrations of utility on the basis of various examples. Aust J Ecotoxicol 9:141–156.

van den Brink PJ, van Donk E, Gylstra R, Crum SJH, Brock TCM. 1995. Effects of chronic low concentrations of the pesticides chlorpyrifos and atrazine in indoor freshwater microcosms. Chemosphere 31:3181–3200.

van den Brink PJ, van Wijngaarden RPA, Lucassen WGH, Brock TCM, Leeuwangh P. 1996. Effects of the insecticide Dursban® 4E (active ingredient chlorpyrifos) in outdoor experimental ditches: II. Invertebrate community responses and recovery. Environ Toxicol Chem 15:1143–1153.

van den Brink PJ, Verboom J, Baveco JM, Heimbach F. 2007. An individual-based approach to model spatial population dynamics of invertebrates in aquatic ecosystems after pesticide contamination. Environ Toxicol Chem 26:2226–2236.

van der Geest HG. 2001. Insects in polluted rivers: an experimental analysis [PhD thesis]. Amsterdam (The Netherlands): University of Amsterdam, 152 p.

van der Geest HG, Greve GD, Boivin ME, Kraak MHS, van Gestel CAM. 2000. Mixture toxicity of copper and diazinon to larvae of the mayfly (Ephoron virgo) judging additivity at different effect levels. Environ Toxicol Chem 19:2900–2905.

van der Hoeven N, Gerritsen AAM. 1997. Effects of chlorpyrifos on individuals and populations of *Daphnia pulex* in the laboratory and field. Environ Toxicol Chem 16:2438–2447.

van der Kooij LA, van de Meent D, van Leeuwen CJ, Bruggeman WA. 1991. Deriving quality criteria for water and sediment from the results of aquatic toxicity tests and products standards: application of the equilibrium partitioning method. Water Res 25:697–705.

van Geest GJ, Zwaardemaker NG, van Wijngaarden RPA, Cuppen JGM. 1999. Effects of a pulsed treatment with the herbicide Afalon (active ingredient linuron) on macrophyte-dominated mesocosms. II structural responses. Environ Toxicol Chem 18:2866–2874.

van Gestel CAM, Hensbergen PJ. 1997. Interaction of Cd and Zn toxicity for Folsomia candida Willem (Collembola: Isotomidae) in relation to bioavailability in soil. Environ Toxicol Chem 16:1177–1186.

van Leeuwen CJ, Hermens JLM. 1995. Risk assessment of chemicals: an introduction. Dordrecht (The Netherlands): Kluwer Academic. 374 p.

van Leeuwen CJ, van der Zandt PTJ, Altenberg T, Verhaar HJM, Hermans JLM. 1991. The application of (Q)SARs, extrapolation and equilibrium partitioning in aquatic effects assessment for narcotic pollutants. Sci Tot Environ 109–110:681–690.

vannote RL, Minshall GW, Cumming KW, Sedell JR, Cushing CE. 1980. The river continuum concept. Can J Fish Aquat Sci 37:130–137.

van Straalen NM. 1992. Ecolosical receptors of environmentally hazards compounds. In augural address, vrye universiteit, Amsterdam.

van Straalen NM. 2002. Threshold models for species sensitivity distributions applied to aquatic risk assessment for zinc. Environ Toxicol Pharmacol 11:167–172.

van Straalen NM, Bergema WF. 1995. Ecological risks of increased bioavailability of metals under soil acidification. Pedobiologia 39:1–9.

van Straalen NM, Hoffmann AA. 2000. Review of experimental evidence for physiological costs of tolerance to toxicants. In: Kammenga JE, Laskowski R, editors. Demography in ecotoxicology. Chichester (UK): John Wiley, p 147–161.

van Straalen NM, van Leeuwen CJ. 2002. European history of species sensitivity distributions. In: Posthuma L, Suter GW, Traas TP, editors. Species sensitivity distributions in ecotoxicology. Boca Raton (FL): CRC Press, p 211–254.

van Vlaardingen PLA, Traas TP, Wintersen AM, Aldenberg T. 2004. ETX 2.0: a program to calculate risk limits and fraction affected, based on normal species sensitivity distributions. No. 601501028/2004. Bilthoven (The Netherlands): National Institute of Public Health and the Environment (RIVM), 49 p.

van Wijngaarden R, Leeuwangh P, Lucassen WGH, Romijn K, Ronday R, van der Velde R, Willigenburg W. 1993. Acute toxicity of chlorpyrifos to fish, a newt, and aquatic invertebrates. Bull Environ Contam Toxicol 51:716–723.

van Wijngaarden RPA, Brock TCM. 1999. Population and community responses in pesticide-stressed freshwater ecosystems. In: Trevisan M, editor, Human and environmental exposure to xenobiotics, proceedings of the XI Symposium in Pesticide Chemistry, September 11–15, 1999, Cremona, Italy, p 19.

van Wijngaarden RPA, Brock TCM, Douglas MT. 2005a. Effects of chlorpyrifos in freshwater model ecosystems: Do experimental conditions change ecotoxicological threshold levels? Pest Manag Sci 61:923–935.

van Wijngaarden RPA, Brock TCM, van den Brink PJ. 2005b. Threshold levels of insecticides in freshwater ecosystems: a review. Ecotoxicol 14:353–378.

van Wijngaarden RPA, Brock TCM, van den Brink PJ, Gylstra R, Maund SJ. 2006. Ecological effects of spring and late summer applications of lambda-cyhalothrin on freshwater microcosms. Arch Environ Contam Toxicol 50:220–239.

van Wijngaarden RPA, Crum SJH, Decraene K, Hattink J, van Kammen A. 1998. Toxicity of Derosal (active ingredient carbendazim) to aquatic invertebrates. Chemosphere 37:673–683.

van Wijngaarden RPA, Cuppen JGM, Arts GHP, Crum SJH, van den Hoorn MW, van den Brink PJ, Brock TCM. 2004. Aquatic risk assessment of a realistic exposure to pesticides used in bulb crops: a microcosm study. Environ Toxicol Chem 23:1479–1498.

van Wijngaarden RPA, van den Brink PJ, Crum SJH, Oude Voshaar JHO, Brock JH, Leeuwangh P. 1996. Effects of the insecticide Dursban 4E (active ingredient chlorpyrifos) in outdoor experimental ditches I. Comparison of short-term toxicity between laboratory and field. Environ Toxicol Chem 15:1133–1142.

van Wijngaarden RPA, van den Brink PJ, Oude Voshaar JH, Leeuwangh P. 1995. Ordination techniques for analyzing response of biological communities to toxic stress in experimental ecosystems. Ecotoxicol 4:61–77.

Verdonck FAM, Aldenberg T, Jaworska J, Vanrolleghem VA. 2003. Limitations of current risk characterization methods in probabilistic risk assessment. Environ Toxicol Chem 22:2209–2213.

Verdonck FAM, Jaworska J, Thas O, Vanrolleghem PA. 2001. Determining environmental standards using bootstrapping, Bayesian and maximum likelihood techniques: a comparative study. Anal Chim Acta 446:429–438.

Verhaar HJM, van Leeuwen CJ, Hermans JLM. 1992. Classifying environmental pollutants 1: structure-activity relationships for prediction of aquatic inherent toxicity. Chemosphere 25:471–491.

Vermeire TG, Jager DT, Bussian B, Devillers J, den Haan K, Hansen B, Lundberg I, Niessen H, Robertson S, Tyle H, van der Zandt PTJ. 1997. European Union system for the evaluation of substances (EUSES): principles and structure. Chemosphere 34:1823–1836.

Versteeg DJ, Belanger SE, Carr GJ. 1999. Understanding single-species and model ecosystem sensitivity: data-based comparison. Environ Toxicol Chem 18:1329–1346.

Vijver MG. 2004. The ins and outs of bioaccumulation: metal bioaccumulation kinetics in soil invertebrates in relation to availability and physiology [PhD thesis]. Amsterdam (The Netherlands): Vrije Universiteit, 179 p.

Vijver MG, Vink JPM, Miermans CJH, van Gestel CAM. 2003. Oral sealing using glue: a new method to distinguish between intestinal and dermal uptake of metals in earthworms. Soil Biol Biochem 35:125–132.

Vinson MR, Hawkins CP. 2003. Broad-scale geographical patterns in local stream insect genera richness. Ecography 26:751–767.

Voinov A, Voinov H, Costamza R. 1999. Landscape modelling of surface water flow flow 2: Patuxent watershed case study. Ecol Model 119:211–230.

Volosin JS, Cardwell RD. 2002. Relationships between aquatic hazard quotients and probabilistic risk estimates: what is the significance of a hazard quotient > 1? Human Ecol Risk Assess 8:355–368.

von der Ohe PC, Liess M. 2004. Relative sensitivity distribution of aquatic invertebrates to organic and metal compounds. Environ Toxicol Chem 23:150–156.

Vouk VV, Butler GC, Upton AC, Parke DV, Asher SC, editors. 1987. Methods for assessing the effects of mixtures of chemicals. New York (NY): John Wiley, 894 p.

Wagner C, Løkke H. 1991. Estimation of ecotoxicological protection levels from NOEC toxicity data. Water Res 25:1237–1242.

Walker B. 1991. Biodiversity and ecological redundancy. Conserv Biol 6:12–23.

Walker CH, Hopkin SP, Sibley RM, Peakall DB. 2001. Principles of ecotoxicology. New York (NY): Taylor & Francis, 309 p.

Walker JD. 2003. (Q)SARs promote more efficient use of chemical testing resources: carpe diem. Environ Toxicol Chem 22:1651.

Walker JD, Carlsen L, Jaworska J. 2003a. Improving opportunities for regulatory acceptance of (Q)SARs: the importance of model domain, uncertainty, validity and predictability. (Q)SAR & Combinatorial Science 22:346–350.

Walker JD, Jaworska J, Comber MHI, Schultz TW, Dearden JC. 2003b. Guidelines for developing and using quantitative structure-activity relationships. Environ Toxicol Chem 22:1653–1665.

Walter H, Consolaro F, Gramatica P, Scholze M, Altenburger R. 2002. Mixture toxicity of priority pollutants at no observed effect concentrations (NOECs). Ecotoxicol 11:299–310.

Walthall WK, Stark JD. 1997a. A comparison of acute mortality and population growth rate as endpoints of toxicological effect. Ecotoxicol Environ Safety 37:45–57.

Walthall WK, Stark JD. 1997b. Comparison of two population-level ecostoxicological end-points: the intrinsic (rm) and instantaneous (ri) rates of increase. Environ Toxicol Chem 16:1068–1073.

Wang H. 1987. Factors affecting metal toxicity to (and bioaccumulation by) aquatic organisms: overview. Environ Int 13:437–457.

Ward S, Arthington AH, Pusey BJ. 1995. The effects of a chronic application of chlorpyrifos on the macroinvertebrate fauna in an outdoor artificial stream system: species responses. Ecotoxicol Environ Safety 30:2–23.

Warne MSJ. 2003. A review of the ecotoxicity of mixtures, approaches to, and recommendations for, their management. In: Kennedy B, editor. Fifth National Workshop on the Assessment of Site Contamination. Adelaide (Australia): National Environment Protection Council Service Corporation, p 253–276.

Warne MSJ, Hawker DW. 1995. The number of components in a mixture determines whether synergistic, antagonistic or additive toxicity predominate: the Funnell hypothesis. Ecotoxicol Environ Safety 31:23–28.

Warren-Hicks W, Parkhurst B. 2003. Whole effluent toxicity tests: using Bayesian methods to calculate model-based variability. SETAC 24th Annual Meeting, Austin, Texas.

Webber EC, Bayne DR, Seesock WC. 1992. Ecosystem-level testing of a synthetic pyrethroid insecticide in aquatic mesocosms. Environ Toxicol Chem 11:87–105.

Weis JS. 2002. Tolerance to environmental contaminants in the mummichog, *Fundulus heteroclitus*. Human Ecol Risk Assess 8:933–953.

Wellman P, Ratte H-T, Heimbach F. 1998. Primary and secondary effects of methabenzthiazuron on plankton communities in aquatic outdoor microcosms. Aquat Ecol 32:125–134.

Weltje L. 1998. Mixture toxicity and tissue interactions of Cd, Cu, Pb and Zn in earthworms (Oligochaeta) in laboratory and field soils: a critical evaluation of data. Chemosphere 36:2643–2660.

Wendt-Rasch L, Friberg-Jensen U, Woin P, Christoffersen K. 2003. Effects of the pyrethroid insecticide cypermethrin on a freshwater community studied under field conditions. II. Direct and indirect effects on the species composition. Aquat Toxicol 63:373–389.

Weston DP, Penry DL, Gulmann LK. 2000. The role of ingestion as a route of contaminant bioaccumulation in a deposit-feeding polychaete. Arch Environ Contam Toxicol 38:446–454.

Wheeler JR, Grist EPM, Leung KMY, Morritt D, Crane M. 2002a. Species sensitivity distributions: data and model choice. Mar Poll Bull 45:192–202.

Wheeler JR, Leung KMY, Morritt D, Sorokin N, Rogers H, Toy R, Holt M, Whitehouse P, Crane M. 2002b. Freshwater to saltwater toxicity extrapolation using species sensitivity distributions. Environ Toxicol Chem 21:2459–2467.

Widianarko B. 1997. Urban ecotoxicology: spatial and temporal heterogeneity of pollution [PhD thesis]. Amsterdam The Netherlands): Free University, 138 p.

Williams EK, Hall JA. 1999. Seasonal and geographic variability in toxicant sensitivity of *Mytilus galloprovincialis* larvae. Aust J Ecotoxicol 5:1–10.

Williams P, Whitfield M, Biggs J, Bray S, Fox G, Nicolet P, Sear D. 2003. Comparative biodiversity of rivers, streams, ditches and ponds in an agricultural landscape in southern England. Biol Conserv 115:329–341.

Williams P, Whitfield M, Biggs J, Fox G, Nicolet P, Shillabeer N, Sherratt T, Heneghan P, Jepson P, Maund S. 2002. How realistic are outdoor microcosms? A comparison of the biota of microcosms and natural ponds. Environ Toxicol Chem 21:143–150.

Williams PH, Gaston KJ, Humphries CJ. 1997. Mapping biodiversity value worldwide: combining higher-taxon richness from different groups. Proc R Soc Lond B Biol Sci 264:141–148.

Willis KJ, van den Brink PJ, Green JD. 2004. Seasonal variation in plankton community responses of mesocosms dosed with pentachlorophenol. Ecotoxicol 13:707–720.

Wintersen A, Posthuma L, De Zwart D] National Institute of Public Health and the Environment (RIVM). 2004. The RIVM e-toxBase: a database for storage, retrieval and export of ecotoxicity data. Bilthoven (The Netherlands): National Institute of Public Health and the Environment (RIVM). http://www.ru.nl/environmentalscience/research/risk_assessment/e-toxbase?mode=print&popup=normal (accessed December 28, 2007).

Wong DCL, Maltby L, Whittle D, Warren P, Dorn PB. 2004. Spatial and temporal variability in the structure of invertebrate assemblages in control stream mesocosms. Water Res 38:128–138.

Woodbury PB. 2003. Do's and don'ts of spatially explicit ecological risk assessments. Environ Toxicol Chem 22:977–982.

Woolard CD, Linder PW. 1999. Modelling of the cation binding properties of fulvic acids: an extension of the random algorithm to include nitrogen and sulphur donor sites. Sci Tot Environ 226:35–46.

World Conservation Monitoring Centre. 1992. Global biodiversity: status of the earth's living resource. London (UK): Chapman & Hall, 585 p.

World Health Organization International Program on Chemical Safety. 1999. Principles for the assessment of risks to human health from exposure to chemicals. Environmental Health Criteria No. 20. Geneva (Switzerland): World Health Organization.

Xu FL, Dawson RW, Tao S, Li BG, Cao J. 2002. System-level responses of lake ecosystems to chemical stress using exergy and structural exergy as ecological indicators. Chemosphere 46:173–185.

Xu S, Nirmalakhandan N. 1998. Use of (Q)SAR models in predicting joint effects in multi-component mixtures of organic chemicals. Water Res 32:2391–2399.

Yachi S, Loreau M. 1999. Biodiversity and ecosystem productivity in a fluctuating environment: the insurance hypothesis. Proc Nat Acad Sci US 96:1463–1468.

Yang RSH. 1994a. Introduction to the toxicology of chemical mixtures. In: Yang RSH, editor. Toxicology of chemical mixtures: case studies, mechanisms, and novel approaches. San Diego (CA): Academic Press, p 1–10.

Yang RSH. 1994b. Toxicology of chemical mixtures: case studies, mechanisms, and novel approaches. San Diego (CA): Academic Press, 720 p.

Yang RSH, Thomas RS, Gustafson DL, Campain J, Benjamin SA, Verhaar HJM, Mumtaz MM. 1998. Approaches to developing alternative and predictive toxicology based on PBPK/PD and (Q)SAR modeling. Environ Hlth Perspect 106:1385–1393.

Yoder CO, Rankin ET. 1995. Biological response signatures and the area of degradation value: new tools for interpreting multi-metric data. In: Davis WS, Simon TP, editors. Biological assessment and criteria: tools for water resource planning and decision making. Boca Raton (FL): Lewis Publishers, p 263–286.

Zeeman M, Auer CM, Clements RG, Nabholz JV, Boethling RS. 1995. USEPA regulatory perspectives on the use of (Q)SAR for new and existing chemical evaluations. SAR (Q)SAR Env Res 3:179–201.

Zeeman M, Gilford J. 1993. Ecological hazard evaluation and risk assessment under EPA's Toxic Substances Control Act (INV): an introduction. In: Lewis MA, editor. Environmental toxicology and risk assessment. Philadelphia (PA): American Society for Testing and Materials, p 7–21.

Zhang J, Jorgensen SE, Tan CO, Beklioglu M. 2003. A structurally dynamic modeling: Lake Mogan, Turkey as a case study. Ecol Model 164:103–120.

Index

Other Titles from the Society of Environmental Toxicology and Chemistry (SETAC)

SETAC

A Professional Society for Environmental Scientists and Engineers and Related Disciplines Concerned with Environmental Quality

The Society of Environmental Toxicology and Chemistry (SETAC), with offices currently in North America and Europe, is a nonprofit, professional society established to provide a forum for individuals and institutions engaged in the study of environmental problems, management and regulation of natural resources, education, research and development, and manufacturing and distribution.

Specific goals of the society are

- Promote research, education, and training in the environmental sciences.
- Promote the systematic application of all relevant scientific disciplines to the evaluation of chemical hazards.
- Participate in the scientific interpretation of issues concerned with hazard assessment and risk analysis.
- Support the development of ecologically acceptable practices and principles.
- Provide a forum (meetings and publications) for communication among professionals in government, business, academia, and other segments of society involved in the use, protection, and management of our environment.

These goals are pursued through the conduct of numerous activities, which include:

- Hold annual meetings with study and workshop sessions, platform and poster papers, and achievement and merit awards.
- Sponsor a monthly scientific journal, a newsletter, and special technical publications.
- Provide funds for education and training through the SETAC Scholarship/Fellowship Program.
- Organize and sponsor chapters to provide a forum for the presentation of scientific data and for the interchange and study of information about local concerns.
- Provide advice and counsel to technical and nontechnical persons through a number of standing and ad hoc committees.

SETAC membership currently is composed of more than 5000 individuals from government, academia, business, and public-interest groups with technical backgrounds in chemistry, toxicology, biology, ecology, atmospheric sciences, health sciences, earth sciences, and engineering.

If you have training in these or related disciplines and are engaged in the study, use, or management of environmental resources, SETAC can fulfill your professional affiliation needs.

All members receive a newsletter highlighting environmental topics and SETAC activities, and reduced fees for the Annual Meeting and SETAC special publications.

All members except Students and Senior Active Members receive monthly issues of Environmental Toxicology and Chemistry (ET&C) and Integrated Environmental Assessment and Management (IEAM), peer-reviewed journals of the Society. Student and Senior Active Members may subscribe to the journal. Members may hold office and, with the Emeritus Members, constitute the voting membership.

If you desire further information, contact the appropriate SETAC Office.

1010 North 12th Avenue	Avenue de la Toison d'Or 67
Pensacola, Florida 32501-3367 USA	B-1060 Brussels, Belgium
T 850 469 1500 F 850 469 9778	T 32 2 772 72 81 F 32 2 770 53 86
E setac@setac.org	E setac@setaceu.org

www.setac.org
Environmental Quality Through Science®